Biotherapy - History, Principles and Practice

Martin Grassberger • Ronald A. Sherman
Olga S. Gileva • Christopher M.H. Kim
Kosta Y. Mumcuoglu

Editors

Biotherapy - History, Principles and Practice

A Practical Guide to the Diagnosis
and Treatment of Disease using
Living Organisms

Springer

Editors
Martin Grassberger
Institute of Pathology and Microbiology
Rudolfstiftung Hospital
   and Semmelweis Clinic
Vienna, Austria

Olga S. Gileva
Department of Preclinic Dentistry
   and Physiotherapy
Perm State Academy of Medicine
   Perm, Russia

Kosta Y. Mumcuoglu
Department of Microbiology
   and Molecular Genetics
Hebrew University – Hadassah
   Medical School, Jerusalem, Israel

Ronald A. Sherman
BioTherapeutics, Education
   & Research Foundation
Irvine, CA, USA

Christopher M.H. Kim
Graduate School of Integrated Medicine
CHA University
Seongnam, Gyeongghido
Korea, Republic of (South Korea)

ISBN 978-94-007-6584-9   ISBN 978-94-007-6585-6 (eBook)
DOI 10.1007/978-94-007-6585-6
Springer Dordrecht Heidelberg New York London

Library of Congress Control Number: 2013940774

Printed on acid-free paper

Springer is part of Springer Science+Business Media (www.springer.com)

Preface

Biotherapy – the use of living organisms for the treatment of human and animal illness – is a practice known since antiquity. But it is not antiquated! Thanks to modern scientific methods and the dedication of many clinicians, biologists, biochemists, and patient advocates, biotherapy today is a rapidly advancing multi – disciplinary field of medicine. The story of biotherapy is a story of life and evolution, a story of human history, a story of scientific discovery, a story of deadly diseases and miraculous cures. The story of biotherapy, as revealed in these pages, is the story of life itself, and the story of man's will and capacity to harness the power of life. The story of biotherapy teaches us that all living beings on this planet are interrelated. Occasionally, those relationships can be beneficial to both parties. Biotherapists have found or created such mutually beneficial relationships, in their efforts to tackle illness and disability. This is their story, too.

The body of scientific publications concerning biotherapy has grown astonishingly large, especially during the past several decades. Much of the credit for biotherapeutic discoveries must go to open-minded and observant clinicians and scientists. The observations of biotherapists provide strong evidence of the concept and power of evolution. It is reasonable to expect that two organisms living together, one parasitic on the other, would eventually (if given enough time) alter their biology or behaviors in such a way that they could both thrive together. After all, if the host cannot survive invasion by the parasite, its species will eventually come to an end. And if the parasite kills its host, then another host must be found (a major waste of time and energy). It is mutually beneficial when both organisms can coexist. Evolution is not simply about "survival of the fittest." It is about adapting to one's condition, so as to become more fit. Biotherapists have recognized many such adaptations that evolved around problems of illness. They have also adapted other inter-species relationships in order to apply them to unmet health needs. This treatise is filled with examples of both.

Take, for example, the blow fly maggot, which evolved to live in the most hostile environments: decaying carcasses, teeming with highly pathogenic bacteria. Clinician researchers, observing that blow fly larvae can also infest wounds in live hosts – and remove the dead tissue from those wounds, without harming the live tissue – began intentionally applying maggots to the non-healing wounds of their patients.

Today, maggot therapy is one of the most rapidly advancing fields within biotherapy, having already gained acceptance by the medical establishment and by health care regulatory authorities.

Need often drives innovation. Better treatments for infectious diseases remain a powerful need these days, because so many of our antibiotics are now useless against the microbes which developed resistance to them (another example of adaptation). This need for better defense against infection is helping to open doors for maggot therapy and other biotherapeutic modalities (e.g., phage therapy). Corpse-dwelling maggots coevolved with bacteria, and now their antimicrobials are proving to be effective in hospitals against the microbes, which have become resistant to man's antibiotics.

Maggot therapy epitomizes some of the common benefits of biotherapy: effective, relatively safe, and low-cost treatment of serious medical problems (in this case, diabetic foot ulcers, pressure ulcers, and other problematic or recalcitrant wounds). When performed responsibly, biotherapeutic modalities have little or no impact on the environment and can often be administered by paramedical personnel rather than highly trained medical specialists.

Hirudotherapy – the medical use of leeches – is one of the oldest practices in all of medicine. Over the years, hirudotherapy has evolved from a simple bloodletting procedure to a scientifically based physiologic process with rationally defined clinical applications. During the Middle Ages, the golden era for bloodletting, leeches were used by nearly every physician to cure anything from headaches to hemorrhoids. In Russia, hirudotherapy reached its zenith in the late eighteenth and early nineteenth centuries, when leech harvesting and leech therapy netted the country an annual six million silver rubles. By the end of the nineteenth century, leeching fell out of favor and became associated with medical quackery in most countries. However, the past 25 years have seen a renaissance in leech therapy, primarily because of its newfound value in reconstructive, transplant, and microvascular surgery. Nowadays, leech therapy is a standard treatment for postoperative venous congestion and has become an integral part of the armamentarium to salvage vascularly compromised flaps or replants. In the twenty-first century, several clinical studies were performed in Germany and Russia demonstrating the efficacy of leeches in relieving osteoarthritis pain. In Asia, Russia, and parts of Eastern Europe, hirudotherapy is officially recognized as a classic alternative treatment for diseases such as phlebitis, osteoarthritis, hypertension, and glaucoma.

Therapy with honey bee venom (HBV) is a bio-therapeutic treatment that utilizes the venom of honeybees. Physicians dating back to Hippocrates used HBV to treat a variety of illnesses. Today, physicians are using bee venom to treat patients with chronic pain disorders such as rheumatism and arthritis, and to combat many inflammatory and degenerative connective tissue diseases. Neurological disorders such as migraines, peripheral neuritis, and chronic back pain are also being treated successfully with HBV. In the case of autoimmune disorders such as multiple sclerosis and lupus, bee venom restores movement and mobility by strengthening the body's natural defensive mechanism. In addition, dermatological conditions such as eczema and psoriasis, and some infectious diseases such as herpes and some urinary tract infections, have been effectively treated.

Apitherapy is the science and art of the use of honeybee products to regain and maintain health. In the past, products of the hive, that is, honey, pollen, propolis, and royal jelly, were frequently used as natural remedies for health maintenance, while bee venom was used for the treatment of illness. More recently, the products of the hive have been incorporated into Western medical practice, where the focus of attention is mainly on illness and its prevention.

Helminth therapy – the therapeutic introduction of helminths (parasitic worms) into the body – provides another example of host-parasite evolution. Over the course of millions of years, helminths and their human hosts have developed mechanisms to co-exist relatively well. In order to survive our immune system, many parasitic worms developed mechanisms to modulate (alter) their hosts' normal immune defense mechanisms. At the same time, mammals optimized their immune responses so as to prevent excessive and potentially lethal invasion by helminths. There are more than 80 different autoimmune diseases, which afflict people in highly developed industrialized countries, but they are rarely seen in tropical countries, where helminth exposure is common. As countries eradicated helminth infections over the past century, autoimmune and inflammatory diseases rose significantly. With exposure to helminths again, many patients with autoimmune and inflammatory diseases have experienced significant remission of their symptoms. Two such helminths – the porcine whipworm, *Trichuris suis*, and the hookworm, *Necator americanus* – were recently evaluated for potential medical application, with very encouraging results. Diseases currently being studied for treatment by helminths include Crohn's disease, ulcerative colitis, multiple sclerosis, celiac disease, rheumatoid arthritis, and autism. Furthermore, such observations and experiments have provided new insights into the complex interactions between helminths and their hosts. With these insights has come a better understanding, and novel treatments, for several autoimmune and inflammatory diseases.

Ichthyotherapy is defined as the treatment of skin diseases such as psoriasis and ichthyosis with the so-called doctor fish of Kangal, *Garra rufa*. Pilot clinical studies and numerous anecdotal reports indicate that ichthyotherapy is a promising treatment for these conditions and deserves further study.

Phage therapy is the use of bacteriophages – viruses that can only infect bacteria – to treat bacterial infections. In some parts of the world, phages have been used therapeutically since the 1930s. Phage therapy was first developed at the Pasteur Institute in Paris early in the twentieth century and soon spread through Europe, the USA, the Soviet Union, and other parts of the world, albeit with mixed success. With the advent of chemical antibiotics in the 1940s, phage therapy found itself largely ignored in the West, though it continued to be used to varying degrees in some other countries, with claims of success. Today, however, the resurgence of bacteria that are resistant to most or all available antibiotics is precipitating a major health crisis, and interest is growing in the potential use of phages to complement antibiotics as a way to fight infection. In 2006, the US FDA and the EU both approved phage preparations targeting *Listeria monocytogenes* on ready-to-eat foods. Currently, phage therapy is part of standard medical practice in the Republic of Georgia and is fairly readily available in Russia, Poland, and other Eastern European countries.

The psychological benefits of human interaction with higher animals like dogs and horses might be explained by our recent coexistence and shared cultural history. The human connection with horses can be traced back even to prehistoric cave paintings. This connection that human beings still subconsciously seek with other living beings has been termed "biophilia" by sociobiologist Edward O. Wilson.

Human-dog partnerships have traditionally provided a service for clients with disabilities, such as the service dogs, which assist people with visual or other physical impairments. Since the first report of a dog as "cotherapist" by American child psychiatrist Boris Levinson in the 1960s, there have been numerous publications outlining the medical and psychological benefits of the human-animal bond. A growing number of practitioners have been integrating animals into their practice, wherein the animal plays an integral role in assisting with the mental health, speech, occupational therapy, or physical therapy goals. When a horse or a donkey is part of the treatment team, it is generally referred to as equine-assisted therapy. Although animal-assisted therapy is well established in many countries, there are numerous exciting avenues still open for research.

A relatively recent example of using animals to help solve medical problems is the application of animals' exquisite olfactory senses to detect illnesses with distinctive odors. For centuries, it has been understood that many human diseases generate characteristic odors. Physicians often recognized the odors associated with conditions such as pneumonia (lung infection), diabetes, and typhoid fever. Dogs have been used by man for their olfactory abilities for many years in the detection of drugs, explosives, banknotes, and other items. Now it has been recognized that dogs may be able to assist in the early detection of human disease, notably various forms of cancer, and hypoglycemic episodes in diabetes.

Each and every biotherapeutic modality discussed in this tome could easily and justifiably be featured in its own volume. Indeed, reviews of several modalities already have been published individually. But biotherapy has seen its rapid advances, in large part, as a result of the multidisciplinary composition of its advocates and researchers. Most biotherapists now recognize that they can learn even more, and advance their own specialties further and faster, by learning about each others' specialties. We biotherapists want to read about the problems that others have faced, and the solutions that others have found. With the publication of this text, the history and current status of the major biotherapeutic modalities finally can be found in one comprehensive yet easily navigated reference book.

The future for biotherapy is bright, exciting, and wide open. Although a lot of work remains to be done, we are confident that all of the modalities described in this volume will become an integral part of conventional medical practice within the near future. The field is advancing quickly, and even as this book is being prepared for print, new microbes and animals are being studied for their potential value as therapeutic agents. We sincerely hope that this volume will stimulate additional research in biotherapy and will help propel the study and practice of biotherapy even faster and further. Join us, and enjoy the journey.

Vienna   Martin Grassberger
Los Angeles   Ronald A. Sherman
Seongnam, Gyeongghido   Christopher M.H. Kim
Perm   Olga S. Gileva
Jerusalem   Kosta Y. Mumcuoglu
February 2013

Acknowledgments

We would like to express our gratitude to Dr. Magdalena Pilz and Dr. Heidrun Grassberger for their additional meticulous proofreading of the manuscripts. Also, the editors are indebted to Springer Science + Business Media, Netherlands, who saw the potential in this project, and our publishing assistant Marleen Moore, who shepherded it to completion.

Contents

Contributors

Zemphira Alavidze LTD Bacteriophage Production Facility, Tbilisi, Georgia

Erin Brewster Bacteriophage Lab, Evergreen State College, Olympia, WA, USA

Theodore Cherbuliez, M.D. Consulting Medical Staff at Spring Harbor Hospital, Maine Medical Center, South Freeport, ME, USA

John C.T. Church, M.D., FRCS. Consultant Orthopaedic Surgeon, Buckinghamshire, UK

Mary Cole, CCC. Cole and Associates Counselling Services, Calgary, Canada

Nina Ekholm Fry, MSSc. Equine-Assisted Mental Health, Prescott College, Prescott, AZ, USA

Associate Faculty, Counseling Psychology, Master of Arts Program, Prescott College, Prescott, AZ, USA

David E. Elliott, M.D., Ph.D. Department of Internal Medicine, Division of Gastroenterology/Hepatology, University of Iowa Carver College of Medicine, Iowa City, IA, USA

Olga S. Gileva, M.D., Ph.D. Department of Preclinic Dentistry and Physiotherapy, Perm State Academy of Medicine, Perm, Russian Federation

Martin Grassberger, M.D., Ph.D. Institute of Pathology and Microbiology, Rudolfstiftung Hospital and Semmelweis Clinic, Vienna, Austria

Dr. Claire Guest, B.Sc., M.Sc. Medical Detection Dogs, Milton Keynes, UK

Guram Gvasalia, M.D. Department of General Surgery, Tbilisi State Medical University, Tbilisi, Georgia

Maureen Howard, M.Sc., R. Psych. Tappen, BC, Canada

Christopher M.H. Kim, M.D. Graduate School of Integrated Medicine, CHA University, Seongnam, Gyeonggi-do, Korea

Elizabeth M. Kutter, Ph.D. PhageBiotics and the Evergreen State College, Olympia, WA, USA

Kosta Y. Mumcuoglu, Ph.D. Department of Microbiology and Molecular Genetics, Hebrew University – Hadassah Medical School, Jerusalem, Israel

David I. Pritchard, Ph.D. Faculty of Science, University of Nottingham, Nottingham, UK

Ronald A. Sherman, M.D., M.Sc. BTER Foundation, Irvine, CA, USA

Tarek I. Tantawi, Ph.D. Department of Zoology, Faculty of Science, Alexandria University, Alexandria, Egypt

Joel V. Weinstock, M.D. Division of Gastroenterology, Tufts Medical Center, Boston, MA, USA

Chapter 1
Biotherapy – An Introduction

John C.T. Church

'Biotherapy' is as old as the hills, in that man has learnt, over the millennia, mostly by trial and error, what the natural world around him has to offer, to alleviate or enhance his condition. By interacting with things, working with them, eating them, or rubbing them on, certain effects will follow, sometimes with dramatic results, including of course, death. As scientists, we divide the natural living world, somewhat arbitrarily, into the plant and animal kingdoms, but these have a massively 'fuzzy' interface. The biological world is diverse, complex, sophisticated and mysterious, changing inexorably over time, so that, whatever we might study or utilise, it is the end result of literally millions, even billions, of years of natural 'research and development'. The 'Bio'- epithet thus is open to a wide range of connotations. A glance at the World Wide Web confirms this diversity of use.

When the International Biotherapy Society (IBS) was founded, at our first conference in May 1996, we defined 'Biotherapy' as 'the use of living organisms in human medicine'. The focus then was primarily on 'Maggot Therapy', 'Hirudothotherapy' and 'Apitherapy'. The emphasis was on utilising the natural abilities, aptitudes and responses of certain organisms, in an environment determined by the practitioner. We were particularly concerned to learn how such organisms would respond to a variety of clinical situations, such as maggots in chronic wounds, and leeches on congested flaps in plastic surgery. Our ideal was to manage the organisms themselves so that they would remain 'happy and hungry', and thereby function optimally, throughout the treatment period, recognising the fact that the environment itself, to which we were subjecting them, might be, or might become, hostile to their wellbeing.

A clinically successful outcome is thus the result of specifically chosen organisms, appropriately prepared for clinical use, introduced to a patient at an opportune time by practitioners who understand the inherent biology of these organisms,

J.C.T. Church, M.D., FRCS (✉)
Consultant Orthopaedic Surgeon, Willand, Abney Court Drive,
Bourne End, Buckinghamshire SL8 5DL, UK
e-mail: jctchurch@doctors.org.uk

M. Grassberger et al. (eds.), *Biotherapy - History, Principles and Practice:*
A Practical Guide to the Diagnosis and Treatment of Disease using Living Organisms,
DOI 10.1007/978-94-007-6585-6_1, © Springer Science+Business Media Dordrecht 2013

and can manage the clinical environment to obtain optimal behaviour. This is challenging, and demanding, and is for many practitioners an entirely new concept, but the clinical results can be dramatic and magnificent.

Modern scientific endeavour is essentially reductionist, whereby objects and phenomena are studied at ever increasing depths of detail. Thus, there are now literally hundreds of known elementary sub-atomic particles, with concepts such as 'String Theory' now well established, albeit yet to be 'captured' and 'looked at'. However, the more that is discovered the more there seems to be open for future discoveries. We may know a lot about 'mass' and 'energy', but 'dark matter' and 'dark energy' still comprise 95 % of all that's there! We may have cracked the human genome, but 'non-coding' or 'Junk DNA' is still 98 %!

By complete contrast with all this, and perhaps aimed at the other end of the spectrum of human enquiry and activity, 'biotherapy' is 'holistic'. The term 'holistic' is yet another with a wide scatter of applications. We use it here to define our aim to engage with naturally occurring phenomena and *healthy intact* organisms, in all their diversity and complexity, introducing them into the clinical arena, with the goal of fully *integrating* biotherapy into modern medicine.

Biotherapy is expanding. Our website (www.biotherapysociety.org) now addresses:

Maggot Therapy: The use of the larvae of the blowfly *Lucilia sericata* as agents of cleansing or débridement, and enhancement of healing of open wounds.

Hirudotherapy: The application of leeches for extraction of blood from congested or inflamed tissues, in a wide range of pathologies.

Apitherapy: The introduction of bee venom, by live bees or by injection, for a wide range of chronic ailments. Bee products such as propolis, honey and Royal Jelly are also efficacious, each in its own spectrum of conditions.

Ichthyotherapy: The use of certain species of small fresh-water fish as scavenging agents for dermatological conditions such as psoriasis.

Helminth Therapy: The use of certain nematode worms as agents for the stimulation of host immunological responses appropriate to the alleviation of certain inflammatory auto-immune bowel diseases such as Crohn's Disease and Ulcerative Colitis. There is the potential use of the schistosomiasis parasite in the prevention of Type 1 insulin-dependant diabetes.

Phagetherapy: The therapeutic use of bacteriophages to treat pathogenic bacterial infections (especially those that do not respond to conventional antibiotics) in human as well as in veterinary medicine.

Animal-assisted therapy: This type of therapy involves higher animals (mainly dogs and other pet animals) for people with physical, psychological, cognitive, social, and behavioural problems. The special term Equine-assisted therapy is applied when a horse is part of the treatment team.

Biodiagnostics: The training of selected dogs to recognise life-threatening medical conditions, such as cancer and diabetic crises, at an early stage, allowing for successful management of the condition by conventional means.

All these aspects of our Society's activities are explained and described in more detail in the appropriate chapters of this book.

Engaging in this way with all these organisms, each with their own particular biology, raises the cogent and attractive question as to whether the active agents, enzymes, cytokines, immuno-stimulants and the rest, that these organisms produce, might be identified and processed by laboratory techniques, to provide us with an enhanced pharmacopoeia. A great deal of study of active agents produced or processed by bees and leeches, has been undertaken. The exo-enzymes secreted by maggots are under active investigation. The volatiles given off by patients suffering from various pathological states, such as cancer and diabetes, are also under scrutiny. 'Sniffer machines' are being developed to recognise specific 'marker' molecules at low concentrations. This must ideally be in collaboration with dogs being trained to identify the chosen 'markers', and the resulting intrinsic specificity fed into the machine programming.

Going down the route towards a 'magical new aspirin' at face value. runs counter to our philosophy of working with intact bio-complexity. A living organism is anything but a tablet. A maggot in a chronic wound will seek out those parts of the wound which provide it with optimal feeding, move relentlessly in the wound in this quest, engage in very efficient group-feeding with other maggots in the wound, produce heat to speed up the wound healing, use oxygen from the air as against from the wound bed, induce the production of cytokines to enhance production of host wound repair cells, produce exo-enzymes appropriate to the type of tissue (skin, fat, muscle) undergoing decomposition, free bacteria from their biofilm, engulf and digest them, thereby increasing its body weight (re-cycling organic waste) 50–80 fold, in 3 days!

There is an aspect of 'biotherapy' not listed above, but which is of fundamental importance to much of modern bio-research and development, and that is the bacterium. Numerous laboratory procedures depend on harnessing selected species of bacteria, and using their natural biology to produce quantities of a given desired product. Their DNA can be altered or added to, to give them further specific laboratory uses. If used in this way to create products of pharmaceutical value, this is a form of biotherapy. But it does not have the *bedside hands-on* aspect that is intrinsic to our other bio-therapeutic protocols.

We mentioned above that biological (natural?) 'research and development' has been active for 'millions of years'. This perforce takes us back to the very first bacterium, with zero antecedents, and enough integrity to withstand a hostile environment perhaps as formidable as that surrounding fumaroles in the deep oceans of today, where certain bacteria are still very much 'at home'. This first bacterium also had to reproduce within hours in order to survive and multiply. From a very early stage, clusters of a mono-culture of these bacteria would then have developed 'quorum sensing' and produced biofilm, with its powerful protective and exclusive attributes. Then such colonies of bacteria would have learnt how to interact with other colonies, themselves protected by their biofilm, and thereby the first multi-cellular organism, of a sort, would have evolved. This phenomenon we now name as a *microbiome*. But, microbiomes interact with other species, for good or ill, and in many ways are controlled by 'host' species, such as insects, that can for instance preserve in an enclosed sac a mono-culture of organisms to provide them with vitamins.

We have lived for decades with the concept that all microbes are bad for us, and demand ever more powerful *Antibiotics*. We, by contrast, should be looking for new '*Probiotic*' mechanisms, that we can harness, and use in the whole arena of '*Biotic*' control (see for example the chapter on phagetherapy in this volume). We do not necessarily need to know the details of how these mechanisms work, any more than for instance we understand how a dog recognises cancer in a human being, or another dog.

But, we need to have the humility to recognise that nature's answers to the challenges of life, as against our mechanistic modern industrial endeavours, out-class us in most aspects of life. In addition, we need to realise that such mechanisms are ubiquitous, and could be harnessed, using our bio-therapeutic principles central to this adventure, to the enhancement of the well-being of our fellow human beings, the prevention and management of disease, and the nurturing of the environment we all live in.

In summary, biotherapy is challenging and demanding, but it is efficacious, relatively safe, low tech, low cost, and eco-friendly, while properly conducted, it is tremendously rewarding.

Chapter 2
Maggot Therapy

Ronald A. Sherman, Kosta Y. Mumcuoglu, Martin Grassberger, and Tarek I. Tantawi

2.1 Introduction

The use and popularity of maggot therapy (MT) – the treatment of wounds with live fly larvae – is increasing rapidly in many countries throughout the world. The advantages of MT, also called larval therapy, maggot debridement therapy (MDT), and biosurgery, include its profound efficacy in debriding necrotic tissue, its relative safety, and its simplicity. These factors, along with other advantages such as its efficiency, its low cost and its effectiveness even in the context of antibiotic-resistant infections, have been responsible for the recent revival in the use of MT.

The end of the twentieth century witnessed the development of antibiotic resistance to some of the most potent antimicrobials yet created. Ironically (or perhaps as a consequence), the end of the century also witnessed the health care community, once again, embracing the maggot – a creature that thrives in the presence of bacteria, putrefaction and "filth."

R.A. Sherman, M.D., M.Sc. (✉)
BTER Foundation, 36 Urey Court, Irvine, CA 92617, USA
e-mail: rsherman@uci.edu

K.Y. Mumcuoglu, Ph.D.
Department of Microbiology and Molecular Genetics, Hebrew University – Hadassah
Medical School, P.O. Box 12272, IL 91120, Jerusalem, Israel
e-mail: kostasm@ekmd.huji.ac.il

M. Grassberger, M.D., Ph.D.
Institute of Pathology and Microbiology, Rudolfstiftung Hospital
and Semmelweis Clinic, Juchgasse 25, 1030 Vienna, Austria
e-mail: martin.grassberger@mac.com

T.I. Tantawi, Ph.D.
Department of Zoology, Faculty of Science, Alexandria University,
Moharrem Bey, Alexandria, Egypt
e-mail: medicinalmaggots@yahoo.com

M. Grassberger et al. (eds.), *Biotherapy - History, Principles and Practice:*
A Practical Guide to the Diagnosis and Treatment of Disease using Living Organisms,
DOI 10.1007/978-94-007-6585-6_2, © Springer Science+Business Media Dordrecht 2013

Maggot therapy is often used when conventional medical and surgical treatments fail to arrest the progressive tissue destruction and heal the wound. The poor blood supply to the deep wound and the consequent inability of immunological mediators and systemic antibiotics to reach the infected area prevent healing. For review see: Thomas et al. (1996), Church (1999), Sherman et al. (2000), Mumcuoglu (2001), Nigam et al. (2006a, b, 2010). Internet sites dealing with the subject of MDT are: BioTherapeutics Education and Research (BTER) Foundation; International Biotherapy Society; and World Wide Wounds.

2.2 History

For centuries, maggot-infested wounds have been associated with decreased infection, faster healing and increased survival. The beneficial effect of fly larvae for wounds was first observed by Ambroise Paré in the sixteenth century. While working with soldiers wounded and left on the battlefield for several days, Baron Larrey (physician-in-chief to Napoleon's armies) and Dr. Joseph Jones (a medical officer during the American Civil war), both described their observations of maggots cleaning the soldiers' wounds without destroying the viable tissue.

J.F. Zacharias, one of the Confederate Surgeons during the American Civil War, may even have facilitated the deposition of fly eggs on his wounded soldiers (Fleischmann et al. 2004). In non-Western civilizations, the intentional application of maggots for wound healing may date back even earlier (Pechter and Sherman 1983; Church 1996; Whitaker et al. 2007).

The earliest first-hand account of fly larvae *intentionally* applied for wound care was by William S. Baer, Chief of Orthopedic Surgery at Johns Hopkins Hospital in Baltimore (Baer 1929). As a military surgeon during World War I, he too witnessed the beneficial effects of maggot-infested wounds. He treated over 100 children, while observing the effects of maggots on chronic osteomyelitis (bone infection) and soft tissue wounds. He developed practical methods for keeping the larvae on wounds, and documented the mechanisms of action involved, i.e. debridement, disinfection and growth stimulation (Baer 1929, 1931). After witnessing serious wound infections, even in maggot-treated wounds, he concluded that medicinal maggots should be disinfected, and therefore developed methods for chemical disinfection (Baer 1931). Soon thereafter, thousands of surgeons were using Baer's maggot treatment and over 90 % were pleased with their results (Robinson 1935a). The pharmaceutical company Lederle Laboratories (Pearl River, NY; taken over by Wyeth in 1994 and by Pfizer in 2009) commercially produced "Surgical Maggots" until the 1940s for those hospitals that did not have their own insectaries (Chernin 1986).

By the mid 1940s, this treatment modality was abandoned, most likely due to the availability of effective antibiotics. Not only were the antibiotics effective in controlling some of the infections that had previously been treated with maggot therapy, but most importantly, use of these new antibiotics pre-empted the bacteremia and local spread of infections that previously led to the soft-tissue complications that the maggots so effectively treated. During the next few decades, rare reports of intentional (Teich and Myers 1986) or accidental myiasis (Horn et al. 1976) illustrated the wound-healing benefits of blow fly larvae, when all else failed.

2.3 Current Status

The end of the twentieth century was the beginning of a new era for maggot therapy, and for that matter, for most of biotherapy: it marks the first controlled, comparative clinical trials of maggot therapy, which led to regulatory oversight and marketing clearance of medicinal maggots in 2004 by the U.S. Food and Drug Administration (FDA), as the first legally marketed "living therapeutic animal."

In early 1990, a prospective clinical study was conducted to evaluate maggot therapy as a treatment for pressure ulcers ("bed sores") in spinal cord injury patients at the Veterans' Affairs Medical Center in Long Beach, CA (Sherman et al. 1995b). Over the next 5 years, the project was expanded to include other patient populations (Sherman 2002b, 2003). Together, these studies demonstrated unequivocally that maggot therapy was associated with faster and more thorough debridement than conventional surgical and non-surgical modalities. Maggot-treated wounds filled with healthy granulation tissue and became smaller more rapidly than those wounds treated with conventional modalities. Wounds scheduled for surgery or amputation but treated first with maggot therapy required fewer surgical interventions, and those that still required surgery had significantly fewer post-operative infections and wound-closure problems (Sherman et al. 2001; Sherman and Shimoda 2004).

In the two following decades, MT has again become accepted as a wound care treatment. Scores of reports, including clinical trials and basic research, were published in the medical and scientific literature since 1995. In 1995, medicinal maggots were produced in the U.S., Israel and the United Kingdom. By the year 2002, they were being produced by over a dozen labs; by 2011, an estimated 50,000 units were produced by at least 24 laboratories, and shipped to patients in over 30 countries. Today, 20 years after its reintroduction, we estimate that more than 80,000 patients were treated by MT (Mumcuoglu et al. 2012).

More recently, MT is taught in medical and surgical training programs, and medical device companies are investing in maggot laboratories, dressings and research. The growing acceptance of MT is due mainly to the growing need for effective, low cost wound care, combined with the growing recognition that maggots can provide exactly that: simple, safe, effective and low-cost wound care (Wayman et al. 2001; Thomas 2006).

2.4   The Fly

Medicinal blowfly maggots are selected for their ability to feed on dead tissue without disturbing viable tissue (Sherman 2002a). The most commonly employed larvae have been those of the green bottle fly, *Lucilia sericata* (or *Phaenicia sericata*, depending on authority) (Fig. 2.1).

Rarely, therapists have used related species, such as *Lucilia cuprina* and *Calliphora vicina* (Table 2.1); but these have not been studied as extensively (Paul et al. 2009; Tantawi et al. 2010; Kingu et al. 2012).

L. sericata is common all over the temperate and tropical regions of the world. It prefers warm and moist climates and accordingly is especially common in coastal regions. The female lays her eggs (up to 200 during her lifetime) on decaying organic material such as feces and animal corpses. Depending on external temperatures, the eggs hatch within 8–24 h, releasing larvae. The larvae will molt twice over the course of 4–7 days (again, depending on temperature). After the third stage (instar) of their larval period (Fig. 2.2), they will wander away from the host and enter their pupal stage. The pupal stage lasts approximately 10–20 days, at the end

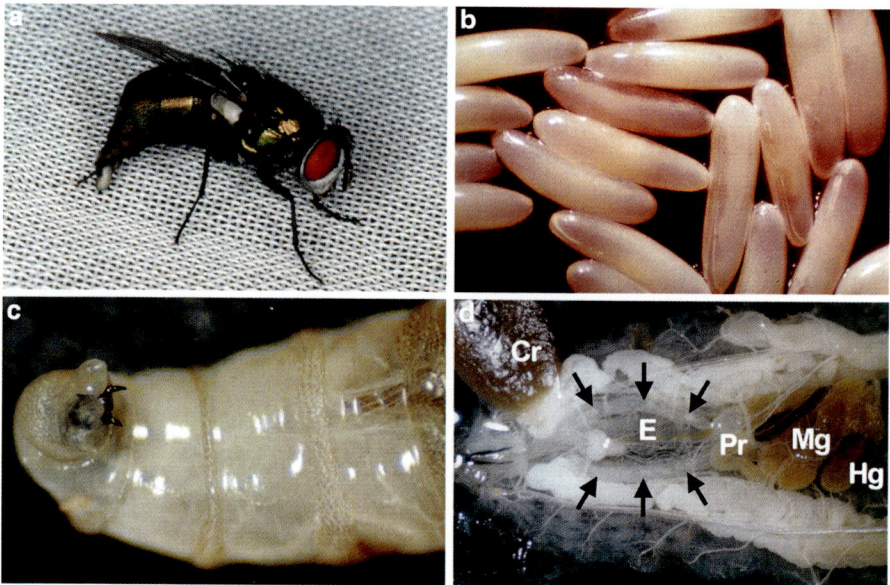

Fig. 2.1 The green bottle-fly, *Lucilia sericata*: (**a**) Female fly with visible ovipositor; (**b**) eggs; (**c**) apical part of a grown third instar larva with the two protruding mouth hooks; (**d**) dissected third instar larva with paired large salivary glands (*arrows*), the crop (*Cr*) rotated to the *top left*, esophagus (*E*), proventriculus (*Pr*), the *light-brown* midgut (*Mg*) and beginning of the darker hindgut (*Hg*). The large *white* masses on the *upper and lower part* of the picture are the larval fat bodies

Table 2.1 Species of flies used in maggot therapy

Family	Species	References
Calliphoridae	*Calliphora vicina*	Teich and Myers (1986)
	Chrysomya rufifacies	
	Lucilia caesar	Baer (1931) and McClellan (1932)
	Lucilia cuprina	Fine and Alexander (1934)
	Lucilia illustris	Leclercq (1990)
	Lucilia sericata	Baer (1931)
	Phormia regina	Baer (1931)
		Horn et al. (1976)
		Horn et al. (1976)
		Robinson (1933)
		Reames et al. (1988)
	Protophormia terraenovae	Leclercq (1990)
Sarcophagidae	*Wohlfahrtia nuba*	Grantham-Hill (1933)
Muscidae	*Musca domestica*	

References indicate the first known use of a specific fly species for this purpose

Fig. 2.2 Third instars of *Lucilia sericata*

of which the organism will have transformed (metamorphosed) and emerged (eclosed) as an adult fly (Greenberg 1973) (Fig. 2.3).

At higher temperatures (37 °C) the newly hatched larvae enter their pre-pupal stage within 48 h.

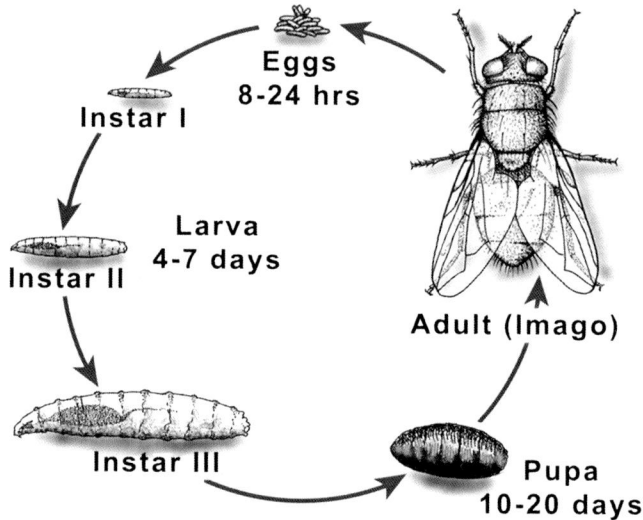

Fig. 2.3 The life-cycle of *Lucilia sericata*

2.5 Mechanisms of Action

The following mechanisms of action have been observed:

2.5.1 Debridement

The most obvious benefit of maggot therapy is the ability of medicinal maggots to effectively debride wounds by removing the sloughy, necrotic tissues. This process is accomplished both physically and chemically. Extracorporeal digestion by the maggots' proteolytic enzymes is one mechanism by which wounds are cleaned. Secreted collagenases and trypsin-like and chymotrypsin-like enzymes have been described (Vistnes et al. 1981; Chambers et al. 2003; Horobin et al. 2003, 2005).

Each maggot is capable of removing 25 mg of necrotic material within just 24 h (Mumcuoglu 2001) (Fig. 2.4).

Maggots secrete their digestive juices directly into their environment, and these proteolytic enzymes are largely responsible for removing the infected, dead tissue from the wound by liquefying the tissue into a nutrient-rich fluid that can be imbibed by the maggots. These proteases may be the vehicle by which the maggots precipitate other wound healing effects, as well. Proteolysis is involved in tissue repair, haemostasis, thrombosis, inflammatory cell activation and tissue reconstruction. Proteinases (mainly matrix metalloproteinases, such as the serine proteases)

Fig. 2.4 Increase in weight of the *Lucilia sericata* larvae at 37 °C

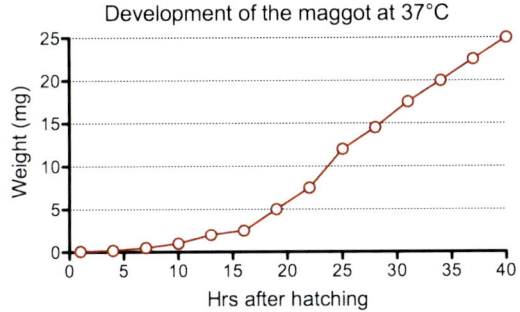

Development of the maggot at 37°C

are involved in collagen degradation, keratinocyte migration, and activation of endothelial cells, fibroblasts, keratinocytes and platelets, through proteinase-activated receptors (Chambers et al. 2003; Brown et al. 2012; Pritchard et al. 2012; Telford et al. 2012).

In addition to the serine proteases, two other classes of proteolytic enzymes are also present in *L. sericata* larval secretions (Chambers et al. 2003); but the predominant activity belongs to the trypsin-like and chymotrypsin-like serine proteases. All together, the larval proteinases are active across a wide pH range (pH 5.0 to pH 10.0).

In addition to enzymatic debridement, maggots also exert a physical debridement over the wound bed (Barnard 1977; Thomas et al. 2002). Blow fly larvae are covered by spines and two mouth-hooks, which aid in locomotion. As the maggots crawl about the wound, these rough structures loosen debris just like a surgeon's "rasper."

2.5.2 Disinfection

Since the natural habitats of blow flies are corpses, excrement, wounds, and similar decaying organic matter, it is obvious that they must be resistant to microbial attack. With the work of William Baer (1931), the antimicrobial activity of maggots began to be appreciated. Early theories were that the maggots killed bacteria through ingestion (Livingston and Prince 1932; Robinson and Norwood 1933, 1934). Greenberg (1968) demonstrated bacterial killing within the maggot gut, and provided evidence that the activity was due, at least in part, to metabolic products of *Proteus mirabilis* (a commensal gram-negative bacterium within the larval gut. Erdmann and Khalil (1986) went on to identify and isolate two antibacterial substances (phenylacetic acid and phenylacetaldehyde) produced by *P. mirabilis* that they isolated from the gut of screwworm larvae (*Cochliomyia hominivorax*).

Mumcuoglu et al. (2001) added to our understanding of alimentary disinfection, and demonstrated that *P. mirabilis* is not required for the process, by feeding green fluorescent protein-producing *Escherichia coli* to *L. sericata*, and using a laser

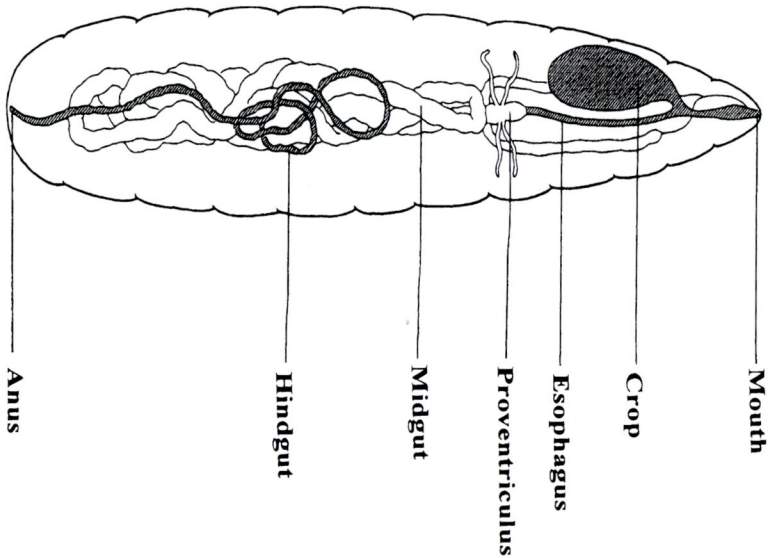

Fig. 2.5 The digestive tract of maggots from *Lucilia sericata*

Fig. 2.6 The crop of the maggot packed with large numbers of fluorescent *Escherichia coli* bacteria

scanning confocal microscope to detect the fluorescence as the bacteria traveled through the maggot's gut (Fig. 2.5). The number of bacteria, very high in the crop (Fig. 2.6), decreased significantly in the mid- and hindgut, with essentially no bacteria reaching the end of the gut (anus) (Fig. 2.7).

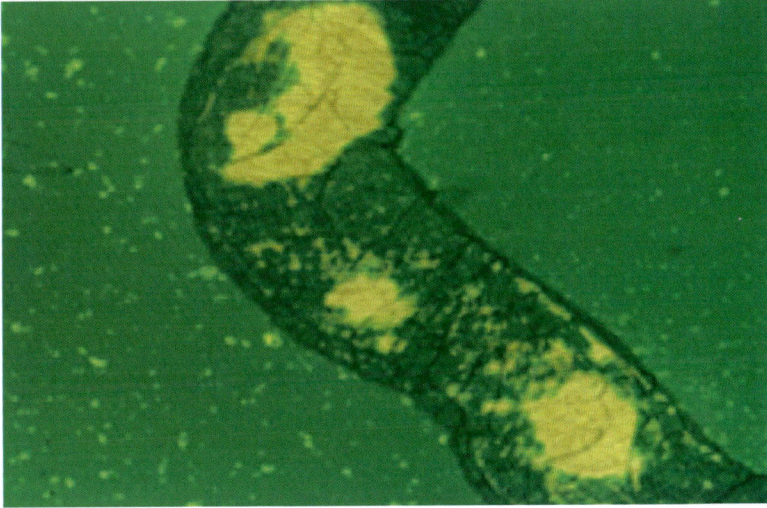

Fig. 2.7 The hindgut of the maggot with decreased numbers of fluorescent *Escherichia coli* bacteria

Antimicrobial killing also occurs outside the maggot gut in the wound bed itself. Some early researchers pointed to the fact that the profuse exudates might be washing the bacteria out of the wounds. Changes in the alkalinity of the wound were also felt to be a mechanism of bacterial killing (Baer 1931) as well as the ammonium products themselves (Messer and McClellan 1935; Robinson and Baker 1939).

Simmons (1935) and Pavillard and Wright (1957) demonstrated that excretion of the antimicrobial gut contents was probably the major mechanism by which the maggots were killing bacteria even before ingesting them.

A preliminary study demonstrated that sterile *L. sericata* secretions exhibit marked antimicrobial activity against liquid cultures of the Gram positive *Streptococcus* sp., *Staphylococcus aureus* and a clinical strain of methicillin-resistant *S. aureus* (MRSA) (Thomas et al. 1999).

Bexfield et al. (2004) and Kerridge et al. (2005) isolated two antibacterial fractions from maggots of *L. sericata*; one with 0.5–3.5 kDa and the second <500 Da. It was shown that these antibacterials are active against a range of bacteria, including the Gram positive *S. aureus*, both methicillin-resistant *S. aureus* (MRSA) and methicillin-sensitive *S. aureus* (MSSA), *Streptococcus pyogenes* and, to a lesser extent, the Gram negative *Pseudomonas aeruginosa*.

Huberman et al. (2007b) isolated three molecules with a molecular weight (MW) of 194, 152 and 138 Da, having antibacterial activity. The antibacterial activity was observed in extracts of whole body, haemolymph and in the excretions of the maggots, and it was effective against a large number of pathogenic and non-pathogenic bacteria (Huberman et al. 2007a). By examining the influence of an infected environment and physical injury on maggots, it was found that the activity was higher in non-sterile reared maggots and even more so in injured maggots when compared with sterile ones, showing that maggots produce higher quantities of antibacterials in the presence of bacteria. Higher antibacterial levels were observed in *L. sericata* maggots removed from chronic wounds of patients. The low MW molecules found in this study have an antibacterial activity against Gram-positive bacteria including methicillin susceptible Staphylococcus aureus (MSSA) and MRSA and Gram-negative bacteria such as *P. aeruginosa, Serratia marcescens, E. coli* and *Klebsiella pneumoniae*, which are also found in chronic wounds. This is in agreement with the findings of other authors, who showed that maggot excretions/secretions are capable of neutralising bacteria such as *E. coli, P. aeruginosa*, MSSA and MRSA. The destruction of the bacteria by the low molecular fractions started after 2 min, and over 90 % of the bacteria were destroyed within 15 min. The quick influx of K^+ and changes in the membrane's potential showed that the low molecular fraction caused bacterial lysis, which was also demonstrated by scanning electron microscopy (SEM) (Huberman et al. 2007a, b).

Apparently, these molecules with low MW are a first-line defense mechanism against the intrusion of bacteria. Unlike antibacterial peptides, they exist already in sterile insects and can be recruited immediately upon invasion of any bacteria. The antibacterial peptides which take longer to be synthesized (Kerridge et al. 2005), strengthen the activity of the former.

Kawabata et al. (2010) applied sterile larvae of *L. sericata* to a test tube containing a bacterial suspension of *S. aureus* or *P. aeruginosa*. To collect the larval extracts, the incubated larvae were cut into multiple pieces with scissors, transferred to a test tube containing PBS and centrifuged. The supernatant was used to test for antibacterials activities. The results showed that infected larvae had better antibacterial capacities than sterile larvae. Antibacterial activities were induced by pretreatment with a single bacterial species, *S. aureus* or *P. aeruginosa*, within 24 and 12 h, respectively, and disappeared after 36 h. The activities were effective against *S. aureus*, but not against *P. aeruginosa*.

Kruglikova and Chernysh (2011) isolated two groups of antibacterial compounds from the excretion: polypeptides with molecular masses ranging from 6.5 to 9 kDa and small molecules with molecular masses ranging from 130 to 700 Da. The polypeptides characterized by the masses of 8.9 and 9 kDa and showing selective activity against Gram negative bacteria correspond well to dioptericins, antimicrobial peptides previously found in the hemolymph of Calliphoridae maggots and known to be part of the immune response to bacterial pathogens.

Ceřovský et al. (2010) were the first to completely sequence a 40-residue defensin-like antimicrobial peptide from *L. sericata*, which they called "lucifensin." Subsequently, the researchers managed to synthesize the molecule by several different methods (Ceřovský et al. 2011). Research continues to isolate other lucifensin-like molecules, since, in other animals where one defensin has been found, the discovery of other related molecules soon followed.

Other high-molecular compounds with masses 6.5, 6.6, 5.8 and 8.6 Da have no clear analogs among antimicrobial peptides present in the hemolymph. The nature of small molecules present in the excretion awaits further study. Thus, the diversity of antimicrobial compounds discovered in *Lucilia* excretion demonstrates a sophisticated strategy that helps the maggots to fight bacteria and other microorganisms settling their environment. The strategy combines secretion of a set of antibacterial peptides involved in insect immune response as well as molecules which function outside the host organism.

Barnes et al. (2010) studied the antibacterial activities of excretions/secretions (ES) from larvae of a carrion feeding beetle, *Dermestes maculatus*, a detritus feeding beetle, *Tenebrio molitor*, and those of *Calliphora vicina* and *L. sericata*. Viable counts were used to assess time-kill of ES against five bacterial species, *S. aureus, E. coli, Bacillus cereus, P. aeruginosa* and *Proteus mirabilis*. The two blowflies were more effective in controlling a wider range of both Gram-positive and Gram-negative bacteria.

Maggot secretions and excretions contain molecules which not only kill bacteria directly, but also molecules which dissolve biofilm and inhibit the growth of new biofilm (Cazander et al. 2009, 2010). At least two types of biofilm have been tested: *S. aureus* and *P. aeruginosa*. Anti-biofilm activity is a very important discovery because biofilm appears to be an important mechanism by which bacteria evade both the body's immune system and also antibiotic activity.

Today, we have not only laboratory evidence for antimicrobial killing, but also clinical evidence that the maggots are disinfecting the wounds. In a study conducted by Bowling et al. (2007), it was demonstrated that MRSA was eliminated from all but one of 13 MRSA-colonized diabetic foot ulcers (92 %) after a mean of three maggot therapy applications (within an average of 19 days). Tantawi et al. (2007) and Contreras-Ruiz et al. (2005) demonstrated a significant decrease in the number of microbial organisms and species after maggot therapy. Armstrong et al. (2005) demonstrated the clinical relevance of this by establishing a case-controlled study of maggot therapy for lower extremity wounds in hospice patients: maggot therapy resulted in fewer clinical assessments of infection, less use of antibiotics, and fewer amputations.

2.5.3 Growth Stimulation

William Baer observed that wounds debrided with maggots heal normally thereafter. But if maggot therapy is continued beyond the point of complete debridement, then the wounds heal even faster than expected (Baer 1931). Maggot-induced wound healing has fascinated therapists and researchers ever since.

Many theories have arisen to explain the rapid growth of granulation tissue and the rapid wound closure associated with maggot therapy. Some researchers speculated that the maggots facilitated normal healing simply by removing the debris and infection that was impairing wound healing (Robinson and Norwood 1934). Others speculated that the simple action of maggots crawling over the wounds stimulated the healing process (Buchman and Blair 1932). This is a reasonable explanation, as we now know that such physical stimulation could very well cause the release of growth factors from the wound bed. Yet, for decades, researchers have believed there were other mechanisms at play, and the search to explain the rapid healing of maggot-treated wounds has continued through the present day.

Clinically, maggot therapy continues to be associated with increased rates of granulation tissue formation (Sherman 2002b, 2003) and faster wound closure (Markevich et al. 2000; Sherman 2002b, 2003; Armstrong et al. 2005) in all but one controlled study (Dumville et al. 2009). But in the latter study, maggot therapy was used only for debridement, and then withdrawn instead of continued; so perhaps it simply proved Baer's observations about normal healing rates following maggot debridement.

Laboratory investigations to explain these observations of enhanced wound healing have also continued ever since Baer's early work. In the 1930s, the alkalinity of maggot-treated wounds, along with the isolated allantoin and urea-containing compounds were believed to be responsible for wound-healing (Robinson 1935b). In fact, today, allantoin and urea are still components of many cosmetics.

More recently, maggot secretions have been shown to stimulate the proliferation of fibroblasts (Prete 1997) and endothelial tissue, in tissue culture. Biopsies of treated wounds reveal profound angiogenesis and luscious granulation tissue (Sherman 2002a). Wollina et al. (2002) demonstrated increased vascular perfusion and tissue oxygenation in patients treated with maggot therapy, using remittance spectroscopy to evaluate the patients before and after maggot therapy.

In 3-dimensional gel-based wound models, maggot secretions significantly enhanced fibroblast migration over the wound surface (Horobin et al. 2003, 2005), apparently as a result of proteolytic cleavage of fibronectin and other molecules which otherwise anchor the fibroblasts to each other and to the wound edges. These migrating fibroblasts are then free to replicate in all directions (not just from the margins of the wound towards the center). Since fragmented fibronectin may itself be bioactive in the wound-closure process, this, too, may be another mechanism by which the maggots stimulate wound healing.

2.6   Clinical Use of Maggot Therapy

2.6.1   Indications

In the U.S., the production of medicinal maggots is regulated by the FDA. The FDA-cleared indications are for debriding non-healing necrotic skin and soft tissue wounds, including pressure ulcers, venous stasis ulcers, neuropathic foot ulcers, and non-healing traumatic or post surgical wounds (hence the acronym MDT). Maggot therapy has also been used to treat necrotic/sloughy wounds associated with a wide variety of pathology (including burns, arterial ulcers, Burger's disease, cellulitis, lymphostasis, neuropathies, osteomyelitis, mastoiditis, thalassemia, polycythemia, and necrotic tumors), and to stimulate and close clean but non-healing wounds (Mumcuoglu et al. 1997, 1999; Angel et al. 2000; Namias et al. 2000; Wollina et al. 2002; Bowling et al. 2007; Sherman et al. 2007a; Gilead et al. 2012).

2.6.2   Contraindications and Adverse Events

Maggot therapy is relatively safe, but complications ("adverse events") are possible. The most common complaint is pain, occurring in 6–40 % of reported patients. When first applied, medicinal maggots are usually too small to be felt, however as they grow larger (generally after the first 24 h of therapy), the movements of their rough exoskeleton and their two hook-like teeth, which are used for locomotion, can be appreciated by those with sensate wounds. Patients most likely to feel discomfort or pain with maggot therapy are easily identified, as they are the patients with wound pain prior to MDT. Once the larvae are satiated (at about 48–72 h), they will spend all of their time trying to escape, by prying the dressing off the skin or wound bed. Thus, MDT-associated pain is also associated with leaving the dressings on the wound for too long (beyond the point when the maggots have finished feeding (Dumville et al. 2009)). Pain is usually well controlled with analgesics; but if not, removal of the dressing and release of the maggots will immediately put an end to the discomfort.

In Israel, 30–35 % of the patients with superficial, painful wounds complain of increased pain during treatment with maggots, which can be treated successfully with analgesics (Mumcuoglu 2001; Mumcuoglu et al. 1998, 2012). Sherman et al. (2000) reported that during MDT, the most common patient complaint was the physical discomfort. Although treatment-associated pain has been reported in only 6 % of nursing home patients (n = 113), pain was reported in 38.1 % (n = 21) of ambulatory patients. Wollina et al. (2002) used MDT to debride the wounds of 30 patients with chronic leg ulcers of mixed origin. Twelve patients (40 %) reported temporary pain, but only two needed analgesic treatment. Wolff and Hansson (2003) treated 74 patients with necrotic or sloughy chronic ulcers of different etiologies and found that maggots effectively debrided 86 % of the necrotic ulcers, and a single application was clinically beneficial in two-thirds of the patients. One-quarter of the study group experienced less pain during treatment, while 41 % felt no difference in pain. Although 34 % noted an increase in pain, most of these patients wanted to continue the treatment because of subjective and objective improvement of the wound.

Steenvoorde et al. (2005a) determined pain levels in patients treated with MDT using a visual analogue scale. Paracetamol (acetaminophen; 1 g/3× daily) and fentanyl patch (25 μg/h every 3 days and 50 μg/h a day before the maggot challenge) were administered for pain relief in the outpatient clinic. Diabetic patients experienced the same amount of pain before and during MDT, while 8 out of 20 non-diabetic patients experienced more pain during MDT than before. The difference between diabetic and non-diabetic patients was statistically significant for all applications combined. In 78 % (n = 37) of patients, pain was adequately treated with analgesic therapy. The authors concluded that a standardized but individually tailored pain management protocol is mandatory.

MDT associated pain is probably the result of the maggots' movements over the wound surface, especially because the maggots use their two mouth-hooks to pull their body forwards, and because the cuticular layer of their body is covered with thorn-like hairs. In addition, the secretion/excretion products of the maggots, which include proteolytic enzymes, might have an influence on the exposed nerves and nerve endings of an open wound (Mumcuoglu et al. 2012).

Since patients prone to MDT-associated pain can readily be identified (they have pain before maggot therapy is even initiated), several interventions are generally undertaken to prevent or minimize additional discomfort during therapy: application of the maggots for shorter periods of time, e.g. 6–8 h, mainly during the daytime hours, use of maggot containment bags (which restrict the maggots' access to the wound bed), and application of smaller maggots and/or fewer maggots.

The second most common problem associated with maggot therapy is late delivery of the maggots. Delivery delays and exposure to extreme temperatures decrease the survival of these young, starving larvae. The shipping container may restrict access to oxygen, especially when designed to be an air-tight thermal insulator. Even though medicinal maggots are generally shipped by overnight courier (and timed to be used within 24 h of arrival), post-marketing studies in the United States demonstrate that 1–2 % still arrive late or dead (unpublished data, RAS).

Occasionally the maggot dressings may come loose, especially if left in place for more than 48 h. Larvae which manage to escape often will pupate under furniture or between mattresses, and emerge from their hiding places 1–2 weeks later as adult flies. Although these flies are not yet mature enough to lay eggs, they can be a nuisance in the clinic or hospital. Moreover, "used maggots" and flies are essentially mobile fomites. There is no data on the frequency of this complication, but clearly it is related to the experience of the therapist in making secure dressings, and the length of time that those dressings are left in place (considering that once the maggots are satiated, they will instinctively attempt to break out of the dressing, leave the wound, and find a secluded place to pupate).

Patient anxiety (the "yuck factor") is more frequently discussed than encountered. In studies of patients with chronic wounds, most patients were quite accepting of maggot therapy (Sherman and Shimoda 2004; Steenvoorde et al. 2005b).

Maggot therapy should not be performed on patients with allergies to the maggots, media or dressing materials used. Media and dressings may need to be specially adapted for those patients. No allergic reactions related to maggots have been described. Maggot therapy should not be performed in patients whose wounds are aggressively advancing, or need around-the-clock inspection (such as anaerobic or mixed aerobic-anaerobic soft tissue infections and fasciitis). The appropriate treatment for these life- and limb-threatening infections is immediate surgical drainage and/or resection. MDT can be a useful ancillary treatment, post-operatively; but should not be administered until all emergency surgical interventions have been delivered.

Maggot therapy is not recommended for use in cases where there is an opening to internal organs (open peritoneum or chest cavity), although successful use of maggot therapy in these situations has certainly been reported (Sherman et al. 2007a).

Therapists may reduce the anxiety of ambulatory patients by providing 24 h/day telephone access to medical assistance.

Escaping maggots could be unpleasant for patients, their relatives and medical staff. Care should be taken to restrict the maggots to the area of the wound, using appropriate dressings. In cases where escaping maggots could cause damage to sensitive structures (i.e. wounds near the eyes or mouth) or where confinement dressings are particularly difficult due to excessive perspiration or soiling (i.e. coccygeal ulcers), use of containment dressings could be helpful.

The digestive enzymes of the maggots can cause erythema or cellulitis. Therefore, the maggots should be restricted to the wound bed, and the periphery of the wound should be protected by the plaster or hydrocolloid dressing. Although maggots will always look for necrotic tissue and slough as a food source, they should not be left on a completely debrided tissue, since they could damage living tissue in their attempt to escape or scavenge for necrotic tissue.

A special device should be built for ambulatory patients with plantar wounds in order to prevent them from squashing the maggots. It is possible that ammonium salts produced by maggots (Robinson and Baker 1939), if not adequately absorbed by the maggot dressing, could precipitate an increase in the patient's body temperature. In rare cases, bleeding of the wound has been observed during maggot therapy, especially if surgical debridement was performed shortly before. If non-sterile maggots are used, there is a danger of bacteremias and/or sepsis (Nuesch et al. 2002).

Fetor has been noted by a few patients and therapists, and probably results from the liquefaction of necrotic tissue, followed by the vaporization of volatile products of putrefaction. Ammonium salts excreted by maggots and the odor of bacteria such as *Pseudomonas* might contribute to fetor, which can be minimized or eliminated by frequently changing the absorbent gauze covering the maggot dressing (Mumcuoglu 2001).

Although the standard of care for bone infection is surgical resection, when surgery is not feasible, maggot debridement may be a reasonable option.

Maggot therapy will dissolve the necrotic soft tissue, even when it surrounds or involves the artery wall. If this occurs, blood vessels – especially those under pressure, such as arteries – may leak blood or perforate. Coaggulopathy (natural or induced) increases the risk of bleeding even further (Steenvoorde and Oskam 2005). Often, the preferred treatment of a necrotic wound involving major vessels is surgery, because it allows direct visualization of the vessels during debridement. However, when this is not feasible, maggot therapy can be performed in the context of potential rupture of a major vessel as long as the patient remains under close observation, i.e., in the hospital or intensive care unit, if necessary.

Without adequate blood flow, even the best-cleaned (debrided) wound will not heal. In fact, it is often argued that a wound that no longer is "protected" by its shell of hard, leathery dead tissue (eschar) is more susceptible to microbial invasion and spread. However, the precise definition or assessment of that critical level of blood flow remains elusive. In published studies of wounds with arterial insufficiency or for any other reason scheduled for amputation, when maggot debridement was performed as a last resort, 40–60 % of wounds healed, and the patients avoided amputation (Mumcuoglu et al. 1998; Sherman et al. 2001; Jukema et al. 2002;

Sherman and Shimoda 2004; Armstrong et al. 2005). Therefore, it is well justified to offer a trial of maggot therapy, even in patients with severe arterial insufficiency, as long as they can be appropriately monitored and treated if maggot therapy is not successful. Such patients may require peri-debridement antibiotic therapy to avoid bacteremia or cellulitis or even surgical intervention.

Maggot therapy should not be left to unproven or inadequately disinfected larvae. Strain differences may affect efficacy and safety, and even experienced laboratories have witnessed infectious complications in their patients as a result of inadequate processing and disinfection (Nuesch et al. 2002).

The administration of systemic antibiotics is not a contraindication for maggot therapy (Sherman et al. 1995a; Peck and Kirkup 2012). Antibiotics may be needed to prevent serious infections from spreading to the neighboring skin (cellulitis) or bloodstream (bacteremia). Topical antibiotic ointments should not be used, however, as these may coat and plug the maggots' spiracles, through which they obtain oxygen.

Therapists should always read and follow the packaging information accompanying the disinfected maggots. Like any other treatment, the risks associated with maggot therapy must be weighed against its potential benefits and against the risks of alternative strategies.

2.6.3   How to Apply Medicinal Maggots

Maggot dressings are intended to maintain the medicinal maggots on the wound in a manner that will be acceptable to the patient and therapist. Accordingly, the dressing should keep the maggots from wandering off, prevent the wound drainage from pooling on the skin (which could lead to maceration, inflammation or dermatitis), and prevent the maggots from drowning, suffocating, or otherwise becoming injured during treatment. The dressing must allow oxygen to reach the maggots, must allow the egress of fluid from the wound, should be relatively inexpensive to produce and simple to maintain, and should prevent the premature escape of maggots (Sherman 1997).

There are two general types of maggot dressings:

Confinement Dressings. Confinement dressings restrict the maggots to the wound area, but allow them complete and free access to the wound tissue. They are sometimes referred to as "cage dressings," or "free range" maggot dressings. The dressings usually are comprised of a foundation or "fence" on the skin that surrounds the wound, topped by a porous polyester net. The maggots are placed inside (at a dose of 5–10 larvae/cm^2) along with some light wet gauze, before sealing the netted roof of this dressing cage (Fig. 2.8). An absorbent pad (i.e. gauze pad) is then held in place just over the porous net, to collect the drainage. This absorbent pad should be changed at least once daily, but more frequently if it becomes soiled

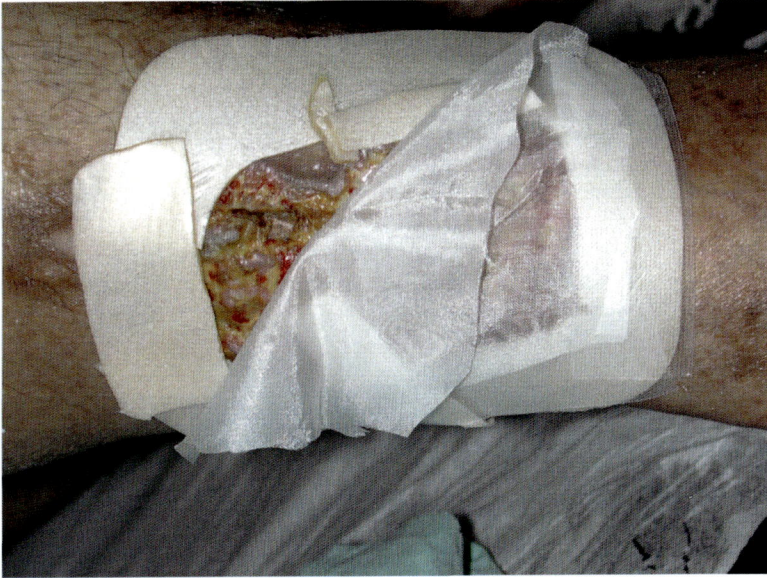

Fig. 2.8 Confinement dressing for maggot therapy

(always follow manufacturer guidelines). The cage dressing itself, however, should be left in place until the maggots are satiated, at about 28–72 h. At that time, the maggots and their cage dressings are all removed. Confinement dressings can be constructed at the bedside, from readily available materials (Sherman et al. 1996; Sherman 1997; AMCICHAC 2010) or they can be purchased ready-made, specifically manufactured for maggot therapy (LeFlap and LeSoc, by Monarch Labs [Irvine, CA]; AgilPad by Agiltera [Dormagen, Germany]).

Containment Dressings. Containment dressings completely contain or envelop the maggots (Grassberger and Fleischmann 2002). These bags may be constructed from simple polyester net (e.g. Biobag by BioMonde [Bridgend, Wales]) or they can be constructed from polyvinyl alcohol foam (Vitapad, by BioMonde). The sealed maggot-bags are then placed on the ulcer bed, covered with a light gauze wrap, and left in place for 3–5 days.

Each dressing style has its advantages and disadvantages (Table 2.2). Limited clinical (Steenvoorde et al. 2005c) and laboratory (Thomas et al. 2002; Blake et al. 2007) comparisons have been made, but only once have confinement and containment dressings been compared in a prospective clinical trial (Dumville et al. 2009). The results of that study demonstrated no significant differences other than faster debridement with the confinement dressings (average of 14 days) compared to bagged maggots (debridement achieved in an average of 28 days).

Table 2.2 Advantages and disadvantages of confinement and containment dressings

Characteristics	Confinement dressings	Containment dressings
Efficacy	Able to access and more efficiently debride undermined areas, sinus tracks, and other crevices	Valuable for wounds near eyes, mouth, or other sensitive cites where it is imperative to avoid escapes
Debridement efficiency	+++	++
Wound pain	More painful	Less painful
Escaping	Inexperienced therapists report more escapes from confinement dressings	Experienced therapists report escapes equally between the two dressing types
Aesthetic acceptability	Less acceptable	More acceptable
Application time	24–72 h	72–96 h
Cost	Less	More

2.6.4 Case Reports

Patient #1. A 68-years old female with venous stasis had a painful chronic wound on her right leg for 24 months, during which she was treated with several conventional debridement methods (Fig. 2.9a). Complete debridement of the wound was achieved within 2 weeks, with six treatment cycles (each for 24 h) of 50–300 confined maggots (Fig. 2.9b).

Patient #2. A 66-years-old male with venous stasis presented to the Department of Dermatology of the Hadassah Hospital in Jerusalem with a chronic wound, which first appeared 6 months earlier (Fig. 2.10a). The wound was completely debrided after four maggot cycles within a week (Fig. 2.10b), each lasting 24–48 h.

2.7 Maggot Therapy in the Veterinary Medicine

Many small animals succumb to complications of serious wounds. Sometimes infection and sepsis overwhelm the animal, while sometimes the costs of intensive care overwhelm the owner. In a study conducted by Sherman et al. (2007b), all eight US veterinarians who had been provided with medicinal maggots were surveyed to determine if this treatment was being used for small animals, and for what indications. At least two dogs, four cats and one rabbit were treated with maggot therapy between 1997 and 2003. The most common indications for using maggot therapy were to effect debridement and control infection, especially if the wound failed to respond to conventional medical and/or surgical therapy. Between 1997 and 2003, 13 horses were treated by 8 veterinarians who used MDT to control infection or debride wounds, which could not easily be reached surgically or were not responding to conventional therapy. Seven animals were lame, and six were expected to require

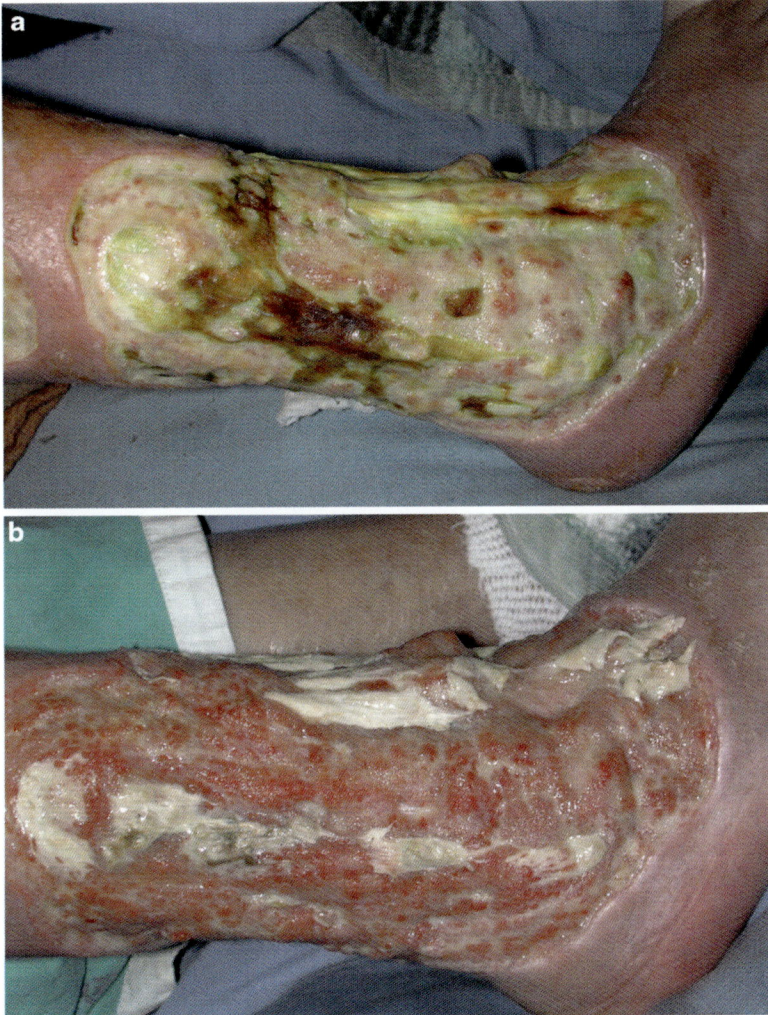

Fig. 2.9 (a) The wound of patient #1, with venous stasis before maggot therapy was initiated. (b) The wound of the same patient after six cycles of maggot therapy. The *whitish* areas are remaining necrotic tendons

euthanasia. Following maggot therapy, all infections were eradicated or controlled, and only one horse had to be euthanized. No adverse events were attributed to maggot therapy for any of these cases, other than presumed discomfort during therapy (Sherman et al. 2007c). The data collected suggested that maggot therapy was useful for treating some serious equine hoof and leg wounds. Practitioners reported the treatment to be safe and often beneficial. Amputation and euthanasia may have been avoided. It was concluded that maggot therapy may have utility for small animals, and should be evaluated further.

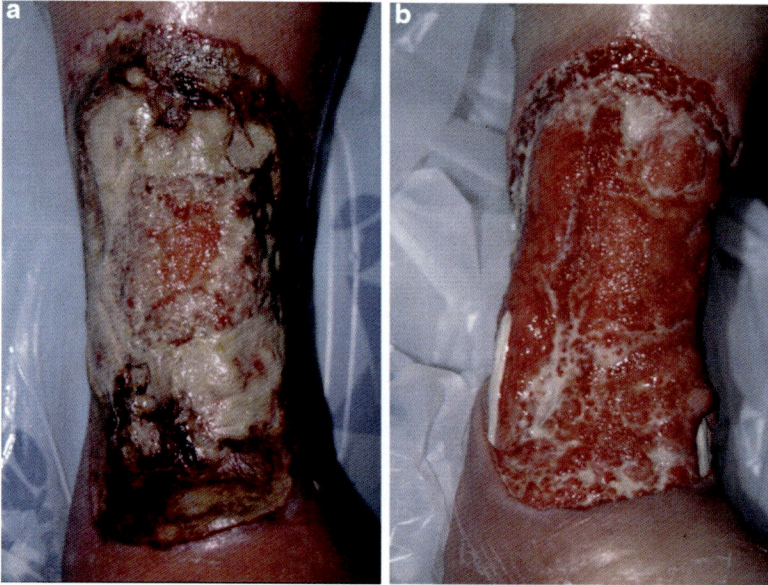

Fig. 2.10 The wound of the patient #2, before (**a**) and after (**b**) maggot treatment

2.8 Future of Maggot Therapy

Over the past 75 years, maggot therapy has saved countless lives and limbs. For just as long, researchers have attempted to isolate the therapeutic molecules in hopes of being able to provide "maggot therapy" without the maggots. These efforts are valuable because they will yield a better understanding of the mechanisms of maggot therapy, as well as a better understanding of wound healing itself. Still, the efficacy of maggot therapy does not lie with a single molecule, nor even a whole family of molecules, but rather a family of molecules in the ideal relative concentrations, combined with the maggot's anatomy and ambulatory behavior that bring these molecules into the recesses of the wound while the maggots physically debride the wound. Perhaps, someday, maggot therapy will be replaced by maggot-derived pharmaceuticals. However, in the short run, it is more likely that we will continue to improve the delivery of maggot therapy by extending their shelf-life, shortening their delivery time, preventing or mitigating the occasional discomfort, and simplifying dressing application. Additional comparative, clinical studies are necessary to further define the advantages of maggot therapy in general, and to investigate the medical utility of species other than *L. sericata*.

In the near future, maggot therapy will continue to spread around the world, as more and more countries recognize the benefits (Contreras-Ruiz et al. 2005). For now, thousands of patients each year are already singing praises and gratitude for the lowly maggot.

References

Angel K, Grassberger M, Huemer F, Stackl W (2000) Maggot therapy in Fournier's gangrene – first results with a new form of treatment. Aktuelle Urologie 31:440–443

Armstrong DG, Salas P, Short B, Martin BR, Kimbriel HR, Nixon BP, Boulton AJ (2005) Maggot therapy in "lower-extremity hospice" wound care: fewer amputations and more antibiotic-free days. J Am Podiatr Med Assoc 95:254–257

Asociación Mexicana para el Cuidado Integral y Cicatrización de Heridas A.C. [AMCICHAC – Mexican Association for Wound Care and Healing] (2010) Clinical practice guidelines for the treatment of acute and chronic wounds with maggot debridement therapy. http://aawconline.org/wp-content/uploads/2011/09/GPC_larvatherapy.pdf. Last accessed 26 Aug 2012

Baer WS (1929) Sarco-iliac joint-arthritis deformans-viable antiseptic in chronic osteomyelitis. Proc Int Assemb Inter-state Postgrad Med Assoc North Am 371:365–372

Baer WS (1931) The treatment of osteomyelitis with the maggot (larva of the blowfly). J Bone Joint Surg 13:438–475

Barnard DR (1977) Skeletal-muscular mechanisms of the larva of Lucilia sericata (Meigen) in relation to feeding habit. Pan-Pac Entomol 53:223–229

Barnes KM, Gennard DE, Dixon RA (2010) An assessment of the antibacterial activity in larval excretion/secretion of four species of insects recorded in association with corpses, using Lucilia sericata Meigen as the marker species. Bull Entomol Res 100:635–640

Bexfield A, Nigam Y, Thomas S, Ratcliffe NA (2004) Detection and partial characterization of two antibacterial factors from the excretions/secretions of the medicinal maggot Lucilia sericata and their activity against methicillin-resistant Staphylococcus aureus (MRSA). Microbes Infect 6:1297–1304

Blake FA, Abromeit N, Bubenheim M, Li L, Schmelzle R (2007) The biosurgical wound debridement: experimental investigation of efficiency and practicability. Wound Repair Regen 15:756–761

Bowling FL, Salgami EV, Boulton AJ (2007) Larval therapy: a novel treatment in eliminating methicillin-resistant Staphylococcus aureus from diabetic foot ulcers. Diabetes Care 30:370–371

Brown A, Horobin A, Blount DG, Hill PJ, English J, Rich A, Williams PM, Pritchard DI (2012) Blow fly Lucilia sericata nuclease digests DNA associated with wound slough/eschar and with Pseudomonas aeruginosa biofilm. Med Vet Entomol 26(4):432–439

Buchman J, Blair JE (1932) Maggots and their use in the treatment of chronic osteomyelitis. Surg Gynecol Obstet 55:177–190

Cazander G, van Veen KE, Bouwman LH, Bernards AT, Jukema GN (2009) The influence of maggot excretions on PAO1 biofilm formation on different biomaterials. Clin Orthop Relat Res 467:536–545

Cazander G, van de Veerdonk MC, Vandenbroucke-Grauls CM, Schreurs MW, Jukema GN (2010) Maggot excretions inhibit biofilm formation on biomaterials. Clin Orthop Relat Res 468: 2789–2796

Ceřovský V, Zdárek J, Fucík V, Monincová L, Voburka Z, Bém R (2010) Lucifensin, the long-sought antimicrobial factor of medicinal maggots of the blowfly Lucilia sericata. Cell Mol Life Sci 67:455–466

Ceřovský V, Slaninová J, Fučík V, Monincová L, Bednárová L, Maloň P, Stokrová J (2011) Lucifensin, a novel insect defensin of medicinal maggots: synthesis and structural study. Chembiochem 12:1352–1361

Chambers L, Woodrow S, Brown AP, Harris PD, Phillipes D, Hall M, Church JCT, Pritchard DI (2003) Degradation of extracellular matrix components by defined proteinases from the greenbottle larva Lucilia sericata used for the clinical debridement of non-healing wounds. Br J Dermatol 148:14–23

Chernin E (1986) Surgical maggots. South Med J 79:1143–1145

Church JCT (1996) The traditional use of maggots in wound healing, and the use of larva therapy (biosurgery) in modern medicine. J Altern Complem Med 2:525–527

Church JCT (1999) Larva therapy in modern wound care: a review. Prim Intent 7:63–68

Contreras-Ruiz J, Fuentes-Suarez A, Karam-Orantes M, MdL E-M, Dominguez-Cherit J (2005) Larval debridement therapy in Mexico. Wound Care Can 3:42–45

Dumville JC, Worthy G, Jm B, Cullum N, Dowson C, Iglesias C, Mitchell JL, Nelson EA, Soares MO, Torgerson DJ, on behalf of the VenUS II team (2009) Larval therapy for leg ulcers (VenUS II): randomised controlled trial. Br Med J 338:1047–1050

Erdmann GR, Khalil SKW (1986) Isolation and identification of two antibacterial agents produced by a strain of *Proteus mirabilis* isolated from larvae of the screwworm (*Cochliomyia hominivorax*) (Diptera: Calliphoridae). J Med Entomol 23:208–211

Fine A, Alexander H (1934) Maggot therapy— technique and clinical application. J Bone Joint Surg 16:572–582

Fleischmann W, Grassberger M, Sherman R (2004) Maggot therapy: a handbook of maggot-assisted wound healing, 1st edn. Thieme, Stuttgart

Gilead L, Mumcuoglu K, Ingber A (2012) The use of maggot debridement therapy in the treatment of chronic wounds in hospitalised and ambulatory patients. J Wound Care 21:78–85

Grantham-Hill C (1933) Preliminary note on the treatment of infected wounds with the larva of *Wohlfahrtia nuba*. Trans R Soc Trop Med Hyg 27:93–98

Grassberger M, Fleischmann W (2002) The biobag – a new device for the application of medicinal maggots. Dermatology 204:306

Greenberg B (1968) Model for destruction of bacteria in the midgut of blow fly maggots. J Med Entomol 5:31–38

Greenberg B (1973) Flies and disease, vol. 2. Biology and disease transmission. Princeton University Press, Princeton

Horn KL, Cobb AH Jr, Gates GA (1976) Maggot therapy for subacute mastoiditis. Arch Otolaryngol 102:377–379

Horobin AJ, Shakesheff KM, Woodrow S, Robinson C, Pritchard DI (2003) Maggots and wound healing: an investigation of the effects of secretions from *Lucilia sericata* larvae upon interactions between human dermal fibroblasts and extracellular matrix components. Br J Dermatol 148:923–933

Horobin AJ, Shakesheff KM, Pritchard DI (2005) Maggots and wound healing: an investigation of the effects of secretions from *Lucilia sericata* larvae upon the migration of human dermal fibroblasts over a fibronectin-coated surface. Wound Repair Regen 13:422–433

Huberman L, Gollop N, Mumcuoglu KY, Block C, Galun R (2007a) Antibacterial properties of whole body extracts and haemolymph of *Lucilia sericata* maggots. J Wound Care 16:123–127

Huberman L, Gollop N, Mumcuoglu KY, Breuer E, Bhusare SR, Shai Y, Galun R (2007b) Antibacterial substances of low molecular weight isolated from the blowfly, *Lucilia sericata*. Med Vet Entomol 21:127–131

Jukema GN, Menon AG, Bernards AT, Steenvoorde P, Rastegar AT, van Dissel JT (2002) Amputation-sparing treatment by nature: "surgical" maggots revisited. Clin Infect Dis 35:1566–1571

Kawabata T, Mitsui H, Yokota K, Shino KI, Guma KO, Sano S (2010) Induction of antibacterial activity in larvae of the blowfly *Lucilia sericata* by an infected environment. Med Vet Entomol 24:375–381

Kerridge A, Lappin-Scott H, Stevens JR (2005) Antibacterial properties of larval secretions of the blowfly, *Lucilia sericata*. Med Vet Entomol 19:333–337

Kingu HJ, Kuria SK, Villet MH, Mkhize JN, Dhaffala A, Iisa JM (2012) Cutaneous myiasis: is *Lucilia cuprina* safe and acceptable for maggot debridement therapy? J Cosmet Dermatol Sci Appl 2:79–82

Kruglikova AA, Chernysh SI (2011) Antimicrobial compounds from the excretions of surgical maggots, *Lucilia sericata* (Meigen) (Diptera, Calliphoridae). Entomol Rev 91:813–819

Leclercq M (1990) Utilisation de larves de Dipteres – maggot therapy – en medicine: historique et actualite. Bull Annls Soc belge Entomol 126:41–50

Livingston SK, Prince LH (1932) The treatment of chronic osteomyelitis with special reference to the use of the maggot active principle. J Am Med Assoc 98:1143–1149

Markevich YO, McLeod-Roberts J, Mousley M, Melloy E (2000) Maggot therapy for diabetic neuropathic foot wounds: a randomized study. In: Abstract of the 59th European association for the study of diabetes, Jerusalem, 17–21 September 2000

McClellan NW (1932) The maggot treatment of osteomyelitis. Can Med Assoc J 27:256–260

Messer FC, McClellan RH (1935) Surgical maggots. A study of their functions in wound healing. J Lab Clin Med 20:1219–1226

Mumcuoglu KY (2001) Clinical applications for maggots in wound care. Am J Clin Dermatol 2:219–227

Mumcuoglu KY, Lipo M, Ioffe-Uspensky I, Miller J, Galun R (1997) Maggot therapy for gangrene and osteomyelitis (in Hebrew). Harefuah 132:323–325, 382

Mumcuoglu KY, Ingber A, Gilead L, Stessman J, Friedman R, Schulman H, Bichucher H, Ioffe-Uspensky I, Miller J, Galun R, Raz I (1998) Maggot therapy for the treatment of diabetic foot ulcers. Diabetes Care 21:2030–2031

Mumcuoglu KY, Ingber A, Gilead L, Stessman J, Friedmann R, Schulman H, Bichucher H, Ioffe-Uspensky I, Miller J, Galun R, Raz I (1999) Maggot therapy for the treatment of intractable wounds. Int J Dermatol 8:623–627

Mumcuoglu KY, Miller J, Mumcuoglu M, Friger M, Tarshis M (2001) Destruction of bacteria in the digestive tract of the maggot of Lucilia sericata (Diptera: Calliphoridae). J Med Entomol 38:161–166

Mumcuoglu KY, Davidson E, Avidan A, Gilead L (2012) Pain related to maggot debridement therapy. J Wound Care 21:400, 402, 404–405

Namias NN, Varela E, Varas RP, Quintana O, Ward CG (2000) Biodebridement: a case report of maggot therapy for limb salvage after fourth-degree burns. J Burn Care Rehabil 21:254–257

Nigam Y, Bexfield A, Thomas S, Ratcliffe NA (2006a) Maggot therapy: the science and implication for CAM. Part I—history and bacterial resistance. eCAM 3:223–227

Nigam Y, Bexfield A, Thomas S, Ratcliffe NA (2006b) Maggot therapy: the science and implication for CAM. Part II—maggots combat infection. eCAM 3:303–308

Nigam Y, Dudley E, Bexfield A, Bond AE, Evans J, James J (2010) The physiology of wound healing by the medicinal maggot, Lucilia sericata. Adv Insect Physiol 39:39–81

Nuesch R, Rahm G, Rudin W, Steffen I, Frei R, Rufli T, Zimmerli W (2002) Clustering of bloodstream infections during maggot debridement therapy using contaminated larvae of Protophormia terraenovae. Infection 30:306–309

Paul AG, Ahmad NW, Lee HL, Ariff AM, Saranum M, Naicker AS, Osman Z (2009) Maggot debridement therapy with Lucilia cuprina: a comparison with conventional debridement in diabetic foot ulcers. Int Wound J 6:39–46

Pavillard ER, Wright EA (1957) An antibiotic from maggots. Nature 180(4592):916–917

Pechter EA, Sherman RA (1983) Maggot therapy: the surgical metamorphosis. Plast Reconstr Surg 72:567–570

Peck GW, Kirkup BC (2012) Biocompatibility of antimicrobials to maggot debridement therapy: medical maggots Lucilia sericata (Diptera: Calliphoridae) exhibit tolerance to clinical maximum doses of antimicrobials. J Med Entomol 49:1137–1143

Prete PE (1997) Growth effects of Phaenicia sericata larval extracts on fibroblasts: mechanism for wound healing by maggot therapy. Life Sci 60:505–510

Pritchard DI, Telford G, Diab M, Low W (2012) Expression of a cGMP compatible Lucilia sericata insect serine proteinase debridement enzyme. Biotechnol Prog 28:567–572

Reames MK, Christensen C, Luce EA (1988) The use of maggots in wound debridement. Ann Plast Surg 21:388–391

Robinson W (1933) The use of blowfly larvae in the treatment of infected wounds. Ann Entomol Soc Am 26:270–276

Robinson W (1935a) Progress of maggot therapy in the United States and Canada in the treatment of suppurative diseases. Am J Surg 29:67–71

Robinson W (1935b) Stimulation of healing in non-healing wounds: by allantoin occurring in maggot secretions and of wide biological distribution. J Bone Joint Surg Am 17:267–271

Robinson W, Baker FL (1939) The enzyme urease and occurrence of ammonia in maggot infected wounds. J Parasitol 25:149–155

Robinson W, Norwood VH (1933) The role of surgical maggots in the disinfection of osteomyelitis and other infected wounds. J Bone Joint Surg 15:409–412

Robinson W, Norwood VH (1934) Destruction of pyogenic bacteria in the alimentary tract of surgical maggots implanted in infected wounds. J Lab Clin Med 19:581–586

Sherman RA (1997) A new dressing design for use with maggot therapy. Plast Reconstr Surg 100:451–456

Sherman RA (2002a) Maggot therapy for foot and leg wounds. Int J Low Extrem Wounds 1:135–142

Sherman RA (2002b) Maggot versus conservative debridement therapy for the treatment of pressure ulcers. Wound Repair Regen 10:208–214

Sherman RA (2003) Maggot therapy for treating diabetic foot ulcers unresponsive to conventional therapy. Diabetes Care 26:446–451

Sherman RA, Shimoda KJ (2004) Presurgical maggot debridement of soft tissue wounds is associated with decreased rates of postoperative infection. Clin Infect Dis 39:1067–1070

Sherman RA, Wyle FA, Thrupp L (1995a) Effects of seven antibiotics on the growth and development of *Phaenicia sericata* (Diptera: Calliphoridae) larvae. J Med Entomol 32:646–649

Sherman RA, Wyle F, Vulpe M (1995b) Maggot therapy for treating pressure ulcers in spinal cord injury patients. J Spinal Cord Med 18:71–74

Sherman RA, Tran JM-T, Sullivan R (1996) Maggot therapy for venous stasis ulcers. Arch Dermatol 132:254–256

Sherman RA, Hall MJ, Thomas S (2000) Medicinal maggots: an ancient remedy for some contemporary afflictions. Annu Rev Entomol 45:55–81

Sherman RA, Sherman JM-T, Gilead L, Lipo M, Mumcuoglu KY (2001) Maggot debridement therapy in outpatients. Arch Phys Med Rehabil 82:1226–1229

Sherman RA, Morrison S, Ng D (2007a) Maggot debridement therapy for serious horse wounds – a survey of practitioners. Vet J 174:86–91

Sherman RA, Shapiro CE, Yang RM (2007b) Maggot therapy for problematic wounds: uncommon and off-label applications. Adv Skin Wound Care 20:602–610

Sherman RA, Stevens H, Ng D, Iversen E (2007c) Treating wounds in small animals with maggot debridement therapy: a survey of practitioners. Vet J 173:138–143

Simmons SW (1935) A bactericidal principle in excretions of surgical maggots which destroys important etiological agents of pyogenic infections. J Bacteriol 30:253–267

Steenvoorde P, Oskam J (2005) Bleeding complications in patients treated with maggot debridement therapy. Int J Low Extrem Wounds 4:57–58

Steenvoorde P, Budding T, Oskam J (2005a) Determining pain levels in patients treated with maggot debridement therapy. J Wound Care 14:485–488. Erratum in J Wound Care 2006 15:71

Steenvoorde P, Budding TJ, van Engeland A, Oskam J (2005b) Maggot therapy and the "Yuk" factor: an issue for the patient? Wound Repair Regen 13:350–352

Steenvoorde P, Jacobi CE, Oskam J (2005c) Maggot debridement therapy: free-range or contained? An in-vivo study. Adv Skin Wound Care 18:430–435

Tantawi TI, Gohar YM, Kotb MM, Beshara FM, El-Naggar MM (2007) Clinical and microbiological efficacy of MDT in the treatment of diabetic foot ulcers. J Wound Care 16:379–383

Tantawi TI, Williams KA, Villet MH (2010) An accidental but safe and effective use of *Lucilia cuprina* (Diptera: Calliphoridae) in maggot debridement therapy in Alexandria, Egypt. J Med Entomol 47:491–494

Teich S, Myers RAM (1986) Maggot therapy for severe skin infections. South Med J 79:1153–1155

Telford G, Brown AP, Rich A, English JS, Pritchard DI (2012) Wound debridement potential of glycosidases of the wound-healing maggot, *Lucilia sericata*. Med Vet Entomol 26:291–299

Thomas S (2006) Costs of managing chronic wounds in the UK, with particular emphasis on maggot debridement therapy. J Wound Care 15:465–469

Thomas S, Jones M, Shutler S, Jones S (1996) Using larvae in modern wound management. J Wound Care 5:60–69

Thomas S, Andrews AM, Hay NP, Bourgoise S (1999) The anti-microbial activity of maggot secretions: results of a preliminary study. J Tissue Viability 9:127–132

Thomas S, Wynn K, Fowler T, Jones M (2002) The effect of containment on the properties of sterile maggots. Br J Nurs 11:S21–S28

Vistnes LM, Lee R, Ksander GA (1981) Proteolytic activity of blowfly larvae secretions in experimental burns. Surgery 90:835–841

Wayman J, Nirojogi V, Walker A, Sowinski A, Walker MA (2001) The cost effectiveness of larval therapy in venous ulcers. J Tissue Viability 10:91–94

Whitaker IS, Twine C, Whitaker MJ, Welck M, Brown CS, Shandall A (2007) Larval therapy from antiquity to the present day: mechanisms of action, clinical applications and future potential. Postgrad Med J 83:409–413

Wolff H, Hansson C (2003) Larval therapy– an effective method of ulcer debridement. Clin Exp Dermatol 28:134–137

Wollina U, Liebold K, Schmidt WD, Hartmann M, Fassler D (2002) Biosurgery supports granulation and debridement in chronic wounds – clinical data and remittance spectroscopy measurement. Int J Dermatol 41:635–639

Chapter 3
Hirudotherapy

Olga S. Gileva and Kosta Y. Mumcuoglu

3.1   Introduction

Hirudotherapy – the use of medicinal leeches for curative purposes – is one of the oldest practices in medicine, dating back to the Stone Age. Over the years, the use of leeches has evolved from a simple blood-letting procedure into a scientifically-based physiologic process with rational defined clinical applications. The Greek physician Nicander of Colophon is considered to have been the pioneer of hirudotherapy. While the Romans were the first to use the name *Hirudo* for leeches in the first century A.D., it was the Swedish physician and zoologist Linnaeus who used the term *Hirudo medicinalis* in 1754 to describe the application of leeches in medical treatment. During the Middle Ages, the golden era for blood-letting, leeches were used by almost all physicians to cure anything from headaches to hemorrhoids. In Russia, hirudotherapy reached its zenith in the late eighteenth and early nineteenth century when leech harvesting and leech therapy netted the country an annual six million silver rubles. By the end of the nineteenth century, with the advent of antibacterial therapy, leeches fell out of favor, and became associated with medicinal quackery in the majority of countries.

Leeches have enjoyed a renaissance in reconstructive microsurgery during the last 25 years. Nowadays, medicinal leech therapy is considered an integral part of the armamentarium to salvage vascularly compromised flaps or replants. In the twenty-first century, several clinical studies were performed in Germany and Russia

O.S. Gileva, M.D., Ph.D. (✉)
Department of Preclinic Dentistry and Physiotherapy, Perm State Academy of Medicine,
26 Petropavlovskaya Street, 614990 Perm, Russian Federation
e-mail: ogileva@rambler.ru

K.Y. Mumcuoglu, Ph.D.
Department of Microbiology and Molecular Genetics, Hebrew
University – Hadassah Medical School, P.O. Box 12272, 91120 Jerusalem, Israel
e-mail: kostam@cc.huji.ac.il

M. Grassberger et al. (eds.), *Biotherapy - History, Principles and Practice:*
A Practical Guide to the Diagnosis and Treatment of Disease using Living Organisms,
DOI 10.1007/978-94-007-6585-6_3, © Springer Science+Business Media Dordrecht 2013

31

to demonstrate the healing effect of leeches in the treatment of painful osteoarthritis. Currently, in some countries of Eastern Europe, Russia and Asia hirudotherapy is officially recognized as a classic alternative therapy of diseases and symptoms such as phlebitis, osteoarthritis, hypertension and glaucoma.

The effectiveness and safety of leech therapy is largely determined by the knowledge of leech maintenance, their application on the skin and mucosae, pre- and post-application handling, as well as preparation of the patients for treatment and post-procedural patient management.

The choice of indications and contraindications, the number of leeches applied, optimal techniques, the regimen of leech therapy, supplemental antibac- terial therapy in high-risk patients as well as monitoring of the patient, are of paramount importance.

Further research on the active components of leech saliva, development of medi- cation based on pharmacologically active components of the leech, the search for novel clinical applications, and comparative, double-blind studies in hirudotherapy are needed.

3.2   History of Hirudotherapy (Leeching)

It's almost impossible to know the exact time of the first use of leeches for blood- letting, although archeologists recently dated the existence of blood-letting tools to the Stone Age. There are numerous reports on paintings depicting the application of leeches inside Egyptian Tombs of the 18th Pharaoh Dynasty, but according to Shipley (1927) these issues on using the leeches for blood-letting in Ancient Egypt are rather controversial.

Apparently, leeching as a practice of blood-letting first appeared in ancient China, India, Greece, Rome and pre-Columbian America. In early medicinal practices, blood-letting was applied based on the so called "humoral theory", first written records of which were found in the Greek philosopher Hippocrates' Collection (*Corpus Hippocraticum*) in the fifth century B.C. Hippocrates – the "Father of Medicine" believed that diseases were caused by imbalance of the four main humors of the body (blood, phlegm, black and yellow bile) that could be stabilized by releasing "impure" blood out of the patients.

The very first citations on the medical use of leeches are credited to the Greek physician Nicander of Colophon (185–138 B.C.) in his medical poem *Alexipharmaca*, and to Themison Laodicea (123–43 B.C., a pupil of Asclepius) who are considered as being the pioneers of hirudotherapy. The practice of leeching called "*jalauka- vacharan*" has been mentioned in Ayurveda since 2.500 B.C. (Kumar et al. 2012).

The Romans were the first who used the name *Hirudo* for leeches. In Roman times, the humoral concept of diseases was expanded by prominent physicians and philosophers of the time. Aelus Galenus (129–200) believed that blood-sucking would eliminate the obnoxious substances in the body and restore the balance of its main humors, which had been altered by disease.

The prominent Persian physician – Avicenna (980–1037) considered the application of leeches to be more useful than wet cupping in "letting of the blood from deeper parts of the body". His famous *"Canon of Medicine"* included several pages of instruction about leeches (Grunner 1930). He was also the first physician to use leeches for treatment of skin diseases.

The Middle Ages were the golden era for blood-letting, and leeches were frequently used by almost all doctors to cure everything from headache to hemorrhoids. It is noteworthy that from medieval times through the Age of Enlightenment leeches were really entwined with medicine and the words "leech" (a deviation of the old Anglo-Saxon "laece" meaning "to heal") and "doctor" were synonymous (Shipley 1927).

Ambroise Paré (1510–1590) – one of the most celebrated surgeons of the Renaissance, devoted an entire chapter of his renowned publication *"Oeuvres"* to leeching. Towards the middle of the sixteenth century, Conrad Gessner of Zurich described the medicinal leech in detail and recommended its use in medical practice. In the seventeenth century, Gerome Negrisoli published a work on using the leeches in gynecology.

In France, under the influence of Francois Joseph Broussais (1772–1832), a surgeon in Napoleon's "Grande Armée", the use of leeches spread rapidly. French physicians prescribed leeches to be applied to newly hospitalized patients even before seeing them (Upshow and O'Leary 2000). Thanks to his activities in medical use of leeches, they became the therapeutic agent par excellence and even inspired fashion. Broussais also gave a tremendous impulse to the leech industry: in 1833 only about 42 million leeches had to be imported from Eastern Europe, Russia and Asia Minor, and their annual consumption approached 100 million specimens.

In Russia, hirudotherapy reached its zenith in the late eighteenth and early nineteenth century and was applied by scores of physicians, including such eminent ones as M. Mudrov, I. Dyadkovsky, G. Zakharyin, F. Pasternatsky and N. Pirogov (Gilyova et al. 1999; Kamenev and Baranovskii 2006; Nikonov 1992). It was the period of intensive medical use of leeches as a remedy of blood-letting in the treatment of a multitude of diseases and disorders, ranging from tuberculosis to epilepsy and rheumatism.

In the United States, leeching did not attain such wide popularity as in Europe due to difficulties in obtaining and breeding the European medicinal leech (*H. medicinalis*), which was more popular than its American counterpart – *Hirudo (Macrobdella) decora*, which consumes smaller amounts of blood. It is interesting that in nineteenth century's America, leeches were not limited to blood-letting by physicians but used as a common home remedy to treat gum diseases and hemorrhoids and to relief the pain of large hematomas (Taylor 1860).

By the end of the nineteenth century, with the advent of the era of antibacterial therapy the medical use of leeches declined significantly. Up to the middle of the twentieth century, hirudotherapy, once used by the physicians of emperors and prominent surgeons, became associated with medicinal quackery in most countries.

However, the scientific interest in leeching had not disappeared completely. In 1884, John B. Haycraft, Professor of Physiology in Birmingham, identified an anticoagulant substance from leech saliva, which was later dubbed "Hirudin" from the Latin "*Hirudo*" (Haycraft 1884). Earlier, in 1809, the Russian physician K.I. Diakonov wrote in his article "*Changes of Human Blood in the Leech*" that "...the lack of blood coagulation of human erythrocytes in the intestinal tract of leeches testifies the presence of some dissolving agents...".

Since the 1970s, hirudotherapy witnessed a resurgence of interest among scientists, physicians and patients. In the mid 1950s, Fritz Markwardt studied the anticoagulant substances from *H. medicinalis*, by isolating hirudin and analyzing its mechanism of action as a thrombin inhibitor (Markwardt 1955). Leech therapy was re-introduced in modern medical practice in the 1960s when two Slovenian surgeons Derganc and Zdravic successfully used leeches for tissue-flap transplantation surgery (Derganc and Zdravic 1960). In 1985, the plastic surgeon Joseph Upton from Harvard published his results on using the leeches for relieving venous congestion after ear replantation in a 5-year-old boy–whose ear had been bitten off by a dog. The first medicinal leech farm, Biopharm, was set up in Swansea (South Wales) in 1981 by Dr. Roy Sawyer.

Leeches are now applied to treat a wide array of diseases, while specific applications vary regionally. For example, U.S. and European practitioners emphasize the value of leeches in microvascular and reconstructive surgery in both the pediatric and adult populations, while in some countries of Eastern Europe, Russian and Asian hirudotherapy is officially recognized as a classic alternative therapy for numerous internal diseases including osteoarthritis, phlebitis, hypertension and glaucoma (Gilyova 2005).

3.3 *Hirudo medicinalis*

Although more than 650 species of leeches are known, but only a few are used as medicinal (therapeutic) leeches. *H. medicinalis*, the European medicinal leech, is one of the most extensively studied annelids (*ringed worms*) and the most frequently used in modern medical practice.

The very first description of leeches was in 1554 by G. Rondelet in his publication *Natural History of Leeches*. Later, in 1758 K. Linnaeus used the term *H. medicinalis* to describe the application of leeches in medicine and classified them in his monograph "Systema Naturae". Today, the species *H. medicinalis* is classified as a member of the kingdom Animalia, phylum Annelida, class Clitellata, order Arhynchobdella, family Hirudinidae and genus *Hirudo*.

The preferred habitats of medicinal leeches are warm, shallow, still ponds with abundant amphibian population. However, by the beginning of the twentieth century the medicinal leech had nearly disappeared from most of its former biotopes in western and southern Europe from the Ural mountains to the countries bordering the eastern Mediterranean (Elliot and Tullet 1984; Sawyer 1986). Nowadays, the

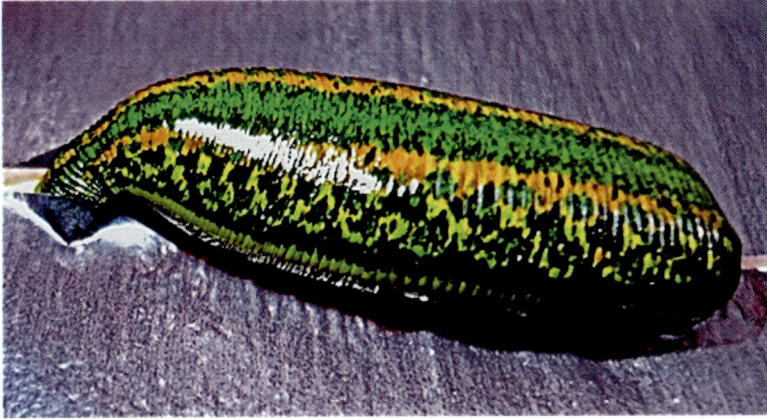

Fig. 3.1 The medicinal leech (*Hirudo medicinalis officinalis*)

European medicinal leech is legally protected by the International Union for Conservation of Nature (IUCN).

Hirudo medicinalis has a segmented elongate and dorso-ventrally flattened body (Fig. 3.1). In extreme extension, the fully mature adults reach 120 mm, with a width of 8–10 mm, while at its maximal contraction it could be 30–35 mm long and 15–18 mm wide. The dorsal surface is usually olive greenish-brown with brownish-black spots and double dotted brown stripes along each lateral side. The ventral surface is usually yellowish-green with black spots, and has a pair of black straight marginal stripes. Leeches are hermaphrodites, i.e., they both exhibit male and female genital organs. In nature *H. medicinalis* breeds once during an annual season, which spans from June through August (Lukin 1976).

Leeches attach themselves to the host by means of their two suckers (anterior and posterior) located at each end of their body. The larger disc-shaped posterior sucker is used for locomotion and attachment whereas the anterior sucker (Fig. 3.2a), consisting of a buccal cavity and jaws, is used for attachment and feeding (Lukin 1976; Sawyer 1986). The leech has three rigid jaws arranged in a tri-radiated configuration. Sixty to 100 pyramidal, sharp teeth on each jaw are used to incise the skin leaving a characteristic Y-shaped bite (Fig. 3.2b). On the side of each tooth, close to the piercing tip, there is an opening through which saliva is injected into the skin (Orevi et al. 2000). After piercing the skin and injecting saliva, the leech starts blood-feeding, gradually releasing saliva chemicals, which are responsible for the therapeutic benefits of hirudotherapy.

The feeding process lasts for about 30–40 min, during this time 10–15 ml of blood are ingested (Sawyer 1986). The blood inside the digestive tract is conserved by endoenzymes, enabling the leech to fast for one and a half years. Digestion of the blood is aided by the presence of endosymbiotic bacteria, such as *Aeromonas hydrophila* and *Pseudomonas hirudinia*, which prevent putrification of the blood and supply enzymes crucial for its digestion.

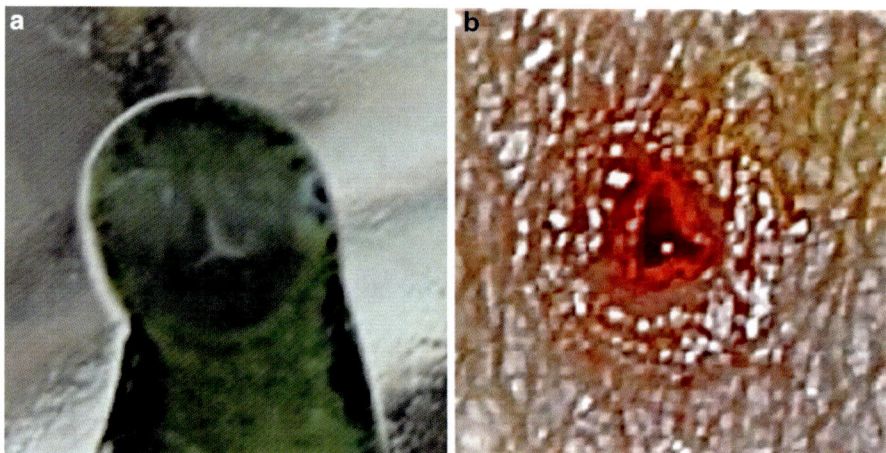

Fig. 3.2 (a) Anterior sucker of the medicinal leech; (b) Y-shaped leech bite on human skin

The excretory system of *H. medicinalis* is formed by 17 pairs of so called nephridia, organs with a function similar to a kidney secreting the urine, which is excreted by nephridiopores. Leeches have no differentiated respiratory system and breathe directly through the epidermis. The nervous system of *H. medicinalis* is composed of a pair of cerebroid ganglions, segmental nerves, sensory papillae, five pairs of eyes and many thermo- and chemoreceptors, helping the leech to orientate in space and react to stimuli (Lagutenko 1981). Leeches appear to be sensitive to water vibration, touch, light, heat, desiccation, sounds and certain chemical substances.

3.4   Mechanism of Action

Hirudotherapy depends on the following main properties of medicinal leeches: the blood-letting action during active suction of blood and passive oozing of the wound, injection of biologically active substances with the saliva into the host, and the neuro-reflex (cuto-visceral) effects of leech applications (Isakhanjan 2003; Zaslavskaya 1940).

The saliva of *H. medicinalis* contains more than 100 bioactive substances, including coagulation inhibitors, platelet aggregation inhibitors, vasodilators, as well as anaesthetizing, antimicrobial and anti-inflammatory agents (Eldor et al. 1996; Fields 1991).

One of the most important ingredients is hirudin, which is the principal anticoagulant responsible for enhanced bleeding and prevention of coagulation. In 1986, hirudin was synthesized through recombinant DNA technology and it is now routinely administered to patients undergoing coronary angioplasty, for the treatment of deep venous thrombosis, and as a substitute for heparin in patients with heparin-induced thrombocytopenia (Eldor et al. 1996; Fields 1991).

In addition to hirudin, leeches secrete two inhibitors of Factor Xa responsible for the conversion of prothrombin to thrombin (Rigbi et al. 1988). The quantity of saliva injected into the skin during the blood feeding of a single leech is enough to prevent coagulation of 50–100 ml of human blood (Heldt 1961).

Furthermore, leech saliva is an effective platelet aggregation inhibitor due to the presence of active ingredients such as calin, apyrase (adenosine diphosphatase), platelet activating factor (PAF)- antagonist, collagenase and prostaglandin. Their main function is to induce secondary bleeding, which can last several hours after the end of each leech session. Calin binds collagen, a natural blood clotting factor that suppresses collagen-induced platelet aggregation (Munro et al. 1991). A mammalian-type collagenase may inhibit collagen-induced platelet aggregation, and increase tissue permeability (Rigbi 1998). Together with plasmin, collagenase helps reduce fibroblast formation in hypertrophic scars and keloids. The PAF-factor antagonist in addition to its role as an aggregation inhibitor can prevent inflammation (Orevi et al. 1992). Apyrase and prostanoid components isolated from leech saliva impede ADP-induced platelet aggregation (Baskova and Nikonov 1987; Rigbi et al. 1996). Destabilase enzyme from the *H. medicinalis* saliva depolymerizes soluble cross-linked fibrin and gives leeches the ability to lyse clots by means of specific fibrinolysis and disaggregation (Baskova and Nikonov 1991). It has been established that large quantities of endo-isopeptides, which are lysed by destabilase enzymes, are detected in affected tissues of patients with atherosclerosis, diabetes mellitus, cataracts and pancreatitis (Gilyova et al. 1988). In addition, destabilase exhibits combined enzymatic and non-enzymatic antibacterial action (Zavalova et al. 2006). The destabilase binds free hirudin and plasma kallikrein inhibitors in a rugged liposome-like destabilase complex that can penetrate through the cell membrane (Nikonov 1996).

Extracts of *H. medicinalis* contain proteinase inhibitors such as bdellins (trypsin-plasmin inhibitors) (Fritz et al. 1969), eglin, inhibitors of α-chymotrypsin, subtilisin, and the granulocytic neutral proteases – elastase and cathepsin G (Seemüller et al. 1977, 1986), responsible for the anti-inflammatory effect of leeching. Eglin is a potential therapeutic agent for the treatment of inflammatory diseases and has been proven effective for the treatment of shock and emphysema in experimental models (Siebeck et al. 1992). Recombinant forms of eglin inhibit NS3-proteinase of hepatatis C virus (Martin et al. 1998).

Medicinal leeches also secrete hirustasin (Hirudo antistasin), which selectively inhibits tissue kallikreins that are largely responsible for maintenance of normal level of blood pressure. Hirustasin can also play a role in the intrinsic coagulation process. The antimetastatic properties of hirustasin, possibly due to inhibition of Factor Xa, are directly activated by tumor cells (Söllner et al. 1994).

The anti-inflammatory and analgesic properties of leeches are subjects of modern hirudobiochemistry and hirudopharmacology and in many respects are associated with the blockage of amidolytic and kininogenase activities of plasma kallikrein, resulting in prevention of pain or pain relief during leech session (Baskova et al. 1992).

It was shown, that experimental rats, which were pre-treated with an aerosol of a leech extract, were less sensitive to pain. The presence of local anesthetics in leech saliva is still controversial, though recently a morphine-like peptide was found in *H. medicinalis* (Salzet 2005). In addition the saliva of *H. medicinalis* exhibits a lipolytic and cholesterol esterase activity that is important in the digestion of the ingested blood (Baskova et al. 1984).

Leeches may also secrete a vasodilative, histamine-like substance, which increases the inflow of blood after a leech bite, and reduces local swelling (Rigbi 1998). Histamine-like biochemicals in leech saliva are also responsible for allergic skin reactions. Accordingly, Leech-Derived Tryptase Inhibitor (LDTI) from leech extracts inhibits proteolytic enzymes of host mast cells, and, possibly is involved in allergic and inflammatory reactions (Sommerhoff et al. 1994). Currently, recombinant forms of tryptase inhibitors are used in clinical practice.

Hyaluronidase (orgelase), known as the "spreading factor", can degrade tissue hyaluronic acid, thus facilitating the infiltration and diffusion of the remaining ingredients of leech saliva into the congested tissue. Tissue permeability, restored with the help of hyaluronidase, promotes the elimination of tissue- and circulatory-hypoxia as well as local swelling (Hobingh and Linker 1999). The role of the combined effect of the salivary components can be summarized as follows: hyaluronidase breaks up the core of proteoglycans and together with collagenase "opens up" the tissue; histamine-like substances act as vasodilatators; kininogenase inhibition and kininase activity prevents pain; coagulation inhibitors and platelet aggregation inhibitors prevent clotting; bdellins and eglins as well as PAF-antagonists prevent inflammation; and antimicrobials promote bacteriostatic and anti-inflammatory effects (Rigbi 1998).

Components of leech secretions, predominantly destabilase and high molecular weight bdellin B, have neurite-stimulating effects in cultures of sensory neurons of chick embryos (Chalisova et al. 2003). Identification of the neurite-stimulating activity of leech saliva provides new therapeutic tools for the treatment of neurodegenerative diseases. The leeches also produced some types of catecholamines (Dopamin, Serotonin and Acetylcholine), affecting the nerve endings. Leech saliva increases the phagocytic activity of blood neutrophils in vitro and possesses anti-complement properties (Baskova et al. 1988).

The application of medicinal leeches to a congested tissue reportedly facilitates blood flow within the affected zone in two ways: directly via active blood-sucking during feeding and indirectly by passive oozing of the blood for as long as 48 h after the leech is removed or detached (Smoot et al. 1990). The persistent bleeding largely potentiates tissue decongestion and leads to loss of blood, relief of capillary net, decrease in venous congestion, decompression of the nerve trunks and endings, increase in lymph flow, positive changes of local hemodynamics, amelioration of hemorheology, increase of oxygen supply, improvement of tissue metabolism, and elimination of tissue ischemia. Powerful anti-edematous and decongestive properties of leeching were shown in a model of acute agarinic inflammation in the feet of Wistar rats, which showed a significant reduction of edema after leech application during all stages of the experimental period, with a more pronounced effect at its earliest stages (3 and 6 h) (Gilyova et al. 2000; Gilyova 2003). Hirudotherapy used on wounds experimentally induced by electric shocks on the skin of Wistar rats, and compared to controls (wounded rats without any treatment), demonstrated clear positive effects on wound healing process by reduction of the severity (size, depth, duration) of local necrotic and inflammatory reactions during the early stages of posttraumatic inflammation; relatively smaller intensity of local intravascular changes in the form of microcirculatory thrombosis, sludge and stasis; acceleration of epithelialization and connective tissue granulation processes; maintenance of the optimal balance of cell number and activity (dominantly of mastocytes and eosino-philes) in the focus of lesion; acceleration of new blood and lymphatic vessel forma-tions; and more rapid restoration of microcirculation and oxygen regimen in affected tissues. Continuous leech applications could have a lymph-stimulating effect, which manifests in the form of increased speed of interstitial humoral transport and lymphatic drainage in the mesentery of experimental rats (Borovaya 2008).

One must take into account the possibility of the neuro-reflex effect of leeching, especially when the leech session is performed in a non-aspirated regimen, i.e., keeping the leech attached on the skin for a short time, during which it injects the first portion of saliva in the wound and takes small quantities of blood. Taking into consideration the existence of skin-visceral connections, Savinov and Chaban (1995), suggested that during leech sucking some part of the active compounds of saliva, reaches the viscera via veins called communicants, promoting the improve-ment of blood circulation of organs, by rendering thrombolytic, antitrombotic, antiaggregative, anti-inflammatory, trophic and vasoactive reflex effects (Savinov and Chaban 1995).

3.5 Legal Aspects Regarding Medicinal Leeches

Since natural resources of medicinal leeches in many countries were practically exhausted, physicians today usually use artificially grown leeches bred under hygienic conditions in the laboratories of special biofactories or cultivated at biofarms. Laws of health care authorities in different countries determine the legal aspects in using medicinal leeches for therapeutic purposes in humans (Michalsen et al. 2007). In 2004, the Food and Drug Administration (FDA) in the USA allowed the use of leeches for medicinal purposes, particularly for the restoration of venous stasis in cases of skin graft and re-attached body parts such as fingers, noses and ears by classifying them as a live "medical device". In Germany, the health authorities permitted the use of leeches for humans by defining them as "medicinal products" in 2004. According to the Russian State Pharmacopeia, the artificially grown leech *H. medicinalis* may be used as an anticoagulant (Permit number 42-702-97). Accordingly, biofactories breed and sell medical-grade *H. medicinalis*, after having standardized it according to their weight, age and time of last feeding. Usually, leeches weighing 1–2 g, which are fed artificially under laboratory conditions and starved for a period of at least 3 months, are used for treatment.

It is widely believed that the farmed *Hirudo medicinalis officinalis* has the best characteristics regarding the speed of attachment and blood sucking, the amount of blood removed, and the survival rate after blood-sucking (Rassadina 2006). Medical-graded leeches are more mobile and produce more biochemically active and stable salivary secretions than the specimens collected in nature (Kamenev 2001). They can also be used for industrial processing, the pharmaceutical and food industry, as well as for the replenishment of natural populations of medicinal leeches (Kamenev 2007). The artificially grown medicinal leech is successful as a bio-indicator of pollution such as salts of heavy metals and pesticides in aquatic environments (Kamenev et al. 2003).

3.6 Maintenance of Medicinal Leeches

The farmed medicinal leeches are usually delivered in special boxes filled with wet peat. After arrival to the appropriate staff they should be rinsed thoroughly and dead specimens as well as soil particles accompanying the leeches should be removed. Then leeches are transferred to containers from glass or clay, which are half-filled with de-chlorinated water, with low calcium content at 10–12 °C (Fig. 3.3). Usually, up to 50 individuals are placed in a three-liter container. Overcrowding leads to suppression of their vital activities and increases mortality (Rassadina 2006). The container is tightly covered with a piece of dense linen fabric attached with an elastic

Fig. 3.3 Glass container suitable for storage of leeches

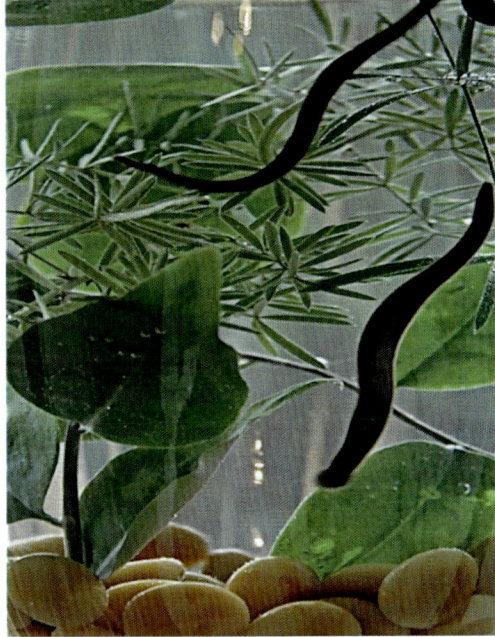

Fig. 3.4 Suitable environment for leeches to periodically shed their skin

band or tight-fitting plastic lid with small holes, which should allow the necessary ventilation. If distilled water is used some commercially available salt additives such as Hirudo Salt™ (Leeches USA) should be added.

The water in the container should be changed once or twice a week, while the container itself should be thoroughly cleaned monthly, without the use of chemicals. As the leech sheds its skin every 2–3 days, it is important to place a stone with a rough surface or plants such as *Trifolium fibrinum*, *Equisetum palustre* or *Potamogeton natans* into container, which would facilitate molting (Fig. 3.4). The container with the leeches should be placed in a dark and cool place, free of sharp odors and vibration. Under such conditions, leeches can survive for up to a year and a half.

3.6.1 Pre-application Handling of the Medicinal Leeches

Leeches, which are to be used for therapeutic purposes should be removed from their container, rinsed with de-chlorinated water and placed into a small plastic container partially filled with water (Fig. 3.5). From there they are transferred into special medicinal devices, i.e., standard or specially shaped glass tubes, which facilitate the application of leeches onto a precise area of the skin or hard-to-reach area of oral or vaginal mucosa (Fig. 3.6). It is also possible to use the barrel of a 5cc plastic syringe for applying the leeches to the treatment site.

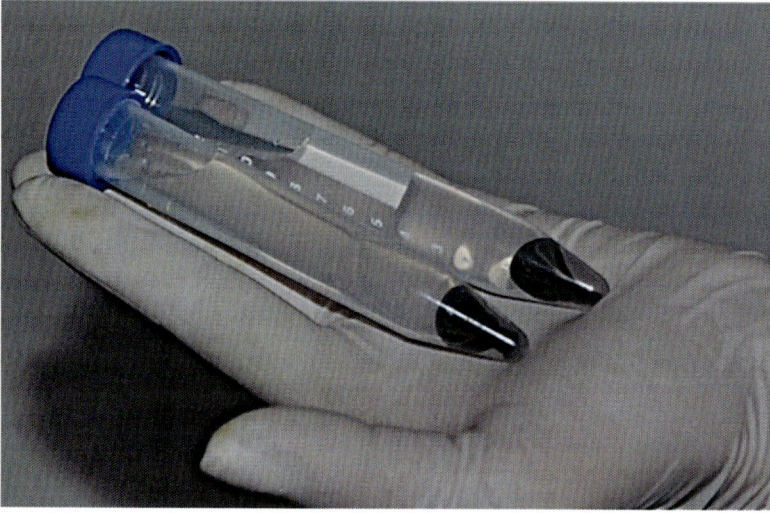

Fig. 3.5 Medicinal leeches used for therapeutic purposes placed into appropriate vials

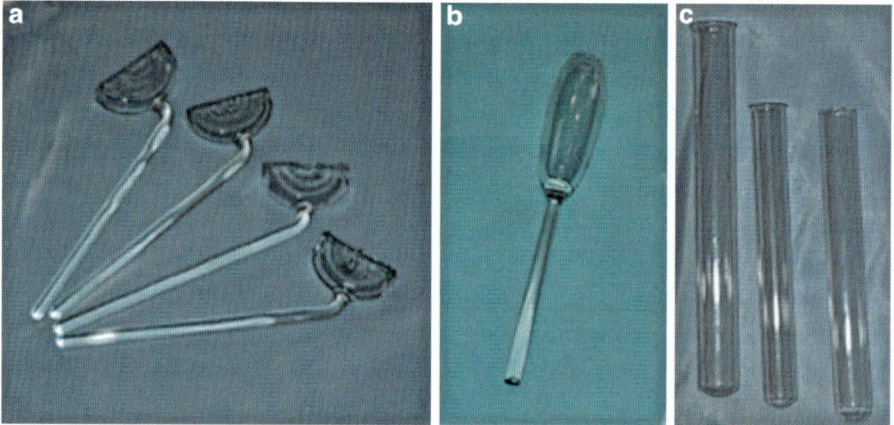

Fig. 3.6 Special medical devices for intraoral (**a**) intravaginal (**b**) and (**c**) on-dermal leech application

The viability (activity) of the leeches can be estimated by their appearance (good motor activity, absence of visible signs of injuries), but also with the help of the **contractile test**: the leech is slightly squeezed with the fingers or inside the palm of the hand, it should show a rapid reflex contraction, by reducing significantly their size (Fig. 3.7). It is not recommended to stimulate the leech to bite by means of pricking, scarification, or scraping of the intact skin, or by application of leech "attractants" such as milk, sugar water, butter or glucose.

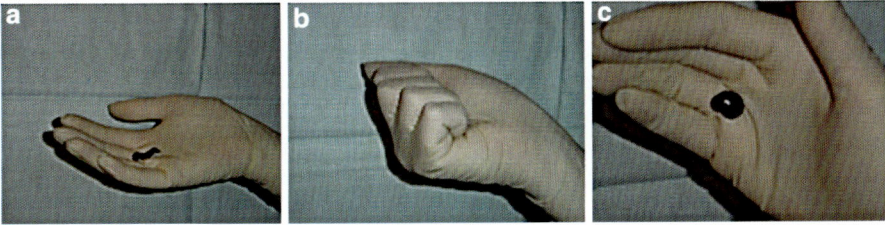

Fig. 3.7 Contractile test for assessment of leech viability

The decision as to when to use hirudotherapy (indications), the number of leeches to be applied, the choice of optimal techniques and regimen of leech therapy, the use of antibacterial therapy in high-risk patients, as well as monitoring the effectiveness of the treatment is determined by a certified physician (surgeon) with a knowledge of leeching. The practical application of leeches is usually done by a trained nurse, who could also give the necessary explanations and psychological support to the patient and his family. Nurses and biotherapists could be responsible for the maintenance of medicinal leeches inside the health institution (Fowler 1999; White et al. 2010).

Medicinal leeches may only be used once for curative purposes. Accordingly, leeches should not be re-used for other patients in order to prevent any cross-infections.

3.6.2 Post-application Handling of the Medicinal Leeches

Leeches removed from the patient should be euthanized by placing them in 70 % alcohol for 15 min and thereafter in a 3–5 % solution of bleach for an hour. Leeches could also be frozen at a temperature of −18 °C for at least 12 h. In some countries such as Germany however, the manufacturer of leeches is ready to take them back after usage and keep them in special ponds, until their natural death.

3.6.3 Preparation of the Patient for Treatment with Leeches and Techniques

Before hirudotherapy, written consent should be obtained from each patient. Patients are treated in a well-ventilated room, while seated or laying comfortably. It is preferable to perform the leeching procedures in the morning hours of the day when leeches are more active, and the patient has an empty stomach or after having a light breakfast. Smoking and the use of perfumes or skin ointments should be avoided before and during the treatment. The use of alcohol, narcotics and benzodiazepines before and during the treatment could result in the reduction of leech sucking activity. Drugs containing acetylsalicylic acid or enzymes should be avoided 2–3 days before the beginning and after the end of the treatment course. Physical examination of the patient and blood tests (complete blood count, CBC) are carried out before the treatment. It is recommended to monitor the blood pressure of the patient before and after leeching.

The area of the skin, which was chosen for leech application, should be cleaned thoroughly with warm water and soap to remove any dirt and cosmetics such as ointments, gels and creams. This skin area is then rubbed vigorously with gauze soaked in warm isotonic sodium chloride solution or distilled, non-chlorinated water to increase the capillary inflow of blood and induce local vasodilatation. These techniques enhance the attachment of leeches and blood-sucking, and also stimulate post-application bleeding from the leech bite.

In order to prevent leeches from leaving the application site, the area should be covered with sterile gauze in the middle of which a hole with a diameter of 1 cm was cut out with scissors. Some commercially available dressings for insulation of the skin during leeching may also be applied. It is preferable to place leeches inside special glass applicators or tubes, which are then placed in a precise area and kept until the leech has attached and started blood feeding. After attachment the tubes are removed, in order not disturb the leech during engorgement. Leeches can also be applied to the skin using plastic gloves.

For intraoral or intravaginal applications, leeches are applied to the mucosa using special glass containers (tubes), by keeping them in place until the end of leeching. In these cases, the path to the oropharynx (intraoral application) or cervix (intravaginal application) is blocked with gauze to prevent leech migration into distal areas in case leeches escape from the applicator. The day before the leeching, the patient undergoing intraoral leech therapy should avoid use of toothpastes containing peppermint or cinnamon flavors, and food products with strong odor such as garlic and horseradish.

The duration of blood feeding is between 30 and 90 min and depends on factors such as the area of application (skin, oral or vaginal mucosa) and its condition (intact, inflamed, ischemic, swollen, congested, erosive etc.), the time of the day, and the size of the leech.

In cases of intraoral or intravaginal leeching, the duration of blood sucking is reduced (half or third of the time), because of the rich blood supply and the abundance of superficial vessels.

During a treatment session, leeches are usually left on the skin or mucosa until their complete engorgement with blood. In some cases however, a limitation of blood sucking time is necessary. For this purpose, leeches should not be forcibly pulled off from the skin with tweezers or hands, but gently removed using chemicals such as 3 % iodine solution, saturated salt solution or table vinegar (Adams 1989; Porshinsky et al. 2011). An applicator soaked with one of these chemicals is brought to the head area of the leech (without touching!), which induces the leech to dislodge from the skin.

Already in 1949, A.S. Abuladze suggested the use of a so-called "leeching without blood extraction" method in hirudotherapy. In his opinion, this method is largely based on the pressure applied by the leech during attachment as a trigger in acupuncture points (Gilyova et al. 1988). Practically, the leech is carefully removed from the bite site with the help of above-mentioned chemicals, immediately after the appearance of the very first peristaltic movements of its body. Within this short period of time, the leech manages to bite the skin, suck the first portions of host blood, and inject its saliva to the skin. In the daily practice, the method is used in case of significant blood loss due to previous leech treatments, as well as for testing individual sensitivity to leeches.

Leech therapy may be used as a monotherapy or in addition to conventional treatment modalities; it could for example be combined with pharmacotherapy, physiotherapy and balneotherapy.

Determination of the optimal number of leeches to be applied simultaneously, the areas of leech application, the duration of blood sucking, the frequency of leeching sessions (daily or every other day), and the duration of the whole treatment course, depends on the indications selected for leech therapy, the severity of the symptoms, the medical status of the patient and his/her response to leech treatment.

3.6.4 Post-procedural Patient Management

The bleeding, following the leech bite is an important component of the treatment, and is an indicator for the success of leeching. Post-procedural bleeding can last up to 48 h. After removal of the engorged leeches from the bite site, the area should be treated with brilliant green solution and covered loosely with a dry bandage for 12–14 h. It is recommended that the patient limits his physical activities to a minimum and enters a sauna not before 2–3 days after the leeching procedure. There is no need to apply any bandage or dressings on the wound of the bite after intraoral and intravaginal leech treatment. In cases of intraoral application, the patient should abstain from eating for the consecutive 2–3 h. After intravaginal applications the vagina is loosely filled with a sterile gauze pad and sexual intercourse should be avoided for few days.

3.7 Hirudotherapy in Modern Medicine

3.7.1 Hirudotherapy in Reconstructive Plastic Surgery

Medicinal leeches have been used in the past 50 years for the salvage of tissue with venous congestion. In 1960, Derganc and Zdravic conducted the first treatment of congested flaps using leeches. Today, especially in the field of reconstructive microsurgery, the medicinal leech therapy enjoys a renaissance.

Leeches are generally used during the critical post-operative period when venous outflow cannot match the arterial inflow, which can lead to venous congestion, clinically identified by the dusky purple appearance of the skin. If this complication is not

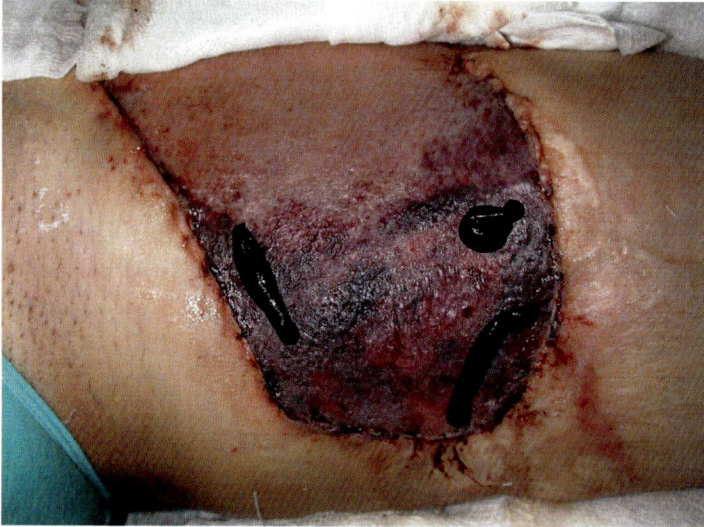

Fig. 3.8 Treatment of a free flap of a patient with a leg injury

corrected, cell death may result and the flap or finger may be lost. Therefore, medicinal leeches are used to salvage compromised microvascular free-tissue transfers, replanted digits, ears, lips and nasal tips the plastic surgery procedures until angiogenesis gradually improves the physiological venous drainage (Hartrampf et al. 1993).

Leech therapy is usually initiated after failure of more conventional treatment modalities such as warming, aspirin, Rheomacrodex (i.v.), immobilization and elevation of the injured area, use of local heparin and vasodilators, to improve venous status.

Successful salvage of such tissues with leeching has been reported in 70–80 % of cases (de Chalain 1996; Durrant et al. 2006; Hayden et al. 1988; Whitaker et al. 2004). In Israel, Eldor et al. (1996) used this technique on 50 patients with congested skin flaps and in one patient with a re-implanted ear, while Shenfeld (1999) used leeches for the treatment of venous insufficiency in a replanted digit. Mumcuoglu et al. (2007) used hirudotherapy in 23 patients, 8–79 years old, presenting with venous congestion of revascularized or replanted fingers and free or local flaps. Of the 15 fingers, 10 fingers were saved (66.7 %) (4 out of 9 replanted fingers and 6 out of 6 revascularized fingers), while 17 out of 18 flaps (94.4 %) were salvaged (3 out of 4 free flaps (Fig. 3.8) and all 14 island and random flaps (Figs. 3.9 and 3.10)). Similar results were reported from countries such as the USA (Chepeha et al. 2002; Koch et al. 2012), UK (Whitaker et al. 2012), Germany (Gröbe et al. 2012), and South Africa (Van Wingerden and Oosthuizen 1997).

As indicated above, in July 2004, the FDA approved leeches as a medical device in the field of plastic and reconstructive surgery. A survey of all 62 plastic surgery units in the United Kingdom and the Republic of Ireland showed that the majority of these units use leeches post-operatively (Whitaker et al. 2004).

Fig. 3.9 Treatment of a skin
flap of the face, after removal
of the skin with scars and
coverage of the area with the
enlarged, adjacent skin

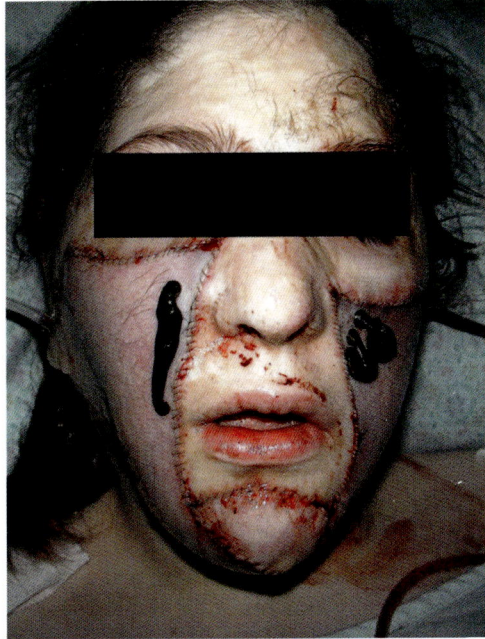

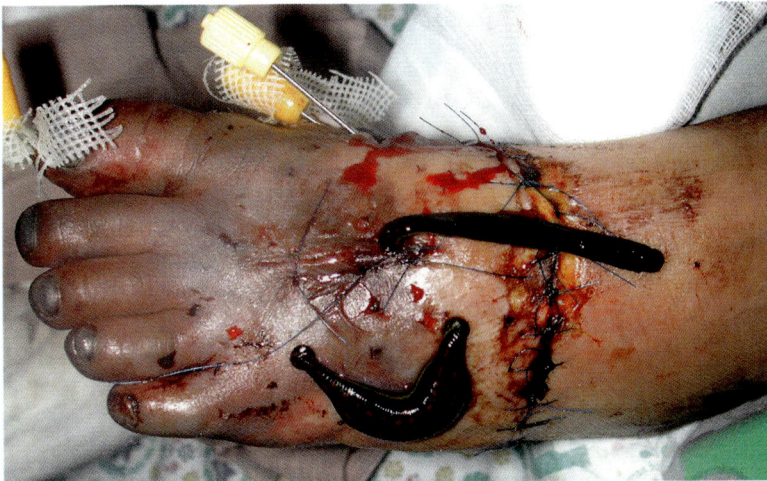

Fig. 3.10 Treatment with leeches of a random skin flap in the leg of patient after an accident

3.7.2 The Medicinal Leech

According to literature in most of the cases, the medicinal leech *Hirudo medicinalis* was used. In some cases, similar results have also been obtained with *Hirudo verbana* and *Hirudo michaelseni* (Van Wingerden and Oosthuizen 1997; Whitaker et al. 2012).

3.7.3 Treatment Procedure

Written informed consent should be obtained from the patients, their parents and/or the legal guardian, before hirudotherapy is initiated.

3.7.4 Method of Application of Leeches

Before application, leeches are thoroughly rinsed with de-ionized water. The area to be exposed to leeches is cleaned with sterile distilled water and ointments (e.g. Doppler gel) are removed. In the case of re-implanted fingers, leeches are placed on the region of the removed nail, whereas in the flaps leeches are deposited on the darker spots. The leeches are usually placed on a given spot of the skin using a 5 ml syringe. For this purpose, the plunger of the syringe is removed using a scissor or scalpel. The leech is placed in the barrel of the syringe and the open proximal end of the syringe is placed on the area to be treated. When the leech starts feeding, the syringe is removed gently (Fig. 3.11).

Leeches normally start feeding immediately, although in some cases the skin has to be punctured with a sterile needle so that oozing blood will stimulate the leeches to feed. When the leech refuses to feed in a given area, the syringe is moved to the neighboring area, until an appropriate place is found as close as possible to the congested area. Depending on the intensity of blood-flow in the area, feeding can last for 30–90 min. During this time, the leeches are monitored by a caregiver. After auto-detachment, the leeches are killed in 70 % ethyl alcohol and are disposed in bags for biological waste.

A plastic adhesive membrane or a thick layer of gauze can be applied around the leech to prevent detached leeches from attaching themselves in other parts of the skin or even under the flap, to fall inside the dressing around the wound or other parts of the patient's body or on the bed, which could be stressful to the patient and not pleasant for others patients in proximity.

Depending on the severity and the size of the congestion, 1–10 leeches are used for each treatment, although some authors recommend higher numbers of leeches. The degree of venous congestion is estimated from the percentage of violaceous color of flap skin paddle, testing capillary refill and color of the blood oozing from the bite site or after having pierced the skin with a needle. At the beginning of the

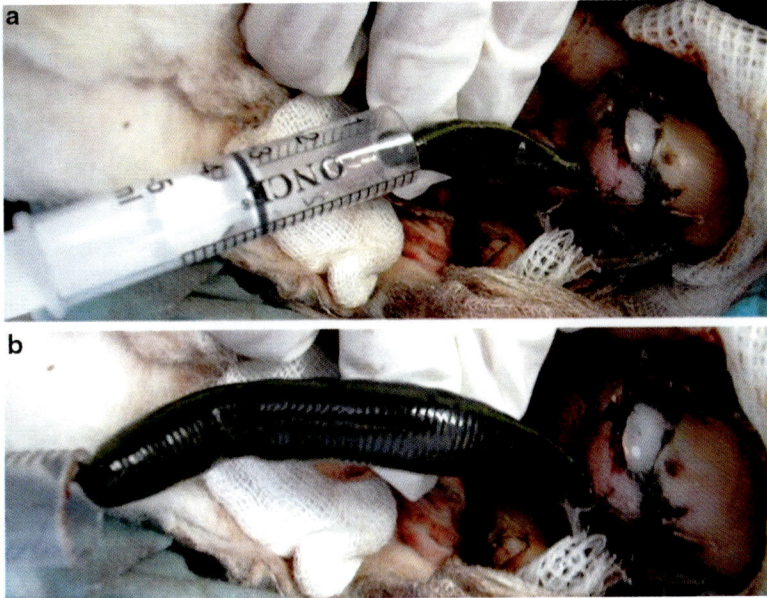

Fig. 3.11 Application of a leech using a syringe (**a**); removal of the syringe after attachment of the leech to skin (**b**)

treatment, the patient might need two treatments per day. Leech therapy is used until venous capillary return is established across the wound border by angiogenesis. Usually the treatment with leeches lasts for 2–6 days (Brody et al. 1989; Chepeha et al. 2002; Henderson et al. 1983; Mutimer et al. 1987). In fact, the decision regarding the duration of the leech treatment is empiric, based on subjective appreciation of the color of the skin, capillary refill, and the color of bleeding after pinprick.

Venous obstruction causes microcirculatory thrombosis, platelet trapping and stasis. Thus, even after successful reanastomosis, secondary changes in the microcirculation can persist and prevent adequate outflow from being reestablished.

In the post-operative period, patients usually receive treatment to reduce blood viscosity and coagulation. The decongestive benefit to the flap or digit comes not only from the initial amount of blood extracted (approximately 5 ml), but from the additional blood loss that typically occurs as a slow ooze over the next 24–48 h (Utley et al. 1998).

As described above, the injected salivary components inhibit both platelet aggregation and the coagulation cascade, accordingly, the flap is decongested initially as the leech extracts blood and is further decongested as the bite wound oozes after the leech detaches. If the transplant necessitates removal of the arterial blood coming to the area, to prevent congestion of the blood, the bite area is cleaned every 3–4 h with a gauze sponge soaked in physiological saline to remove any locally forming clot and with a heparin (5,000 U/ml) soaked gauze, to increase the time of blood oozing.

3.8 Hirudotherapy in Medicine

3.8.1 The Use of Medicinal Leeches in Synosteology

Leech therapy was extensively used in the treatment of joint diseases such as rheumatism, gout and arthritis throughout medical history (Sawyer 1986; Scott 2002). The progress achieved over the past 30 years in understanding the action mechanisms of hirudotherapy has revived the therapeutic interest in leeches, especially in their use in modern synosteology.

Osteoarthritis is a degenerative joint disease affecting the population all over the world and causing pain, loss ability and stiffness. In 2001, Michalsen et al. published the results of a nonrandomized pilot study on the evaluation of using leeches to relieve pain in patients with osteoarthritis of the knee. The study was conducted over a period of 3 months on 16 patients suffering from osteoarthritis of the knee, ten of whom received a single treatment with four locally applied medicinal leeches for a period of 80 min and the other six were given conventional treatment. The patients treated with leeches showed a significant reduction in pain in less than 24 h.

Two years later, in a follow-up randomized controlled study, Michalsen et al. (2003) evaluated the effectiveness of leech therapy in 51 patients with osteoarthritis (II-III stage). Patients in the leech therapy group received a single treatment with 4–6 medicinal leeches applied to the most painful zones of the periarticular soft tissue with unlimited time of blood-letting (after a mean of 70 min). Control group patients received a 28-day topical diclofenac gel treatment. Both treatment regimens resulted in almost similar effective pain relief at day 7. It was however noted that by the end of the study the functional ability and joint stiffness were significantly better only in leech-treated patients. The results of this study once again demonstrated the safety and tolerability of leech therapy. Only minor side effects of leeching such as slight but clinically non-relevant decrease in hemoglobin level and local itching (in more than 70 % of treated patients) were observed.

According to Zhavoronkova (2003), the use of leeches (6–8 leeches per treatment session, left on the skin until complete engorgement with blood; performed twice a week, 8–10 treatments per course) in combination with acupuncture, provided good clinical outcomes in 76 % of patients suffering from Duplay's periarthritis syndrome (periarthritis humeroscapularis).

The earliest studies of Russian leech therapists showed that leech therapy can be well-combined with balneotherapy in patients with osteoarthritis (I-II Stage). In a study conducted in the medical facilities of UST-Katchka Health Resort (Russia, Perm), known for its natural bromine-iodine mineral waters (Gilyova et al. 1988), 79 elderly patients with osteoarthritis of the joints, including knee, shoulder, elbow and wrist, were divided in three groups. Group 1: balneotherapy (5–7, 10 min applications (bandages) per day of concentrated (260–290 g/l) bromine-iodine solution (25–27 °C) on the affected joint, for a period of 4–6 days; group 2: hirudotherapy (2–4 leeches applied to the painful sites of the joints; treatment session lasted up to complete engorgement of the leeches with blood; 5–7 leech treatments in a day);

group 3: balneotherapy combined with hirudotherapy (the alternation of bromine-iodine applications and leech applications on the affected sites of the joint with 2-day intervals between these treatment modalities; 16-day treatment course). Using the WOMAC questionnaire (Western Ontario and McMaster Universities Arthritis Index) (Bellamy et al. 1988), which reflects the severity of joint pain, limitation of physical functioning and stiffness, the best results were obtained in patients receiving a combination of balneo- and leech therapy. Inclusion of leech therapy in the balneotherapy achieved a significant reduction of the main arthritis symptoms (pain, functional disabilities, and join stiffness) for a longer period (up to 6 months) than with the two other treatment regimens. Local itching (in 75 % of subjects) and skin irritation (in 12.5 % of subjects) were observed in patients receiving only leech therapy. Slight local skin irritation was observed in 9.1 % of patients undergoing only bromine-iodine treatment. The absence of such side effects in patients treated with alternating applications of the bromine-iodine rich water and leeches was remarkable and can be explained by the antipruritic, antimicrobial and anti-inflammatory properties of the natural mineral bromine-iodine water.

In 2003, Sulim reported the possibility of using leeches for alleviation of myofascial syndrome symptoms in patients with arthrosis or arthritis of the temporomandibular joint that is very common in elderly patients with secondary edentulous. The treatments consisted by placing 2–3 leeches twice a week on the temporal areas, 5–6 treatments per course. The time of blood extraction by leeches was limited to 10–15 min. The results of the treatment manifested in a reduction in pain (35/41) and joint crunch (33/41), as well as the improvement of the mandibular mobility during functional loads (29/41).

In 2003, Makulova used leeches as a supplementary symptomatic treatment for patients with ankylosing spondylitis or Bechterew's disease, a severe chronic inflammatory disease of the axial skeleton with involvement of peripheral joints and of non-articular structures. The positive effects of a 10-day course of leech therapy (4–6 leeches were applied paravertebral per session; treatments were performed up to complete leech engorgement with blood; the total number of treatments per course were 4–5) were observed in 24 out of 30 patients, and resulted in a decrease in spontaneous spine pain, pain on palpation of the spine, as well as in the increase of spine mobility.

In an open-label study (Salikhov 2004), the effectiveness of leech applications on the knee joints (4–6 leeches were placed on painful sites of the joint until complete engorgement with blood, 6–8 treatments per course) was compared with the intra-articular injections of steroids (Diprospan, 4 mg) course, a "gold standard" therapy for arthritis. It was found that hirudotherapy was as effective as the steroid treatment. Moreover, leech therapy was superior to the steroid treatment regarding its efficacy in coping with pain in patients with advanced forms of rheumatoid arthritis.

A randomized study conducted by Andereya et al. in 2008 focused on the long-term results of repeated use of the leeches in 113 patients with advanced osteoarthritis of the knee. The subjects were randomized into three groups: those receiving a single treatment with leeches (group I), those receiving two leech applications with an interval of 4 weeks (group II) and the control group (group III), which receive a simulated treatment with the help of an "artificial leech". A significant improvement in pain, stiffness and total arthritis symptoms was observed in the treatment groups I and II during the entire follow-up period. The greatest and more lasting improvement was observed in patients, who received two leech sessions.

In a non-randomized study, Zakharova (2008) compared the clinical efficacy of hirudotherapy and its combination with immunosuppressants in 75 patients suffering from rheumatoid arthritis. In one group, the patients received immunosuppressive treatment (methotrexate 7.5 mg/week) supplemented with topical, non-steroid anti-inflammatory drugs, while the second group has received the same pharmaco-therapy, combined with the use of leeches (2–4 leeches applied to the painful zones of the joints; treatment performed up to the complete leech engorgement with blood; a total of five leech sessions). The results were evaluated by means of dynamic estimation of the following indices: The Disease Activity Score (DAS), Ritchie Articular Index (AI), and VAS (visual analogue scale) for the overall assessment of complaints. According to the DAS score the outcome of the combined treatment was significantly better than the use of immunosuppressants only. In reference to AI score, it was found that the use of leeches supplemented with immunosuppresants alleviated the pain symptoms more significantly and in shorter periods of time than the use of basic treatment alone. Only leech-treated patients showed a significant increase in prothrombin index by the end of the treatment course. In addition, the use of leeches demonstrated the ability of a significant increase in the antioxidant activity of the serum.

In a recent non-blinded, randomized controlled trial with outpatients in a crossover design, conducted on 52 subjects suffering from active osteoarthritis of the knee, Stange et al. (2012) evaluated the effects of a single leech therapy with eight leeches (group I) as compared to transcutaneous electrical nerve stimulation (group II) over a period of 9 weeks. Highly significant improvements in means of Lequesne's combined index for pain and function, between days 0 and 21, was observed in patients who received leech therapy, whereas no significant differences were found in patients who had been treated with transcutaneous electrical nerve stimulation. No serious adverse effects were reported by any of the patients.

The anti-inflammatory effect of leech therapy in joint inflammatory diseases was confirmed in an experimental study by testing hirudin as a principal thrombin inhibitor of synovial inflammation in Antigen Induced Arthritis (AIA) (Varisco et al. 2000). Experiments using animal models have shown the ability of hirudin to significantly decrease the severity of AIA measured by synovial histology. It is known, that hirudin could make a significant contribution in patients with arthritis due to its ability to inhibit synovial stimulatory proteins and proteins isolated from fibroblasts. The direct antiphlogistic effect of recombinant polyethylene glycol-hirudin applied subcutaneously was demonstrated histologically in an experimental model of antigen-induced joint inflammation in mice (Marty et al. 2001).

The results obtained with leech therapy in the treatment of inflammatory joint diseases are promising and suggest that leech therapy is an effective treatment modality and is well tolerated by the majority of patients. This method deserves further research, especially regarding the effect of the different leech saliva components and the optimization of the leech therapy protocol for patients with different kinds of arthropathy. Nevertheless, the effectiveness and safety of hirudotherapy in patients with various forms of joint diseases should be evaluated in larger random-ized, blinded and controlled clinical trials with long-term follow-up. Subjective patient-supported treatment outcomes should be confirmed objectively by modern methods of investigation.

3.8.2   The Use of Leeches in Phlebology

Internationally hirudotherapy is recognized as an effective treatment modality for patients with various phlebitic disorders, such as acute, subacute and chronic throm-bophlebitis, varicose veins complicated by phlebitic thrombosis and trophic ulcers.

In 1980, Scheckotov et al. reported their 10-year experience of using the medicinal leech in the treatment of phlebitic disorders of the leg in 487 patients with phlebitis, thrombosis and trophic ulcers. According to the protocol used 8–10 medicinal leeches were applied at once and were left on the skin until complete engorgement with blood. Leeches were applied 2–3 times a week, over a period 2–3 weeks. It was noted that leeches ingested approximately 350–400 ml of congestive, venous blood from the dilated and thrombotic veins. The first subjective signs of improvement were seen after the third leeching procedure. By the end of the first treatment course, a significant reduction of the prothrombin index by 20–25 % and a normalization of the acid-base balance of the blood were observed in the majority of patients. In the long-term observation period after 3–5 years, the results were evaluated as good and acceptable for the majority of patients. It is interesting to note that subjects with relapses usually refused to accept treatments other than leech therapy.

In 1996, Eldor et al. compiled the first Israeli experience of using leech therapy in patients with post-phlebitic syndrome, manifested by swelling, induration and pigmentation of the leg, as well as ulceration in the area of the medial malleus. The following scheme of treatment in 87 patients was used: 10–15 leeches were applied to the intact limb up to their complete saturation with blood, while the treatment lasted for 3–4 weeks. A decrease of the pain in the legs (68/87), a feeling of light-ness in legs (68/87), and the possibility of long walking (33/87) was already observed after the first treatment and lasted for at least 3 weeks. Patients reported that the efficacy of hirudotherapy was superior to previous treatment modalities, including vasodilators and pentoxifylline (a drug with rheological properties). All patients received prophylactic antibiotic therapy (tetracycline 2 g/day) for 2 days. No complications during and after the treatment were observed (even in patients who received more than 20 treatments with leeches). The itching and scarring of the bite site due to leech therapy was reported only rarely.

A team of Indian leech therapists (Bapat et al. 1998) demonstrated the efficacy of medicinal leeches as aids for venous decongestion, the resorption of edema, hyperpigmentation and healing of varicose ulcers. Twenty patients with varicose veins were included in the study and leeches were applied to the perifocal areas of the ulceration, until they were completely engorged with blood. At the end of treatment, all ulcers were healed. The significant reduction in size of edema and the decrease in skin hyperpigmentation were observed in 95 and 75 % of the subjects, respectively. The partial pressure of oxygen in blood, measured in leech-treated patients revealed that leeches preferentially sucked venous blood, thus promoting the healing process. The results of the study showed that hirudotherapy is an effective adjunct method in the management of complicated varicose veins.

The results of using different treatment modalities in 137 subjects with uncomplicated forms of varicose veins of the lower extremities were evaluated in a study conducted by Gilyova et al. (1988). The patients of the main group (n=81) received hirudotherapy, while those of the control group (n=56) were given conventional treatment with heparin ointment and Trental 400. Four to six medicinal leeches were applied along the inflamed vein until they were completely saturated (Fig. 3.12a, b), while additional 1–2 leeches were placed on the sacro-coccygeal area (Fig. 3.12c). Six to eight leeching sessions were performed each day, while the patients were receiving a prophylactic treatment with antibiotics. Physical examination, routine blood test (CBC), blood coagulation tests (spontaneous platelet aggregation, streptase induced plasma lysis, coagulation factor Xa activity), as well as rheovasography were carried out before and after the treatment. By the end of treatment course, 66.7 % of the patients who received hirudotherapy showed diminished levels of pain and tension in the legs, less swelling and induration (Fig. 3.12d), while similar improvement was observed only in 44.6 % of patients receiving conventional drug therapy. The results of the functional tests in leech-treated patients showed an improved vascular tonus, increased local blood flow and increased elasticity of blood vessels. Patients undergoing leech therapy showed significantly better results regarding the rheovasographic indices than those treated with conventional drugs. By the end of the treatment, the blood coagulation parameters showed the reliable changes in patients of both groups, however, in leech-treated patients, the platelet aggregation score decreased more significantly than in the controls.

Hirudotherapy can be successfully used for treatment of complicated varicose disease, accompanied by the formation of poorly healing chronic leg ulcers. In 2000, a non-randomized clinical study focused on the long-term results of using leeches in 20 patients (11 males and 9 females, 54–73 years old) with chronic venous ulcers, was conducted (Gilyova 2000). The clinical results of leech therapy were classified as follows: (a) complete healing of the ulcer; (b) considerable improvement with a significant reduction in size of the ulcer; and (c) no visible improvement. Depending on the size of the ulcer, 5–7 leeches were placed on the ulcer edges, 1–2 cm from the wound bed. In addition, leeches were applied to the coccyx area and along the varicose veins. Broad-spectrum antibiotics (e.g. Ciprofloxacin 250 mg 2–3 times a day) were used for prophylaxis. The vast majority of patients reported a reduction in pain, swelling, and fatigue, immediately after the first leech application. By the end of the first hirudotherapy course the ulcer had healed completely in 15.8 % of the patients, there was considerable improvement in 31.6 % of subjects, improvement in 42.1 % of cases, while in 10.5 % of the patients no changes were observed. The efficiency of

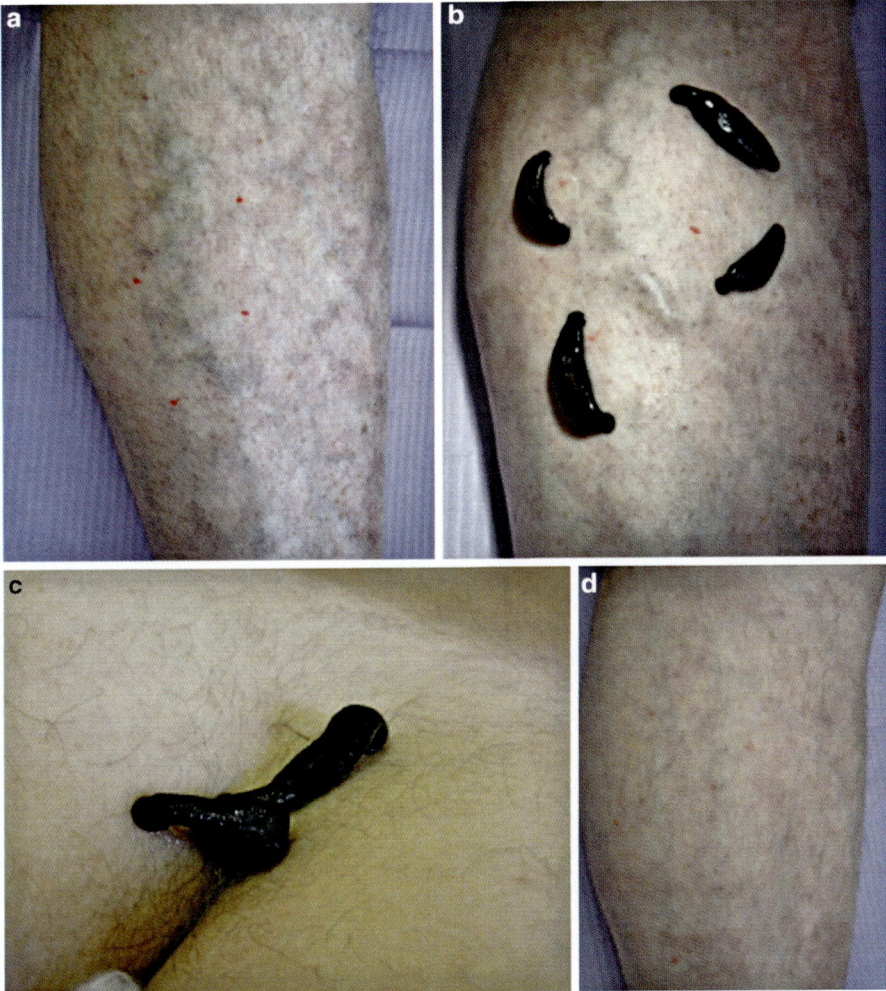

Fig. 3.12 The skin condition of a male patient with varicose veins of lower extremities before hirudotherapy (**a**). Four leeches applied on the right lower leg along the inflamed subcutaneous veins (**b**); two leeches applied on the sacrococcygeal area (**c**). View of the right lower leg 3 weeks after hirudotherapy (**d**)

the treatment increased after the second course of leeching, which was conducted 3–4 months later, when the percentage of patients with completely healed leg ulcers increased to 31.5 %, while all other subjects showed a significant improvement. Leech therapy was generally well tolerated by the majority of the patients, although 42.1 % of them reported localized itching, which lasted for 1–2 days after the first treatment course, and pigmentation of the skin in the sites of the leech bite (10.5 %). It was concluded that patients with chronic varicose require repeated courses of treatment with leeches.

Additional clinical trials on the efficacy of leeches are essential to better define their role in modern phlebology.

3.8.3 The Use of Leeches in Patients with Cardiovascular Disorders

Since ancient times, leeches have been used extensively by physicians for the treatment of various forms of cardiovascular disorders. In the past, leeches were traditionally used for the treatment of arterial hypertension, and were almost the only means to combat the hypertensive crisis. Current understanding of mechanisms of the multifactorial effects of hirudotherapy allow new consideration of the prospects of using medicinal leech in the complex treatment of cardiovascular diseases such as arterial hypertension and its complications, coronary heart disease and postinfarction cardiosclerosis.

The study by Chmeil et al. (1989) points to the possible antihypertensive effects of leeching, which apart from the bloodletting action has also hypovolemic hemodilution, antithrombotic, fibrinolytic, anti-aggregative and hemorheologic properties. Blood hemorheologic tests were performed in 23 patients with different cardiovascular diseases before and immediately after leeching (application of ten leeches on the lumbar area, until complete engorgement with blood) as well as approximately 1 month later. The hemorheologic changes included significant decrease of the blood viscoelasticity and the decrease in aggregation tendency and flexibility of red blood cells. These positive hemorheologic changes were even more marked after 1 month of treatment.

In the Russian medical literature, there are numerous papers on the clinical experience and effectiveness of leech therapy in the complex treatment of arterial hypertension and its complications that account for up to 30–40 % of all diseases of the adult population in the country. In 2001, Gantimurova conducted a study on 114 patients (42 males and 52 females, aged 40–73 years) with an advanced form of arterial hypertension (2nd and 3rd degree of arterial pressure and 3rd degree of cardiovascular risks), receiving leech therapy together with conventional antihypertensive medication (main group), and compared them with those treated with conventional antihypertensive medication alone (control group). In the main group, leeches were placed in the head region (BTE [behind-the-ear] points), neck region (paravertebrally), heart region (5th intercostal space, at the middle of the clavicle), upper quadrant of the liver, and sacro-coccygeal region. Two to four leeches were used per session and 30–40 leeches were applied to each patient during the treatment course. Parameters of central and peripheral hemodynamics (tetrapolar rheography, thoracic leads), as well as some rheological blood parameters, were analyzed before and after treatment. A significant antihypertensive effect was observed in both groups of patients. Taking into consideration the objective clinical and functional parameters, as well as the subjective feelings of patients, the best results were observed when conventional drug treatment was combined with hirudotherapy; considerable improvement was detected twice as often in leech-treated patients as compared to the controls. With regard to hemorheologic parameters, only patients undergoing combined treatment (antihypertensive drugs plus leeches) showed a significant reduction in the ADP-induced platelet aggregation test. Thus, including leech therapy in the conventional treatment of patients with arterial hypertension associated with the greater hemodynamic and hemorheologic improvements, has more therapeutic advantages than the conventional medication alone.

In 2003, Zadorova reported the need for heterogeneous modes of hirudotherapy according to severity of arterial hypertension. He used hirudotherapy for patients with hypertension in addition to the conventional medical treatment. In patients with initial forms of hypertension (1st and 2nd degree of arterial pressure and 1st degree of cardiovascular risk), two leeches were placed daily on the head region (BTE points) with little or no blood extraction, according to Abuladze's method. Twenty to twenty-five leeches were administered to each patient per treatment course. In patients with an advanced form of hypertension, 2–4 medicinal leeches were applied daily to one of the following regions: head (BTE points), posterior neck, heart (5th intercostal space, at the level of the middle of the clavicle and left parasternal point). The leech application points were changed every day. Head and neck regions were treated three times a day, while the number of leeches per treatment course varied from 20 to 25. Sustainable normal blood pressure and disappearance of headaches, dizziness, ear noise, as well as restoring of sleep and appetite was observed at the end of the treatment in 82 % of patients with primary (essential) hypertension. After 3 months, the antihypertensive effect of leech therapy was observed in 60 % of the subjects (15/25), while after 6 month this effect was observed only in 24 % of the patients. The efficacy of the treatment was more pronounced when leeches were used in the early stages of arterial hypertension.

In Russia, a number of balneological treatment modalities are used for the treatment of arterial hypertension. These include sodium chloride and bromine-iodine rich natural mineral waters, obtained from springs in Ust-Katchka Health resort (Perm, Russia). Patients with a mild form of arterial hypertension (I-II stages, first to third degree of cardiovascular risk) are treated with hydrotherapy (baths) with increasing concentrations of bromine-iodine from 6 to 24 g/l (Gilyova et al. 1988; Tuev et al. 2001). In a randomized controlled study performed on 141 patients (59 male and 82 female, aged 40–63 years) with stages I-II arterial hypertension, the effectiveness and safety of hirudotherapy, combined with either conventional medical treatment or with bromine-iodine hydrotherapy, was tested (Korobeinikova 2002; Sidorov et al. 2003). The patients were assigned into two groups: the main group, consisting of two subgroups (leech therapy combined with conventional medication and leech therapy combined with bromine-iodine hydrotherapy). Control subjects were treated with bromine-iodine hydrotherapy alone. By the end of the third week, the following parameters were analyzed: the dynamic of the main clinical symptoms and functional indexes such as complaints, pulse, systolic arterial pressure (SAP), diastolic arterial pressure (DAP), an electrocardiogram (ECG), as well as some blood coagulation parameters (platelet count, rate of spontaneous platelet aggregation, auto-coagulation test, prothrombin index, fibrinogen level, thrombin time of plasma coagulation, antithrombin III level and Hageman-dependent fibrinolysis (HDF).

During each treatment session, 2–4 leeches were applied and left until their full engorgement with blood in one of the following head and neck regions (BTE points, suprascapular ("collar") area, sacrococcygeal area, oral mucosa, and a repeat of BTE points and suprascapular area (Fig. 3.13).

The areas of leech application were changed daily. A full treatment course consisted of six sessions a day. No special psychological problems associated with the intraoral leech application were detected. The clinical experience and manual skills of the leech therapist, as well as special equipment for applying the leeches to oral mucosa, combined with the ability to establish a good relationship with the patient, helped in the success of the treatment. Intraoral, on-mucosal leech application necessitated no special nursing of a bleeding wound at the site of the leech bite; there were no cosmetic problems such as scarring, itching, and pigmentation; it reduced the treatment time by half; and decreased the times for lowering the arterial pressure as compared with on-dermal leech application on the BTE points. Although the arterial pressure decreased in all treatment groups, according to the SAP and DAP levels, the rate of onset of the antihypertensive effect and improvement in subjective symptoms, the best results were obtained in patients who underwent a combination of balneo- and leech therapy. At the end of the third week since the beginning of the treatment, a normalization of SAP was observed in 97 % of the main group of patients and in 95.1 % of subjects in the control group. Normalization of DAP level was achieved in 96.5 % of leech-treated patients and only in 61 % of control group subjects. Patients reported that the treatment with leeches provided a rapid improvement in their general well-being, mood, appetite, sleep, disappearance of headaches, heart pain, dyspnea and dizziness. Overall, 84 % of leech-treated patients in both subgroups of the main group reported a pronounced sedative effect of leech therapy.

ECG examination showed an objective improvement of the myocardial metabolism in 57 % of leech-treated subjects and only in 39 % of control group individuals. The clinical effects of leech therapy in patients with arterial hypertension were supported also by the improvement in hemostasis. The most significant changes related to hemostasis were seen in the platelet aggregation index that was significantly reduced in 56 % of the leech-treated patients (this rate was almost the same in both subgroups) and only in 21.9 % of the controls. After the course of spa treatment combined with hirudotherapy, the coagulation hemostasis scores were improved, demonstrated by the increase in thrombin time in 71 % in leech-treated patients, as compared to an increase of 17 % only in control group patients. The fibrinolytic effect (significant shortening of the HDF time) was observed only in leech-treated patients (in 28 % of subjects of the 1st subgroup and in 24 % of subjects of the 2nd subgroup). Hirudotherapy was usually well tolerated, however, 69 % of the patients suffered from itching (more often after the second or the third on-dermal leech applications). In two cases heavy bleeding after leech application was noted in patients, concomitantly receiving conventional antihypertensive drugs.

Although today the medicinal leeches may not be the first-line medication for hypertension, as there is a plethora of modern antihypertensive drugs, it should be taken into consideration that some patients may have contraindications in the use of

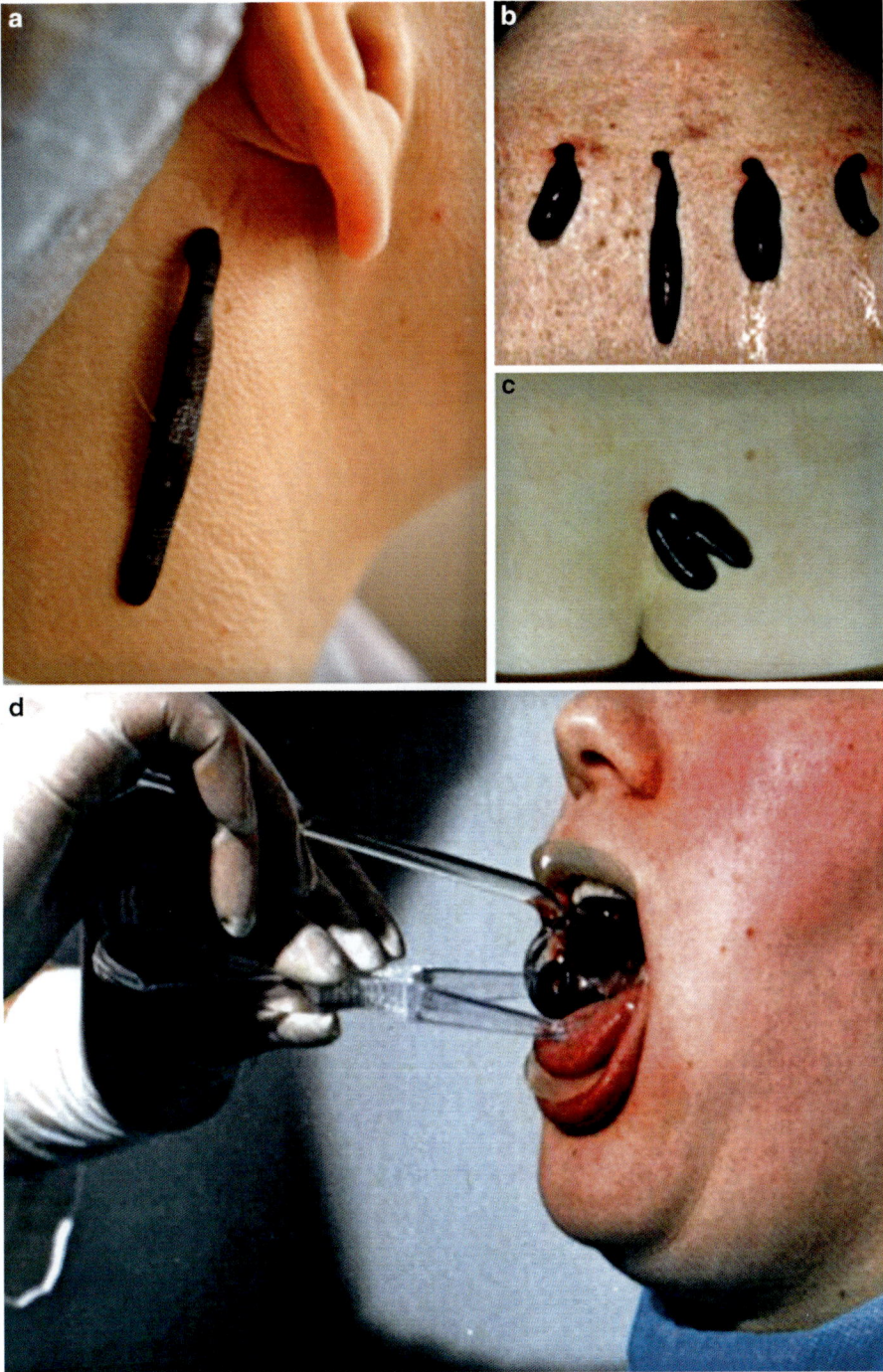

Fig. 3.13 Points of leech application in patients with mild arterial hypertension: BTE-point (**a**), suprascapular ("collar") area (**b**), sacrococcygeal area (**c**), tongue (**d**)

conventional drugs and in these cases, non-drug treatments (e.g. leech therapy) should also be used.

Coronary heart disease (CHD) is the most common type of heart disorder and the number one cause of death for both the male and female population in most of the developed countries in the world. It is known that hirudotherapy has been successfully incorporated in the treatment and rehabilitation of patients with CHD in many clinics, hospitals and health centers of Russia, Armenia, Ukraine, Kazakhstan and Latvia (Baskova 1999; Chemeris and Konirtaeva 1999; Ena 2003; Isakhanyan 1991; Nikonov 1996).

Ptushkin and Lapkes (2003), reported the outcome of a study with 320 patients, aged 43–86 years, suffering from CHD (unstable angina with frequent, drug-resistant heart attacks), who received a course of hirudotherapy concomitantly with conventional medical treatment. The medical history of patients showed that 35 % of them suffered from a myocardial infarction. Leeches were applied for unlimited (until leeches dropped off spontaneously) time of blood-letting to the following areas: heart, posterior neck, parasternum, as well as to BTE points. The amount of leeches per treatment session ranged from 5 to 10. The treatment course varied from 5 to 6 sessions per day. In patients with Angina pectoris who previously had more than two attacks per day after 2 weeks of therapy positive results (reduction in the number of heart attacks, and reduction in the severity of pain) were observed in 60 % of patients. No more analgesics were necessary in 90 % of treated subjects. Side effects such as skin redness and itching in the areas of leech applications have been seen in 8 % of patients and were easily controlled by oral antihistamines.

It was reported (Borovaya 2008) that in elderly persons with increased tonus of the symphathetic nervous system (sympathicotonia), leech therapy leads to excessive parasympathetic shift in autonomic balance, resulting in syncopal states or orthostatic hypotension. Therefore, leech therapy in elderly people should be carried out under strict instructions and with special recommendations, i.e., placing the patient in the supine position, requesting him to get up slowly, and have a long rest after the treatment. Similar side effects were not observed in middle-aged people with CHD. It should also be mentioned that hirudotherapy, applied to patients with hyperlipidemia led to significant improvements of total cholesterol, triglycerides and low-density lipoprotein levels.

A positive effect of hirudotherapy in the treatment of chronic heart failure (CHF) was also observed (Kuznezova et al. 2008). It was shown that after leech therapy more than 50 % of patients (n = 70) with CHF (Functional Class II-III, NYHA) showed a reduction in their dyspnea and peripheral edema and an increase in their physical stress tolerance. In patients suffering from hypertension, the blood pressure decreased significantly while an increase in the ejection fraction was noted. A reduction of the spontaneous platelet aggregation index, and a normalization of fibrinogen and soluble fibrin-monomeric complex levels were observed by the end of the treatment course.

In traditional Chinese medicine, *Hirudo medicinalis nipponia* is recommended for the treatment of CHD, transient disorders of cerebral circulation, as well as for control of hyperlipidemia (Yang et al. 2003).

3.8.4 The Use of Medicinal Leeches in the Treatment of Neurological Disorders

In Russia, for many years, leeches have been used for the symptomatic treatment of different forms of dorsopathy, such as spondylopathy, back pain and intervertebral disc disorders, as well as of peripheral neuropathies (Frolov and Frolova 1999; Gilyova et al. 1988; Mokhov and Zalzman 2003; Muzalevskii et al. 1999; Nikonov 1996; Rashidova 2003). A study showing the positive effect of leech therapy in the relief of pain in patients suffering from sciatic nerve inflammation was first published in 1961 (Kasimov 1961). The therapists reported a significant reduction in pain immediately after the first treatment, when the leeches took a blood meal until they engorged. Paresthesias and numbness in the areas of innervation disappeared after the course of daily applications of 3–4 medicinal leeches, when the total number of treatments varied from 3 to 5. Leeches were placed along the most painful areas on the skin above the inflamed nerve. The use of leeches in patients suffering from acute neuritis is more effective than in the chronic phase of the inflammation. Recent studies showed the effectiveness of leech therapy in relieving pain in patients suffering from neuritis of the facial and trigeminal nerve, as well as of the external femoral cutaneous nerve (Farber 1987; Spizina 1987; Zhavoronkova 1995).

In a study by Krashenjuk and Krashenjuk (2003), the use of leeches in the treatment of infantile cerebral palsy was discussed. Five children (aged 3–6 years) suffering from spastic diplegia of the lower extremities (Little's syndrome) were treated with leeches. Previously, conventional treatment provided only short-term therapeutic effects. The treatment, which lasted for 3–5 months, improved the muscle plasticity of the upper and lower extremities and speech function in all patients, as well as the ability to walk with the support of a stroller (1 of 5 patients). Five months after the initiation of the leech therapy, all children showed an improvement in their psychological and intellectual condition. The inclusion of leech therapy in the treatment of cerebral palsy provided not only a considerable, but also a more prolonged clinical benefit.

The advantages of using medicinal leeches in the treatment of children with perinatal encephalopathy, obsessive-compulsive disorders, neurocirculatory dystonia and progressive myopia were reported by numerous Russian pediatricians and child neurologists (Kolesnikova et al. 1999; Stroganova 1999).

Hirudotherapy can be used parallel to acupuncture, chiropractic, physio- and pharmacotherapy as well as massage in patients suffering from different forms of spondylosis, especially those with frequent exacerbations (Bogdanova et al. 2003). In the course of a treatment session for spondylosis it is recommended to apply 4–8 medicinal leeches to the most painful points and to the intervertebral areas of the affected spine segments (Fig. 3.14). For this purpose, leeches are left on the skin for 7–10 min, while the treatment course consisted of 5–7 sessions performed in 3–4 day intervals. A decrease in pain and muscle tension in the affected areas was observed in the majority of patients already after the first treatment, while an increase in the range of the joint's motion was noted after the second or third leeching session.

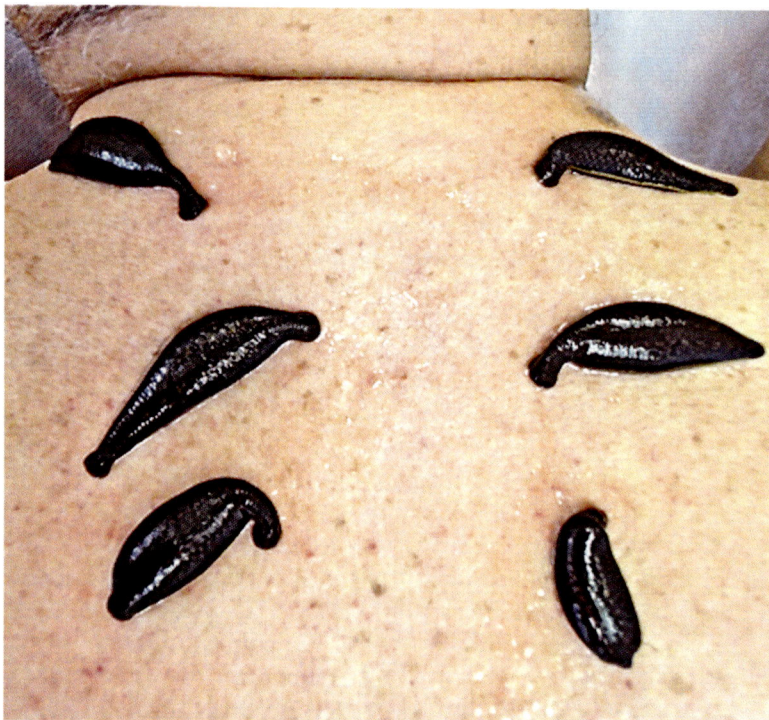

Fig. 3.14 Leeches applied in the area of intravertebral spinous processes (Th1-Th2, Th2-Th3, Th3-Th4) in a patient with spondylosis

According to Muzalevskii et al. (1999), leeches can be used after surgical interventions as well as during the postoperative rehabilitation period in patients with spinal conditions such as spinal disc herniation.

The usefulness of hirudotherapy in the treatment of patients with acute cerebral circulatory disorders such as ischemic stroke has been discussed by several experts (Poprotskii et al. 2001, 2008; Pospelova 2012; Pospelova and Barnaulov 2010; Seselkina 2001; Seselkina et al. 1998). In a study performed on 42 patients (20 males and 22 females, aged 47–65 years) with apoplectic attacks, Seselkina et al. demonstrated the effectiveness of hirudo-reflexotherapy (Seselkina et al. 2003). Nineteen of the patients suffered from basal vertebral system complications, while 23 subjects suffered from disorders of the left and/or right median cerebral arteries. During the treatment, 5–8 leeches were applied on frequently used acupuncture points of the skin, such as BTE points, temporal regions, rhombus of Michaelis and perianal area. The patients were treated 1–3 times a week during 3 weeks and a recovery of functionality was observed in 87 % of patients. The functional condition of oculomotor, facial and hypoglossus peripheral nerves, was estimated, according to which the eye movements and speech recovered by 74 and 78 %, respectively, while the swallowing ability was restored by 42 %. Dopplersonography of cerebral vessels showed significant increase of peak systolic velocity and maximal circulation velocity in the cerebral hemisphere at the site of the apoplectic attack. A significant reduction in rheological parameters of the blood (red blood cell and plasma velocity, red cell aggregation and red cell mechanical fragility) was observed at the end of hirudotherapy. The positive effects of the treatment with leeches were also manifested as a significant increase in the activated plasma recalcification time, significant decrease of prothrombin time, significant reduction of fibrinogen level and level of ADP-induced platelet aggregation.

In an uncontrolled clinical study conducted by Pospelova and Barnaulov in 2003, the following results were obtained after a course of a 10-day hirudotherapy (2–3 leeches were placed on the occipital, thoracic and lumbar-sacral regions, in the vicinity of the vertebral arteries, as well as on the tongue) in 22 patients with chronic vertebrobasilar insufficiency: disappearence of headaches (16/22 ear and head noises (8/12), dizziness (15/22), fatigue (20/22), irritability (5/9) and grief (4/5). In a recent follow-up study conducted on 146 patients with different forms of cerebrovascular disorders such as chronic vertebrobasilar vascular insufficiency, ischemic stroke, high-grade internal artery stenosis or occlusion, and stage 1 of vascular encephalopathy), the positive effects of hirudotherapy on the psycho-emotional status were confirmed. In the course of treatment with leeches a significant decrease in anxiety and depression rates (Beck Depression Scale and Zung Anxiety Scale) was observed in 95.9 and 86.3 % of the patients, respectively (Pospelova and Barnaulov 2006).

3.8.5 The Use of Medicinal Leeches in Patients with Ophthalmologic Disorders

According to the recommendations of Russian ophthalmologists (Semikova and Bondareva 1995), leeches could be used in patients with inflammatory eye diseases, injures of the eye, cataract, and glaucoma.

Leeches can be prescribed for all forms of glaucoma. For this purpose, 2–4 leeches should be placed on the lower temporal and supraorbital regions, until complete saturation with blood. An important reduction in the intraocular pressure could usually be observed in about 3 h after leeching (Semikova and Bondareva 1995). Belezkaya (2007) studied the effectiveness of hirudotherapy in patients in the initial stage of primary open-angle glaucoma and a progressive optic neuropathy, leading to atrophy of the optic nerve. A significant reduction of the intraocular pressure, improvement of hydrodynamic indices of the eye (aqueous outflow easiness rate, moisture chamber production, Becker's coefficient) and visual function, as well as an increase of retinal activity were observed after a course of hirudotherapy (2–6 leeches placed on the lower temporal and supraorbital regions until complete blood saturation, applied daily for 2 weeks). In the terminal stages of glaucoma, when almost all therapeutic and surgical methods of treatment were unsuccessful, leeches were able to prevent complete blindness.

Leeches can be also successfully used in patients with periorbital hematoma (Bunker 1981; Semikova 1992). In such cases it is recommended to apply simultaneously 2–4 leeches on the temporal, supra- or infraorbital and mastoid regions. The time of blood extraction is usually limited to 10 min.

According to Menage and Wright (1991), six leeches applied on the patient's eyelids in cases of severe eye injury enabled an easier examination of the ocular globe and improved the following surgical interventions. In such cases treatment with leeches resulted in a rapid and highly efficient reduction of the periorbital swelling and restoration of the eyesight. Good results were also obtained by using leeches in the treatment of severe post-traumatic exophthalmos. A significant reduction of exophthalmos and a complete recovery of the eyesight were observed in patients after six sessions of two leeches being placed on the temporal region (Semikova and Bondareva 1995). Recommendations for using leeches in the treatment of acute optic neuritis were based on results obtained from a clinical study carried out on 20 patients receiving Heel (Germany) antihomotoxic therapy (Traumeel S, Lymphomyosot, Cerebrum compositum, Hepar compositum, Echinacea compositum) in addition to leeches, and on 11 control patients treated with traditional steroids and antibiotics (Semikova 2011). After combined treatment, the vision of the patient, the conditions of the optic nerve measured by the optical coherence tomography, the critical flicker frequencies, and the central sensitivity, improved significantly. No complications were observed during and after the treatment with leeches. Full recovery of eyesight was achieved twice as fast in leech-treated patients as compared to the control group. The follow up after 1–4 years showed that symptoms did not reappear in the majority of the leech-treated patients.

Medicinal leeches can also be used in ophthalmology as a diagnostic tool for macular edema (Irvine-Gass syndrome) that sometimes appears after cataract surgery (Semikova and Bondareva 1995).

It is well established that the appearance of cataracts is widely connected to hyperproduction of specific ε (γ-Glu)-Lys-isopeptide bonds, which are not hydrolyzed by ordinary proteases. The therapeutic effect of leech therapy in cataract patients could be explained by the fact that leech saliva contains destabilase enzymes that promote highly specific isopeptide activity (Baskova et al. 1988).

3.8.6 The Use of Medicinal Leeches in the Treatment of Periodontal and Oral Mucosal Diseases

The first use of leech therapy in dentistry was reported by Thomas Bell in 1817, who treated a patient with an oroantral fistula and facial swelling using six leeches applied to "the face". In 1845, Harris applied leeches to the gum for drainage of a periodontal abscess. For the last 20 years, Russian dentists accumulated substantial experience in the use of leech therapy for treatment of periodontal and oral mucosal diseases, as well as of stomato-neurological and inflammatory maxillofacial disorders (Gilyova et al. 1999, 2000, 2010; Korobeinikova 2002; Gilyova and Gibadullina 2003).

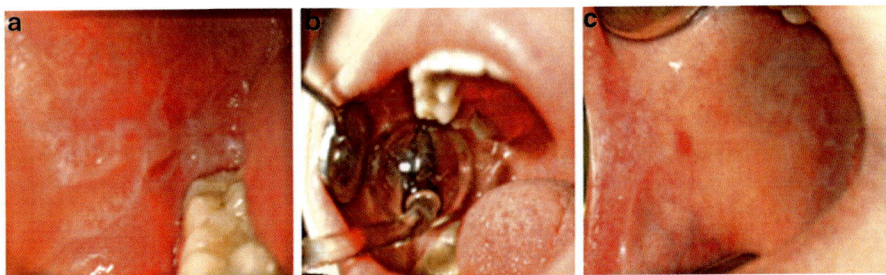

Fig. 3.15 Female patient with erosive oral lichen planus; (**a**) before the treatment; (**b**) application of leeches on the perifocal site of oral ulcer using specially shaped glass containers; (**c**) view of the oral mucosa after hirudotherapy

The number of medicinal leeches prescribed for periodontal treatment depends on the stage and form of gingivitis and periodontitis and varies from 3 to 5 specimens, successively applied to the affected gingiva in cases with a mild gum inflammation, to 5–7 leeches, in severe forms of periodontal disorder such as edematous gingivitis, periodontal abscess and acute periodontitis. In this treatment modality, leeches are left on the gums for 10–15 min using a special application set, which is appropriate for applying leeches to the oral mucosa (Fig. 3.15).

The use of medicinal leeches followed by the application of bio-soluble "Piyavit" (Hirud I.N., Russia) gel, based on extracts of *H. medicinalis* showed local analgesic and anti-inflammatory effects in 89.6 % of patients with chronic, generalized periodontitis. The cupping of the gum inflammation resulted in an improvement of periodontal indices, as well as periodontal photoplethysmogram, which showed that the vasospasm of the gum vessels decreased by 12.9 ± 0.4 %, and the gum blood flow and vascular wall elasticity increased by 24.7 ± 1.6 % and 12.4 ± 0.7 %, respectively (Gilyova et al. 1999).

Medicinal leeches have also been used in patients with oral lichen planus, one of the most common chronic inflammatory oral mucosal disorders that affect oral tissue with or without the involvement of the skin. The advanced forms of oral lichen planus are characterized by the appearance of persistent, hard-to-heal oral ulcers, resistant to traditional anti-inflammatory therapy (Fig. 3.15a). Use of conventional medication, supplemented by daily applications of 1–2 leeches on the perifocal sites of the oral ulcers (Fig. 3.15b) during 7–10 days, resulted in a reduction of the ulcer epithelialization period by 2–3 times as compared with conventional topical steroid therapy only. In 86.4 % of the patients, complete healing of oral ulcers was achieved (Fig. 3.15c). The use of polarography tests at the end of treatment showed a significant improvement in the initial and maximal oxygen supply of the affected oral mucosa (Gilyova et al. 2010).

There are also studies showing that leeches were used in the treatment of patients with Melkersson-Rosenthal syndrome, a rare neurological disorder characterized by recurring facial paralysis, swelling of the lips and desquamative glossitis (Dzekh 2000; Pozharizkaya et al. 2000). According to the recommendations of the dentists,

2–3 specimens should be applied for 10–15 min on the swollen area of the lip, oral mucosa and tongue, daily for 4–5 days. Distinct esthetic and functional improvement was observed just after the first leeching procedures and lasted for long periods in 59 % of patients.

Medicinal leeches can also be used to relief oral and maxillo-facial pain in patients with glossalgia, post-traumatic facial pain and trigeminal neuralgia (Kasimov 1961).

3.8.7   Leech Therapy in Gynecology

Leeches in gynecology are mainly used in cases of parametritis, mastitis and genital endometriosis (Starzeva et al. 2001; Lomaeva 2003; Chaban et al. 2003; Zhivoglyad and Nikonov 1998, 2003).

The effectiveness of hirudotherapy was demonstrated in a study carried out by Lomaeva (2003) on 71 women in childbearing age with hard-to-reach foci of internal retrocervical and rectovaginal endometriosis, which could not be treated by conventional medical and surgical interventions. In these patients endometriosis manifested in form of pain (59 women), hyperpolymenorrhea (43 women), and secondary sterility (16 women). Two to four medicinal leeches were placed to the intravaginal, pubic, perianal and coccygeal zones and left until complete engorgement. A significant reduction in pain, size of retrocervical endometriosis foci, and normalization of the menstruation cycle was observed in 38.2 % of patients after a 10–14 day course of treatment with leeches. 22.7 % of patients with secondary sterility became pregnant after 3 months of hirudotherapy and later successfully gave birth. The positive effect of leeches in patients with endometriosis was also confirmed by ultrasonic examinations.

Hirudotherapy was successfully used in patients with chronic salpingo-oophoritis (Khardikov et al. 2003). The effectiveness of leech therapy was evaluated in 20 women with an average age of 28.6 ± 0.3 years. These patients were treated with 2–4 leeches per session, and 5–8 leech sessions were performed during 1–3 days. The 30 patients of the control group (average age of 27.2 ± 0.4 years) were treated with antibiotics, analgesics, rheologicals vitamins, antihistamines, biostimulators and physiotherapy. A partial reduction in pain was observed in 70 % and a significant reduction in 25 % of the main group of patients, while in the control group these reductions were 50 and 36.7 %, respectively. These improvements were also objectively confirmed by ultrasonic and rheological tests. Thus, hirudotherapy was considered to be an effective adjunct method of treatment for patients with salpingo-oophoritis.

3.8.8   Leech Therapy and Cancer

In 2010, Kalender et al. reported the case of a 62-year old male patient with concurrent renal cell carcinoma and leiomyosarcoma, suffering from severe pain in the lumbar region. During the examinations, the patient reported that he conducted a self-treatment by applying seven leeches to the lumbar region, which resulted in complete control of pain. Apparently, this is the first clinical case showing the possible positive effect of leeches for relief of pain caused by cancer. However, this treatment modality should be further investigated.

3.9 Complications and Possible Side Effects

Modern leech therapy is generally recognized as a relatively safe and well-tolerated treatment modality. Slight localized itching of the bite site persisting for several hours and up to 3 days, is the most common (37.3–75 %) adverse effect of leech therapy (Karadag et al. 2011). Use of 5 % potassium permanganate, cold compresses, 10 % baking soda paste, Golden Star balm or Dimethindene maleate (Fenistil® Gel) on the affected skin areas can be required for more pronounced cases. In severe cases of generalized itching, topical corticosteroids and oral antihistamines should be prescribed. Signs of regional lymphadenitis, slight swelling and painfulness of regional lymph nodes on the side of leech application and sub-febrile temperatures can occur in 6.4–13.4 % of the treated patients and usually appears after 3–4 leech applications (Semikova et al. 1999). Apparently, such adverse reactions never appear when leeches are applied on oral, nasal or vaginal mucosa (Kamenev and Baranovskii 2006). In very rare cases, allergic skin reactions have been observed.

When leeches are applied on intact skin a slight pain is experienced, which ceases after a short period of time and most probably is due to analgesics existing in the saliva of the leech. However, the bite of the leech is not felt when they are placed on flaps of reattached digits. Symbiotic bacteria such as *Aeromonas hydrophila*, *Aeromonas veronii* and *Aeromonas media*, living in the intestinal tract of the leech may cause infections in 4–20 % of the patients, whose flaps or replanted digits are treated with leeches (Lineaweaver et al. 1992; Mercer et al. 1987; Sartor et al. 2002). Accordingly, prophylactic treatment with antibiotics is necessary. *Aeromonas* infections can occur acutely (within 24 h) or delayed (up to 26 days after the beginning of leech therapy). Clinical manifestations of *Aeromonas* infection vary from a minor wound infection to extensive tissue loss. *Aeromonas* species are sensitive to second- and third-generation cephalosporins, ciprofloxacin, sulfamethoxazole-trimethoprim, tetracycline and aminoglycosides (Eldor et al. 1996). Patients undergoing plastic surgery should be treated during each day of leech therapy with 500 mg of ciprofloxacin (Mumcuoglu et al. 2010).

It is important to stress that leech related *Aeromonas* infections more frequently develop in severely sick and immunosuppressed patients. When applied on intact skin, e.g., in patients treated for osteoarthritis, local pain, arterial hypertension and different forms of spondylo- and dorsopathies, *Aeromonas* infections are extremely rare.

The excess bleeding after leeching can be of concern and transfusions may be needed, especially in patients with a tendency to hemorrhage, who suffer from anemia or for those taking anticoagulants or platelet-inhibiting drugs (Eldor et al. 1996).

The hemoglobin level of the patient should be monitored every 4 h and consists of a complete blood cell count, partial thromboplastin time, and serum chemistry. Blood transfusions are given based on a hemoglobin level of <8 g/dL. Usually, 3–6 units of packet blood cells are used to compensate for blood loss (de Chalain 1996; Chepeha et al. 2002).

All measures should be taken to avoid the migration of detached leeches (before or after engorgement) inside body orifices, thus, a special dressing around the wound is necessary (de Chalain 1996).

Smolle et al. (2000) described the appearance of firm, brown-red nodules at the site of the leech bites several weeks later, histologically resembling follicular pseudolymphoma, in a patient with chronic venous insufficiency in both legs. In this case, intralesional corticosteroid treatment was used with success.

In very rare cases, thrombotic microangiopathy and renal failure have been reported when leeches were applied in patients with arterial insufficiency and hirudotherapy is contraindicated in these patients.

Isolated reports described the appearance of syncopal status or orthostatic hypotension (mainly in older persons with reduced sympathicotonus) at the start of or during leech therapy. According to Michalsen et al. (2007) vasovagal attacks can occur in 0.1 % of patients undergoing hirudotherapy, mainly in those with a history of developing vasovagal attacks or syncope during invasive procedures such as venipuncture. It is highly recommended that patients drink plenty of fluids during hirudotherapy. In order to prevent those systemic side effects, hirudotherapy should be performed in a calm atmosphere, under constant blood pressure monitoring and with the patient lying down.

When leeches are applied to the esthetically important areas with thin skin and thin layers of subcutaneous tissue scarring after leech bite could be a cosmetic problem. In some cases, an on-mucosal leech application could avoid these complications (Gilyova et al. 1988).

Hirudotherapy can cause some negative psychological or emotional reactions in patients and even their family. Less than 10 % of patients undergoing leech therapy for osteoarthritis had initial fears before treatment, which usually disappeared after the first treatment course (Michalsen et al. 2007). In fact many patients successfully treated with leeches changed their attitude towards hirudotherapy in a positive way. Nevertheless, it might be necessary to prepare the patients psychologically before the application of leeches. In any case, before undergoing treatment with leeches patients and their relatives must be entirely informed about the benefits and potential risks of the treatment. Informed consent must always be obtained.

3.10 Contraindications to Leech Therapy

Contraindications related to hirudotherapy include hemophilia, hemorrhagic diathesis, hematological malignancies, severe anemia, severe hypotension, sepsis, HIV-infection, decompensated forms of hepatobiliary diseases, any form of cachexia, as well as individual intolerance to leeches. Leech therapy is also not recommended in pregnancy and lactation, in children under 10 years of age, in patients with an unstable medical status, history of allergy to leeches or severe allergic diathesis, disposition to keloid scar formation, severe arterial insufficiency, as well as in those using anticoagulants, immunosuppressants and some vasoactive drugs (e.g., *Ginkgo biloba* products). In addition, leech therapy should not be used in patients who refuse to receive subsequent blood transfusions, e.g., in reconstructive and plastic surgery.

3.11   Conclusions and Future Research

Throughout the long history of medicine, interest in leech therapy has periodically risen and fallen. However, in many countries the past three decades have witnessed the resurgence of interest in leeches as an effective treatment modality among physicians and patients. Leeches have grown in popularity once again in reconstructive and plastic surgery, and traumatology and they have also made a comeback in some fields of internal medicine.

Characteristics of modern health care systems in a given country as well as its medical traditions determine the extent of leech therapy in different fields of medicine. Progress in understanding the action mechanisms of leech therapy, based on scientific research of the past 30 years, has largely stimulated the interest in this unique method of treatment, although as in practically all other treatment modalities, well-designed clinical trials are necessary.

Leeches can be successfully used in the treatment of painful osteoarthritis, disorders of the varicose veins, mild forms of arterial hypertension as well as in patients with contraindications for the use of conventional antihypertensive drugs.

The results obtained in patients with central nervous system and peripheral nervous system disorders, are promising. However, further multi centered, randomized and controlled clinical trials should be also carried out to better define the role of leeches in the treatment of different forms of neuropathology.

The use of leeches could give new perspectives in some fields of modern medicine such as ophthalmology, gynecology, dermatology, otorhinolaryngology and even dentistry. However, more experimental and comparative clinical research should be done to determine the efficacy and advantages of hirudotherapy. The difficulties in conducting blind clinical studies with leeches are obvious. The choice of the most effective and suitable hirudotherapy techniques for each disorder and symptom should be optimized, while novel fields of leech application should be explored.

According to experts, the composition and properties of *H. medicinalis* saliva are still not entirely investigated. Undoubtedly, identification of new physiologically and biochemically active ingredients of leech saliva can deepen our understanding of hirudotherapy, and give impetus to new leech-based medicine. Further research on the biologically active compounds of the saliva of other hematophagous leech species is also warranted. In addition, the resurgence in the use of the medicinal leech in modern medicine might help in its protection in nature.

References

Adams SL (1989) The emergency management of a medicinal leech bite. Ann Emerg Med 18:316–319

Andereya S, Stanzel S, Mans U et al (2008) Assessment of leech therapy for knee osteoarthritis: a randomized study. Acta Orthop 79:235–243

Bapat RD, Acharya BS, Juvekar S, Dahanukar SA (1998) Leech therapy for complicated varicose veins. Indian J Med Res 107:281–284

Baskova IP (1999) Hirudotherapy method in prophylaxis and treatment of cardiovascular diseases. In: Clinical and experimental hirudology in the new millennium. Book of abstracts of the 6th conference of hirudologists of Russia and CIS, Pyatigorsk, pp 4–5

Baskova YP, Nikonov GY (1987) Leeches prostaglandins and the enzyme destabilase as thrombotic agents of the preparation from the medicinal leeches. Thromb Haemost 58:70

Baskova YP, Nikonov GY (1991) Destabilase, the novel isopeptidase with thrombolytic activity. Blood Coagul Fibrinolysis 2:167–172

Baskova IP, Jusupova GI, Nikonov GI (1984) Lipase and choline-esterase activity of salivary secrete of Hirudo medicinalis. Biochimia 49:676–678

Baskova IP, Nikonov GI, Mirkamalova EG et al (1988) The influence of the drug from medicinal leeches (Hirudo medicinalis) on phagocytosis and the complement system. Kaz Med J 69:334–336

Baskova IP, Khalil S, Nartikova VF, Paskhina TS (1992) Inhibition of plasma kallikrein, kininase and kinin-like activities of preparation from the medicinal leech. Thromb Res 67:721–730

Belezkaya GA (2007) Hirudotherapy in patients with early stages of primary open-angle glaucoma. Krasnoyarsk

Bellamy N, Buchanan WW, Goldsmith CH, Campbell J, Stitt LW (1988) Validation study of WOMAC: a health status instrument for measuring clinically important patient relevant outcomes to antirheumatic drug therapy in patients with osteoarthritis of the hip or knee. J Rheumatol 15:1833–1840

Bogdanova II, Demidova OO, Polikarpova EE (2003) Hirudo medicinalis in the treatment of painful muscle syndromes of osteochondrosis. In: Baskova IP (ed) Treatment with leeches and preparations from them. Girudo-Med, Luberzi, pp 51–53

Borovaya EP (2008) Leeches and endoecological rehabilitation in a complex sanatorium treatment of patients with coronary heart disease of middle and old age. Moscow

Brody GA, Maloney WJ, Hentz VR (1989) Digit replantation applying the leech Hirudo medicinalis. Clin Orthop Relat Res 245:133–137

Bunker TD (1981) The contemporary use of the medicinal leech. Injury 12:430–432

Chaban TN, Kapralova VV, Savinov VA (2003) Hirudotherapy in gynecology. In: Baskova IP (ed) Treatment with leeches and preparations from them. Girudo-Med, Luberzi, p 68

Chalisova NI, Pennijajnen VP, Baskova IP, Zavalova LL, Bazanova AV (2003) Neurite-stimulating activity of components of the salivary gland secretion of the medicinal leech in cultures of sensory neurons. Neurosci Behav Physiol 33:411–414

Chemeris AV, Konirtaeva NN (1999) Hirudotherapy in Kazakhstan. In: Clinical and experimental hirudology in the new millennium. Book of abstracts of the 6th conference of hirudologists of Russia and CIS, Pyatigorsk, pp 125–126

Chepeha DB, Nussenbaum B, Bradford CR, Teknos TN (2002) Leech therapy for patients with surgically unsalvageable venous obstruction after revascularized free tissue transfer. Arch Otolaryngol Head Neck Surg 128:960–965

Chmeil H, Anadere I, Moser K (1989) Hemoreological changes under blood leeching. Clin Hemorheol 9:569–576

de Chalain TM (1996) Exploring the use of the medicinal leech: a clinical risk-benefit analysis. J Reconstr Microsurg 12:165–172

Derganc M, Zdravic F (1960) Venous congestion of flaps treated by application of leeches. Br J Plast Surg 13:187–192

Durrant C, Townley WA, Ramkumar S, Khoo CT (2006) Forgotten digital tourniquet: salvage of an ischemic finger by application of medicinal leeches. Ann R Coll Surg Engl 88:462–464

Dzekh SA (2000) The clinic, immunological disorders, hirudotherapy in patients with the Rossolimo-Melkerssona-Rosenthal syndrome. Moscow

Eldor A, Orevi M, Rigbi M (1996) The role of the leech in medical therapeutics. Blood Rev 10:201–209

Eldor A, Orevi M, Rigbi M (1998) Beneficial role of the leech in post-phlebitic syndrome. In: Book of abstracts of the 3rd international conference on biotherapy, Jerusalem, pp 12–13

Elliot JM, Tullet PA (1984) The status of the medicinal leech *Hirudo medicinalis* in Europe and especially in the British Isles. Biol Conserv 29:15–26

Ena YM (2003) Hirudotherapy in complex treatment of patients with hypertension disease. In: Baskova IP (ed) Treatment with leeches and preparations from them. Girudo-Med, Luberzi, pp 24–25

Farber FM (1987) Pathogenic methods of treatment of peripheral palsy of facial nerve. Novosibirsk

Fields WS (1991) The history of leeching and hirudin. Haemostasis 21:3–10

Fowler A (1999) The clinical application of leeches. In: Book of abstracts of the 4th international biotherapy conference "bridging the millenia", Porthcawl, pp 9–10

Fritz H, Oppitz KH, Gebhardt M, Oppitz I, Werle E, Marx R (1969) On the presence of a trypsin-plasmin inhibitor in hirudin. Hoppe Seylers Z Physiol Chem 350:91–92

Frolov VA, Frolova EA (1999) Combined use of hirudotherapy and manual therapy in treatment of scapulohumeral periarthropaty. In: Clinical and experimental hirudology in the new millennium. Book of abstracts of the 6th conference of hirudologists of Russia and CIS, Pyatigorsk, pp 26–27

Gantimurova OG (2001) Leeches in the treatment and rehabilitation of patients with arterial hypertension. In: Practical and experimental hirudology: results in a decade (1991–2001). Book of abstracts of the 7th conference of hirudologists of Russia and CIS, Luberzi, pp 8–10

Gilyova OS (2000) Hirudotherapy and the skin. In: Book of abstracts of the 5th international conference on biotherapy, Wurzburg, pp 9–10

Gilyova OS (2003) Hirudotherapy in modern Russian medicine: clinical and experimental aspects. In: Book of abstracts of the 6th international conference on biotherapy, Sivas, pp 22–23

Gilyova OS (2005) Modern hirudotherapy – a review. The BeTER LeTTER 2(1):1–3

Gilyova OS, Gibadullina NV (2003) Hirudotherapy in treatment of periodontal diseases: clinical and functional indices. In: Book of abstracts of the 6th international conference on biotherapy, Sivas, pp 24–25

Gilyova O, Rogozhnikov G, Barer G et al (1988) Hirudotherapy from the depth of centuries. West-Ural Branch of Russian Academy of Science Publishing House, Perm

Gilyova OS, Korobeinikova GA, Gilyova ES et al (1999) The achievements of modern medicine in systemic hirudotherapy. Russ J Biomech 3:68–77

Gilyova OS, Korobeinikova GA, Gibadullina NV (2000) Experimental studies on the healing and anti-inflammatory effects of leeching. In: Book of abstracts of the 5th international conference on biotherapy, Würzburg, p 10

Gilyova OS, Gibadullina NV, Korobeinikova GA et al (2010) Hirudotherapy in the complex treatment of periodontitis and oral mucosal diseases in patients with systemic cardiovascular pathology. In: Book of abstracts of the 8th international conference on biotherapy, Los Angeles, p 24

Gröbe A, Michalsen A, Hanken H, Schmelzle R, Heiland M, Blessmann M (2012) Leech therapy in reconstructive maxillofacial surgery. J Oral Maxillofac Surg 70:221–227

Grunner OCA (1930) Treatise on the Canon of Medicine of Avicenna incorporating a translation of the first book. Luzac and Co, London

Harris CA (1845) The principles and practice of dental surgery. Lindsay & Blackiston, Philadelphia

Hartrampf CR, Drazan L, Noel RT (1993) A mechanical leech for transverse rectus abdominis musculocutaneous flaps. Ann Plast Surg 31:103–105

Haycraft JB (1884) On the action of secretion obtained from the medicinal leech on coagulation of the blood. Proc R Soc Lond 36:478

Hayden RE, Phillips JG, McLear PW (1988) Leeches. Objective monitoring of altered perfusion in congested flaps. Arch Otolaryngol Head Neck Surg 114:1395–1399

Heldt TJ (1961) Allergy to leeches. Henry Ford Hosp Med Bull 9:498–519

Henderson LP, Kuffler DP, Nicholls J, Zhang R (1983) Structural and functional analysis of synaptic transmission between identified leech neurons in culture. J Physiol 340:347–358

Hobingh P, Linker A (1999) Hyaluronidase activity in leeches (Hirudinea). Comp Biochem Physiol B Biochem Mol Biol 124:319–326

Isakhanjan GS (2003) On the reflex mechanism of action of hirudotherapy. In: Baskova IP (ed) Treatment with leeches and preparations from them, Girudo-Med, Luberzi, pp 22–23

Isakhanyan GS (1991) Hirudotherapy in internal medicine. Girudo-Med, Erevan

Kalender ME, Comez G, Sevinc A, Dirier A, Camci C (2010) Leech therapy for symptomatic relief of cancer pain. Pain Med 11:443–445

Kamenev OJ (2001) Wild and artificially grown medical leech and their application in medical practice. In: Practical and experimental hirudology: results in a decade (1991–2001). Book of abstracts of the 7th conference of hirudologists of Russia and CIS, Luberzi, pp 90–92

Kamenev OJ (2007) Resources of medicinal leeches (Hirudo medicinalis L.) in the waters of the western Caucasus, and their rational use. Krasnodar

Kamenev OJ, Baranovskii AJ (2006) Treatment with leeches. Theory and practice of leeching. Ves', Saint Petersburg

Kamenev OJ, Tonkopii VD, Zagrebin AO (2003) Medicinal leeches as a test object for bioidentification of heavy metals in the aquatic environment. In: Hirudo-2003. Book of abstracts of the 3rd conference of hirudologists of Russia and CIS, Moscow, pp 57–58

Karadag AS, Calka O, Akdeniz N, Cecen I (2011) A case of irritant contact dermatitis with leech. Cutan Ocul Toxicol 30:234–235

Kasimov SS (1961) Clinical and experimental hirudotherapy in treatment of neuritis and neuralgia of sciatic nerve. Izvestia Akademii nauk AzSSR 4:91–95

Khardikov AV, Gazazyan MG, Ivanova OJ (2003) Hirudotherapy in treatment of chronic salpingo-oophoritis. In: Hirudo-2003. Book of abstracts of the 3rd conference of hirudologists of Russia and CIS, Moscow, p 22

Koch CA, Olsen SM, Moore EJ (2012) Use of the medicinal leech for salvage of venous congested microvascular free flaps of the head and neck. Am J Otolaryngol 33:26–30

Kolesnikova EA, Karzeva TV, Romanova GE et al (1999) The efficacy of hirudotherapy in complex treatment of child neuro-circulatory dystonia. In: Clinical and experimental hirudology in the new millennium. Book of abstracts of the 6th conference of hirudologists of Russia and CIS, Pyatigorsk, p 67

Korobeinikova GA (2002) Leeches in complex treatment of patients with essential hypertension and combined oral mucosa lesions in Ust-Katchka Health resort. Perm

Krashenjuk AI, Krashenjuk SV (2003) Hirudotherapy treatment of child cerebral palsy. In: Baskova IP (ed) Treatment with leeches and preparations from them. Girudo-Med, Luberzi, p 45

Kumar VD, Kumar CP, Kumar SA et al (2012) A critical review on historical aspects of jalauka-vacharan (hirudotherapy). IJRAP 3(1), Jan–Feb 2012. http://www.ijrap.net/admin/php/uploads/742_pdf.pdf. Accessed 11 Feb 2012

Kuznezova LP, Lusev VA, Volov VA et al (2008) The place of hirudotherapy in complex treatment of chronic heart failure. Russ Cardiol J 2:28–30

Lagutenko JP (1981) Structural organization of the trunk brain of annelids: evolutionary neuromorphological analysis. Nauka, Leningrad

Lineaweaver WC, Furnas H, Follansbee S, Buncke GH, Whitney TM, Canales F, Bruneteau R, Buncke HJ (1992) Postprandial Aeromonas hydrophila cultures and antibiotic levels of enteric aspirates from medicinal leeches applied to patients receiving antibiotics. Ann Plast Surg 29:245–249

Lomaeva BI (2003) The influence of hirudotherapy on haemostatic indices in women suffering from endometriosis. In: Book of abstracts of the 6th international conference on biotherapy, Sivas, pp 26–27

Lukin, EI (1976) Leeches. In: Fauna USSR. Academy of Science of the USSR, Moscow

Makulova GG (2003) Application of hirudotherapy in patients with Bechterew's disease. In: Hirudo-2003. Book of abstracts of the 3rd conference of hirudologists of Russia and CIS, Moscow, pp 29–30

Markwardt F (1955) Untersuchungen ueber Hirudin. Naturwissenschaften 52:537

Martin F, Dimasi N, Volpari C, Perrera C, Di Marco S, Brunetti M, Steinkühler C, De Francesco R, Sollazzo M (1998) Design of selective eglin inhibitors of HCV NS3 proteinase. Biochemistry 37:11459–11468

Marty I, Péclat V, Kirdaite G, Salvi R, So A, Busso N (2001) Amelioration of collagen-induced arthritis by thrombin inhibition. J Clin Invest 107:631–640

Menage MJ, Wright G (1991) Use of leeches in a case of severe periorbital haemotoma. Br J Ophthalmol 75:755–756

Mercer NS, Beere DM, Bornemisza AJ, Thomas P (1987) Medical leeches as sources of wound infection. Br Med J (Clin Res Ed) 294(6577):93

Michalsen A, Deuse U, Esch T et al (2001) Effect of leech therapy (Hirudo medicinalis) in painful osteoarthritis of the knee: a pilot study. Ann Rheum Dis 60:986

Michalsen A, Klotz S, Lüdtke R, Moebus S, Spahn G, Dobos GJ (2003) Effectiveness of leech therapy in osteoarthritis of the knee: a randomized, controlled trial. Ann Intern Med 139:724–730

Michalsen A, Roth M, Dobos G (2007) Medicinal leech therapy. Thieme Verlag, Stuttgart

Mokhov DE, Zalzman SG (2003) Hirudotherapy in neurology. In: Baskova IP (ed) Treatment with leeches and preparations from them. Girudo-Med, Luberzi, pp 39–40

Mumcuoglu KY, Pidhorz C, Cohen R, Ofek A, Lipton HA (2007) The use of the medicinal leech, Hirudo medicinalis, in the reconstructive plastic surgery. Internet J Plast Surg 4(2). doi:10.5580/3c6

Mumcuoglu KY, Huberman L, Cohen R, Temper V, Adler A, Galun R, Block C (2010) Elimination of symbiotic Aeromonas spp. from the intestinal tract of the medicinal leech, Hirudo medicinalis using ciprofloxacin feeding. Clin Microbiol Infect 16(6):563–567

Munro R, Jones C, Sawer R (1991) Calin – a platelet adhesion inhibitor from saliva of the medicinal leech. Blood Coagul Fibrinol 2:179–184

Mutimer KL, Banis JC, Upton J (1987) Microsurgical reattachment of totally amputated ears. Plast Reconstr Surg 79:535–541

Muzalevskii VM, Pastel VB, Petrova OV (1999) Complex treatment of neuroosteofibrosis foci in osteochondrosis of vertebra. In: Clinical and experimental hirudology in the new millennium. Book of abstracts of the 6th conference of hirudologists of Russia and CIS, Pyatigorsk, pp 24–25

Nikonov GI (1992) Medicinal leech. Yesterday, today, tomorrow. Electronika, Moscow

Nikonov GI (1996) Hirudotherapy and hirudopharmacotherapy. Voenizdat, Moscow

Orevi M, Rigbi M, Hy-Am E, Matzner Y, Eldor A (1992) A potent inhibitor of platelet activating factor from the saliva of the leech Hirudo medicinalis. Prostaglandins 43:483–485

Orevi M, Eldor A, Giguzin Y, Rigbi M (2000) Jaw anatomy of the blood-sucking leeches, Hirudinea Limnatis nilotica and Hirudo medicinalis, and its relationship to their feeding habits. J Zool Lond 250:121–127

Poprotskii AV, Aivazov VP, Khinchagov BP (2001) Differential treatment of different forms of dyscirculatory encephalopathy using hirudotherapy in Essentuki. In: Practical and experimental hirudology: results in a decade (1991–2001). Book of abstracts of the 7th conference of hirudologists of Russia and CIS, Luberzi, pp 30–32

Poprotskii AV, Aivazov VN, Khinchagov BP et al (2008) Efficiency and safety of hirudotherapy in ischemic cerebrovascular disorders. Vopr Kurortol Fizioter Lech Fiz Kult 4:51–55

Porshinsky BS, Saha S, Grossman MD, Beery Ii PR, Stawicki SP (2011) Clinical uses of the medicinal leech: a practical review. J Postgrad Med 57:67–71

Pospelova ML (2012) Overview of pathogenetic mechanisms of hirudotherapy and the rationale for the treatment of patients with cerebrovascular disease. Sovremennie problem nauki I obrazovania 1. www.science-education.ru/101-5291. Accessed 23 Jan 2013

Pospelova ML, Barnaulov OD (2003) Features of hirudotherapy in patients with chronic vertebro-basilar insufficiency. In: Hirudo-2003. Book of abstracts of the 3rd conference of hirudologists of Russia and CIS, Moscow, pp 35–36

Pospelova ML, Barnaulov OD (2006) Hirudotherapy influence on the depression and anxiety rates in cerebrovascular patients. Psychopharmacol Biol Narcol 6:1370–1375

Pospelova ML, Barnaulov OD (2010) Hirudotherapy in the treatment of bilateral internal carotid artery occlusion: case report. Curr Top Neurol Psychiatr Relat Discip 18:51–53

Pozharizkaya MM, Zidra SI, Dzekh SA et al (2000) Hirudotherapy in complex treatment in patients with the Rossolimo-Melkerssona-Rosenthal syndrome. Stomatologiya dlya vsekh 4:4–6

Ptushkin VV, Lapkes TG (2003) Application of hirudotherapy in complex treatment of patients with coronary artery disease and unstable angina. In: Baskova IP (ed) Treatment with leeches and preparations from them. Girudo-Med, Luberzi, pp 15–16

Rashidova ES (2003) The experience of hirudotherapy of neurological diseases. In: Hirudo-2003. Book of abstracts of the 3rd conference of hirudologists of Russia and CIS, Moscow, pp 34–35

Rassadina EV (2006) Environmentally sound reproduction biotechnology of *Hirudo medicinalis* L. in vitro. Uljanovsk

Rigbi M (1998) The antihemostatic armamentarium of leech saliva and its therapeutic potential. In: Book of abstracts of the 3rd international conference on biotherapy, Jerusalem, pp 28–29

Rigbi M, Jackson CM, Latallo ZS (1988) A specific inhibitor of bovine FaX in the saliva of the leech *Hirudo medicinalis*. In the 14th international congress of biochemistry, Prague, 10–15 July, FR 037:53

Rigbi M, Orevi M, Eldor A (1996) Platelet aggregation and coagulation inhibitors in leech saliva and their roles in leech therapy. Semin Thromb Hemost 22:273–278

Salikhov IG (2004) About hirudotherapy in patients with rheumatoid arthritis. Kaz Med J 85:53–55

Salzet M (2005) Neuropeptide-derived antimicrobial peptides from invertebrates for biomedical applications. Curr Med Chem 12:3055–3061

Sartor C, Limouzin-Perotti F, Legré R, Casanova D, Bongrand MC, Sambuc R, Drancourt M (2002) Nosocomial infections with *Aeromonas hydrophila* from leeches. Clin Infect Dis 35:1–5

Savinov VA, Chaban TN (1995) Comparative characteristics of vascular and reflex phenomena in hirudotherapy. In: Book of abstracts of the 4th conference of association of hirudologists, Moscow, p 21

Sawyer RT (1986) Leech biology and behaviour, vol 1–2. Clarendon, Oxford

Scheckotov GM (1980) Use of leeches in varicose veins. Voen Med J 3:68

Scott K (2002) Is hirudin a potential therapeutic agent for arthritis? Ann Rheum Dis 61:561–562

Seemüller U, Meier M, Ohlsson K, Müller HP, Fritz H (1977) Isolation and characterization of a low molecular weight inhibitor (chymotrypsin and human granulocytic elastase and cathepsin G) from leeches. Hoppe Seylers Z Physiol Chem 358:1105–1107

Seemuller U, Dodt J, Fink E et al (1986) Proteinase inhibitors of the leech *Hirudo medicinalis* (hirudins, bdellins, eglins). In: Barettand AJ, Salvesen G (eds) Proteinase inhibitors. Elsevier Science Ltd, New York, pp 337–359

Semikova TS (1992) Hirudotherapy in treatment of subretinal bleeding (clinical case). Ophtalmokhirurgiya 3:63–64

Semikova TS (2011) Complex therapy of acute optic neuritis using antihomotoxic drugs, hirudo-therapy and autohemotherapy. In: Book of abstracts of the 2nd world congress on controversies in ophthalmology, Catalonia Palace of Congresses, Barcelona, p 64

Semikova TS, Bondareva M (1995) Hirudotherapy in ophthalmology: recommendations for practitioners, Girudo-Med, Moscow

Semikova TS, Semikova MV, Jarovoj AJ et al. (1999) Complications and side effects of hirudo-therapy in ophthalmology. In: Clinical and experimental hirudology in the new millennium. Book of abstracts of the 6th conference of hirudologists of Russia and CIS, Pyatigorsk, p 47

Seselkina TN (2001) Leeches as an early replacement therapy in patients with acute ischemic stroke. In: Practical and experimental hirudology: results in a decade (1991–2001). Book of abstracts of the 7th conference of hirudologists of Russia and CIS, Luberzi, pp 14–16

Seselkina TN, Belitskaya RA, Vasilenko GF et al (1998) Hirudoreflexotherapy of acute apoplectic attack. In: Book of abstracts of the 3rd international conference on biotherapy, Jerusalem, p 35

Seselkina TN, Belitskaya RA, Vasilenko GF et al. (2003) Efficacy of hirudoreflexotherapy in treatment of patients with ischemic stroke. In: Baskova IP (ed) Treatment with leeches and preparations from them, Girudo-Med, Luberzi, pp 33–38

Shenfeld OZ (1999) Successful use of the medicinal leech (*Hirudo medicinalis*) for the treatment of venous insufficiency in a replanted digit. Isr Med Assoc J 1:221

Shipley AE (1927) Historical preface. In: Harding WA, Moore JP (eds) The Fauna of British India: Hirudinea. Taylor and Francis, London

Sidorov V, Gilyova OS, Korobeynikova G, Gilyova E, Sidorov D (2003) Hirudotherapy in a combination at UST-Katchka Resort. Sixth international conference on biotherapy, Sivas, p 39

Siebeck M, Hoffmann H, Weipert J, Fritz H (1992) Effect of the elastase inhibitor eglin C in porcine endotoxin shock. Circ Shock 36:174–179

Smolle J, Cerroni L, Kerl H (2000) Multiple pseudolymphomas caused by *Hirudo medicinalis* therapy. J Am Acad Dermatol 43:867–869

Smoot EC 3rd, Debs N, Banducci D, Poole M, Roth A (1990) Leech therapy and bleeding wound techniques to relieve venous congestion. J Reconstr Microsurg 6:245–250

Söllner C, Mentele R, Eckerskorn C, Fritz H, Sommerhoff CP (1994) Isolation and characterization of hirustasin - an antistasin-type serin-proteinase inhibitor from the medicinal leech *Hirudo medicinalis*. Eur J Biochem 219:937–943

Sommerhoff CP, Söllner C, Mentele R, Piechottka GP, Auerswald EA, Fritz H (1994) A Kazal-type inhibitor of human mast cell tryptase: isolation from the medical leech *Hirudo medicinalis*, characterization, and sequence analysis. Biol Chem Hoppe Seyler 375:685–694

Spizina VI (1987) New treatment method of stomalgia with medicinal leeches. Stomatologiya 1:34–36

Stange R, Moser C, Hopfenmueller W (2012) Randomized controlled trial with medical leeches for osteoarthritis of the knee. Complement Ther Med 20:1–7

Starzeva NV, Lomaeva IB, Beda JV et al (2001) Clinical and lab aspects of hirudotherapy of endometriosis. In: Practical and experimental hirudology: results in a decade (1991–2001). Book of abstracts of the 7th conference of hirudologists of Russia and CIS, Luberzi, pp 28–30

Stroganova VA (1999) The experience of hirudotherapy in the practice of pediatrician and child's neurologist. In: Clinical and experimental hirudology in the new millennium. Book of abstracts of the 6th conference of hirudologists of Russia and CIS, Pyatigorsk, pp 62–65

Sulim NI (2003) Hirudotherapy of deforming arthrosis of the temporomandibular joint. In: Hirudo-2003. Book of abstracts of the 3rd conference of hirudologists of Russia and CIS, Moscow, pp 32–33

Taylor HS (1860) The family doctor. John Potter, Philadelphia

Tuev AV, Koryukina IP, Shutov et al (2001) Treatment methods on Ust-Katchka Health resort. Zvezda, Perm

U.S. Food and Drug Administration (FDA) FDA clears medicinal leeches for marketing (2004) FDA Talk Paper. T04-19, Rockville

Upshow J, O'Leary JP (2000) The medicinal leech: past and present. Am Surg 66:313–314

Utley DS, Koch RJ, Goode RL (1998) The failing flap in facial plastic and reconstructive surgery: role of the medicinal leech. Laryngoscope 108:1129–1135

Van Wingerden JJ, Oosthuizen JH (1997) Use of the local leech *Hirudo michaelseni* in reconstructive plastic and hand surgery. S Afr J Surg 35:29–31

Varisco PA, Péclat V, van Ness K, Bischof-Delaloye A, So A, Busso N (2000) Effect of thrombin inhibition on synovial inflammation in antigen induced arthritis. Ann Rheum Dis 59:781–787

Whitaker IS, Izadi D, Oliver DW, Monteath G, Butler PE (2004) *Hirudo medicinalis* and the plastic surgeon. Br J Plast Surg 57:348–353

Whitaker IS, Oboumarzouk O, Rozen WM, Naderi N, Balasubramanian SP, Azzopardi EA, Kon M (2012) The efficacy of medicinal leeches in plastic and reconstructive surgery: a systematic review of 277 reported clinical cases. Microsurgery 32:240–250

White RL, Fries CM, Wells MM (2010) Leech therapy: new applications for an old treatment. Nurse. http://ce.nurse.com/ce193-60/leech-therapy-new-applications-for-an-old-treatment/. Accessed 23 Jan 2013

Yang T, Peng JR, Zhang HZ (2003) Medicinal uses of a blood-sucking leech, Hirudo nipponia in China. In: Book of abstracts of the 6th international conference on biotherapy, Sivas, p 47

Zadorova EV (2003) Hirudotherapy experience in patients with essential hypertension and coronary heart disease. In: Baskova IP (ed) Treatment with leeches and preparations from them. Girudo-Med, Luberzi, pp 16–18

Zakharova OA (2008) Pathogenetic substantiation of hirudotherapy in patients with articular form of rheumatoid arthritis. Chita

Zaslavskaya SD (1940) Capillary blood-letting as a method of the deep decongestion of internals. Vrach 9:613–616

Zavalova LL, Yudina TG, Artamonova II, Baskova IP (2006) Antibacterial non-glycosidase activity of invertebrate destabilase-lysozyme and of its helical amphipat'hic peptides. ∞ 52:158–160

Zhavoronkova NM (1995) Hirudotherapy in complex treatment of Bernhardt-Roth syndrome. In: Book of abstracts of the 4th conference of association of hirudologists. Moscow, p 10

Zhavoronkova NM (2003) Hirudotherapy in complex treatment of Dupley's syndrome. In: Baskova IP (ed) Treatment with leeches and preparations from them. Girudo-Med, Luberzi, p 57

Zhivoglyad RN, Nikonov GI (1998) Hirudotherapy of uterine lesions. In: Book of abstracts of the 3rd international conference on biotherapy. Jerusalem, pp 39–40

Zhivoglyad RN, Nikonov GI (2003) Comparative analysis of hirudo- and antibiotic therapy in hormone-induced and inflammatory diseases of female genital mutilation. In: Baskova IP (ed) Treatment with leeches and preparations from them. Girudo-Med, Luberzi, p 69

Chapter 4
Apitherapy – Bee Venom Therapy

Christopher M.H. Kim

4.1 Introduction

Bee Venom Therapy (BVT) is a bio-therapeutic medical treatment that utilizes the venom of the honeybee for the treatment of diseases.

Physicians dating back to Hippocrates used honeybee venom (HBV) to treat a variety of illnesses. Today, physicians are still using HBV to treat patients worldwide. Clinical trials and rigorous testing under certified licensed physicians have proven that HBV is an effective treatment modality. The benefits of this drug have proven to be remarkable all over the world from Russia to the United States.

The proponents of bee venom are extensive. In the case of chronic pain disorders such as rheumatism and arthritis, bee venom is used to combat inflammation and the degeneration of connective tissue. Neurological disorders such as migraine, peripheral neuritis and chronic back pain have also been treated successfully. In the case of autoimmune disorders such as multiple sclerosis and lupus, it restores movement and mobility by strengthening the body's natural defense mechanism. In addition, dermatological conditions such as eczema, psoriasis, herpes can be effectively treated. Most recently, bee venom is being investigated for treatment of cancerous tumors as well.

Bee venom therapy has been proven to be both safe and effective through extensive regulation and standards set under federal guidelines. Formerly known as a complementary alternative medicine (CAM), bee venom is gaining a larger audience in international scientific communities. As evidenced throughout medical history to be a traditional miracle treatment, bee venom is nature's organic solution to human affliction.

C.M.H. Kim, M.D. (✉)
Graduate School of Integrated Medicine, CHA University,
Yatapdong, Bundanggu 222, 463-836 Seongnam, Gyeonggi-do, Korea
e-mail: cmk@apimeds.com

M. Grassberger et al. (eds.), *Biotherapy - History, Principles and Practice:*
A Practical Guide to the Diagnosis and Treatment of Disease using Living Organisms,
DOI 10.1007/978-94-007-6585-6_4, © Springer Science+Business Media Dordrecht 2013

4.2   History of Bee Venom Use

From ancient times onwards it is recorded, that topical application of bee stings has been used for a wide range of ailments and diseases, all through Africa, Asia, Europe and the Americas. Hippocrates (ca. 460–370 BC), an ancient Greek physician and father of western medicine, was the first referring to apitherapy. According to Galen (129 – ca. 200 AD), a prominent Roman physician as well as surgeon and philosopher (of Greek ethnicity) apitherapy was understood as a common form of medical treatment for the sick. Famous rulers such as Charlemagne (Carolus Magnus, 747–814 AD) and Ivan the Terrible (1530–1584 AD) applied bee stings to treat illnesses such as gout. The intrinsic benefits of bee venom is also referred to in the Koran: *"There proceeded from their bellies a liquor wherein is a medicine for men"* (Kim 1986).

In the late 1800s, scientific studies and articles were in progress in Western Europe and Russia. The French physician Dr. Desjardins (1859) published his experiments on treating rheumatism and skin cancer with bee stings in the journal *"Abeille Medical"* (Medical Bee Journal). Dr. Langer (1897) reported his inspiring progressive scientific studies. In Austria, Dr. Philip Terc (1904) developed the first systematic application of bee stings with over 25 years of expertise and research in rheumatic disorders.

In the 1900s, medical understanding of bee venom treatment progressed with Dr. Rudolph Tertsch (1912) who published several research papers on the treatment of rheumatic fevers and arthritis as well as a book (*"Das Bienengift im Dienste der Medizin"*) describing his father's research. Dr. Franz Kretsky (1928) from Austria developed the first injectable form of bee venom. Studies performed by Neumann et al. (1952) indicated that crude bee venom was a complex compound characterized by a unique biochemical composition. Decades later, in 2003, an American physician Dr. Christopher Kim would patent the first standardized form of injectable honeybee venom branded as Apitoxin in South Korea. The apitherapy (Bee Venom Therapy) timeline is summarized in Table 4.1.

4.2.1   Modern Bee Venom Therapy

Modern-day apitherapy is known as bee sting therapy (BST) or bee venom therapy (BVT). BST utilizes the actual honeybee with its stinger, whereas BVT involves injection of venom, which is extracted from honeybees in their hive environment without any harm to their colony. BVT injections can treat a wide range of disorders such as multiple sclerosis, lupus, chronic inflammation and pain, and as well as rheumatoid arthritis, neurological disorders such as migraines and peripheral neuritis. Furthermore, dermatological conditions such as eczema, psoriasis, herpes, urinary tract infections, and other viral infections are combated as well (NIH 1995). In South Korea, BVT injections are Korean FDA certified as a proven bio-therapeutic drug, and in the United States, it is currently awaiting FDA approval.

Table 4.1 Brief history of apitherapy

460–370 BC	Hippocrates utilized bee stings on his patients for treatment of disease
23–79 AD	Pliny the Elder, Roman naturalist and Commander of the Roman Army and Navy, prescribed bee venom and cited the beneficial uses of it in his "Natural History"
129–199 AD	Galen, the "Prince of Physicians" and also the "Father of Experimental Physiology," mentioned the uses of bee venom in his 500 treatises on medicine
742–814 AD	Charlemagne, the "King of Franks," who built the largest empire in Western Europe since Roman times was treated with bee stings. At this point in time, bee stings were used to cure almost all maladies and illnesses
1530–1584	Ivan the Terrible, Ivan IV of Russia, who suffered from gout was cured with bee stings
1600–1634	Monfat, another known naturalist, prescribed bee sting venom for reducing kidney stones, as well as strengthening the urinary tract in treating infections and other ailments
1859	Dr. Desjardins, a French physician, published the first scientific paper on the successful treatment and curative properties of bee venom for rheumatic diseases in the journal "*Abeille Medical*" (Medical Bee Journal). He also reported to have cured two individual cases of skin cancer
1888	Dr. Philip Terc, an Austrian physician, one of the foremost pioneers of bee venom therapy, applied over 39,000 bee stings to over 500 rheumatic patients during a 25 year period. He was the first to apply bee stings systematically and published his first article, "Report About a Peculiar Connection Between the Bee Stings and Rheumatism" in the *Vienna Medical Press*
1894	Dr. M.I. Lukomsky, a Russian professor at Saint-Petersburg Forestry, published his successes of bee venom therapy in treating rheumatic fever, gout, neuralgia, and other diseases
1897	Dr. Lyubarsky, a Russian military surgeon, concluded from his many years of expertise with bee venom that it was a successful treatment and remedy for rheumatic fever. He published these findings in his article entitled, "Bee Venom As A Curative Agent"
1912	Viennese Dr. Rudolph Tertsch, the "Father of Modern Apitherapy," published scientific research studies and several publications on the treatment of rheumatic fever with bee venom. Out of his 660 rheumatic arthritis patients, his findings were staggering: 82 % were completely cured (544 persons); 15 % showed improvement (99 persons); and only 3 % (17 persons) failed to show symptoms of recovery
1928	Dr. Franz Kretsky, an Austrian physician, was the first ever to invent an injectable form of bee venom
1932	Yoannovoitch and Chahovitch both treated experimental cancerous tumors with bee venom and published their findings in the *Bulletin de l'Academie de Medicine*
1935	Dr. Bodog F. Beck, Hungarian-American physician, coined the phrase "bee venom therapy" for the first time ever in history, and published his work on the effects of bee venom entitled, "Bee Venom, Its Nature, and Its Effects on Arthritis and Rheumatoid Conditions" in New York, USA
2003	Dr. Christopher Kim, Korean-American physician, patents the first standardized and federal regulated injectable form of honeybee venom known as *Apitoxin (purified Apis mellifera toxin)* in Korea; this bio-pharmaceutical drug is awaiting United States FDA approval branded as *Apitox*

The list of renowned apitherapists in the United States include Bodog F. Beck, M.D. (New York), Raymond Carey, M.D. (California), P.H. O'Connell, M.D. (Connecticut), Joseph Broadman, M.D. (New York), L.A. Doyle, M.D. (Iowa), Joseph Saine, M.D. (Montreal) and Charles Mraz, Master Beekeeper (Vermont).

4.2.2 American Apitherapy Society (AAS)

In 1977 the American Apitherapy Society was originally organized as the North American Apitherapy Society (NAAS) in Washington, D.C., but was reconstituted as the American Apitherapy Society (AAS) in 1988 as a non-profit organization (Kim 2011). The Society is primarily responsible for informing the general public and the medical profession in matters relating to apitherapy. They collect research data and information on apitherapy approaches and methods, and also conduct seminar workshops on a yearly basis.

4.3 Venom Source (*Apis mellifera*)

The genus *Apis* or the honeybee, has at least five separate species with several sub-species. All species thrive in large colonies set up in their natural home environment called the hive. Since honeybees are social insects, their hives are polymorphistic in nature, where the inter-relationship and structural mechanism of the colony thrives on the unique existential relationship between its queen, drones, and worker bees (Fig. 4.1).

Larvae fed on royal jelly only will develop to future queens. The queen, upon birth, must fight for the crown and for this purpose kill her sisters. The newly established queen, which will live for a period up to 7 years, will be fed with the royal food and will grow three times larger and twice her length to be able to oviposite large numbers of eggs. She will lay about 1,000–2,000 fertilized eggs per day via the ovipositor. The worker bees have a modified ovipositor called a stinger with which they can sting to defend the hive against outside intruders.

The male bees are called drones and their primary function is to mate with the queen in order to fertilize the eggs. Like the queen, drones cannot procure food and are entirely dependent on the worker bees. The worker bees are females only, and they take care of all the hive's needs, e.g., collect nectar, pollen, and water, construct combs, produce honey, and feed the brood. The worker bees also produce wax and propolis and perform maintenance functions such as hive cleaning and ventilation of the hive with the help of their wings and muscles in order to maintain the hive's temperature at a constant level.

When intruders such as wasps penetrate the hive, they can kill the bees and use the different developmental stages of the bee as food. In such cases, the bees fight back and use their stinger at the expense of their lives. If a worker bee uses her

Fig. 4.1 Honeybees: queen, drone and worker

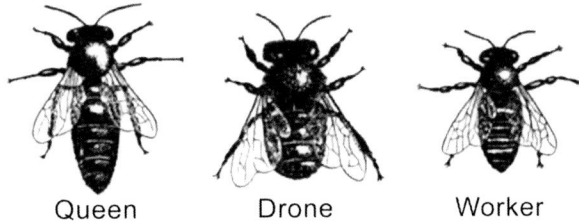

Queen   Drone   Worker

Table 4.2 Important honeybee species

Species name	Description
Apis cerana	This species is found in Asia and is known as the Indian honeybee. Bees of this species are spread throughout most of the Asian continent
Apis dorsata	This species is the largest among honeybees, and is known to have the most powerful and strongest sting out of the five species
Apis florea	This species (also called dwarf honeybee) is quite rare and the smallest of the colony numbers. In Asia, the honey from this particular species commands a premium price due to its rarity
Apis dorsata laboriosa	This species is the world largest honeybee and is morphologically quite close to Apis dorsata and is found only in the Himalayas
Apis mellifera	This species of honeybee is worldwide the most commercially used one and is used for Bee Sting Therapy and Bee Venom Therapy for approved bio-therapeutic drugs such as Apitoxin/Apitox

stinger, she loses the entire sting apparatus, which is deadly to the bee. Hence, honeybees only attack as a last resort and are usually not harmful, unless they are threatened with annihilation.

It is important to note that throughout history, humans have cultivated honeybee hives usually without danger. Despite the fact that honeybees have a powerful sting, these bees only use it when they know they are being attacked. The history of apitherapy, hive cultivation, and farm cultivation prove thast people and honeybees can coexist together in harmonic balance and subsistence with one another in nature. Important honeybee species are listed in Table 4.2.

4.4 Application Methods

4.4.1 Bee Venom Therapy (BVT)

Bee Venom Therapy (BVT) is the use of *Apis mellifera* toxin with the market name *Apitoxin* and *Apitox*. This remedy is applied intradermally, and never intravenously. Up to 2.0 mL of this solution can be injected subcutaneously, however the therapeutic effects are less satisfactory than intradermal delivery. Only licensed medical doctors and government approved acupuncture physicians can deliver bee venom injections,

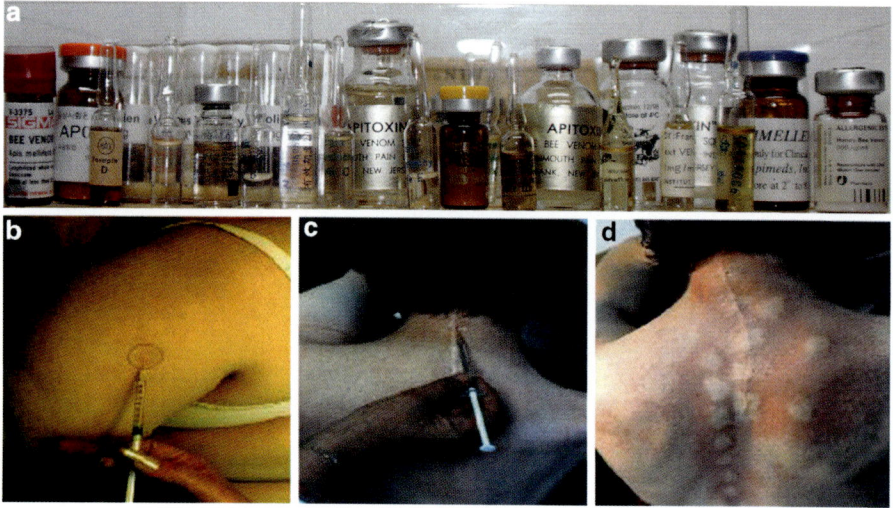

Fig. 4.2 (**a**) samples of bee venom for injection; (**b**) injection site; (**c**) before injection; (**d**) after several injections

since it must be regulated and monitored. BVT is used for the treatment of a variety of autoimmune diseases, neurological disorders, chronic ailments and inflammations. Figure 4.2 shows samples of bee venom and the local reactions before and after injection of bee venom.

4.4.2 Bee Sting Therapy (BST)

Bee Sting Therapy (BST) or Apipuncture is a method which is used by licensed practitioners and licensed acupuncturists to treat ailments such as skin disease and arthritis using the sting of a live bee. The treatment involves placing a live bee, held by tweezers, on the affected area of the body and simply allowing it to sting the patient.

Before the initiation of the treatment, the patient should be tested for allergies to bee venom. The test is usually applied on the forearm by forcing a bee to sting the skin and removing the stinger after a minute. Since most people have localized skin reactions to bee venom, the practitioner will determine the level of severity depending on the patient's condition. If the patient does not show any abnormal reactions to the test after a period of 10–15 min, the treatment session can begin. The therapist must record the patient's history, symptoms, and the results of treatment. It is safe to start with only 1–2 stings in the first session when the patient's condition and reaction time to sting is recorded.

The bee stings appear to work best when applied to so-called "trigger" points that align with certain acupressure locations on the body. Once the point is identified and the sting applied, the initial effect will be minimal. However, as time goes by, the area will redden, swell, and feel warm (Fig. 4.2d). The patient may also experience localized pain. This indicates that an immune response has been triggered

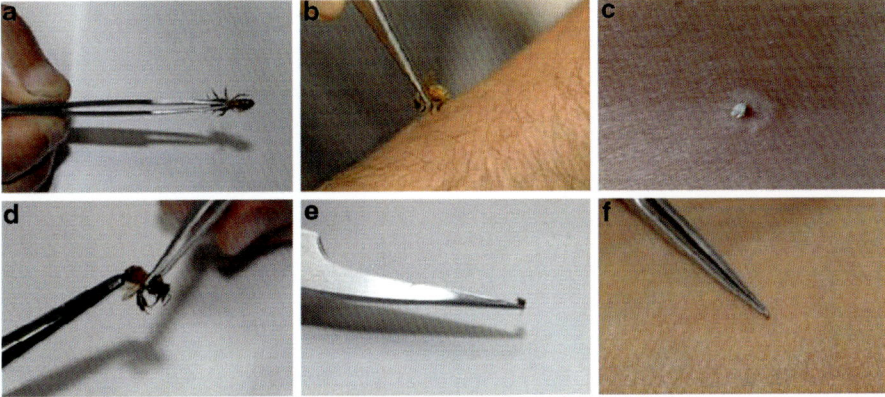

Fig. 4.3 The two methods of bee sting therapy. (**a-c**) direct application; (**d-f**) indirect application

and potential healing processes are occurring. Over the next several weeks, the patient will receive stings at regular intervals. As the treatment progresses, the patient will become desensitized to the bee venom and will stop showing a swelling when stung. At this point, the treatment is usually discontinued.

There are two ways to apply Bee Sting Therapy:

The direct application – catching a bee with tweezers or fine forceps, and grasp a bee by the head or thorax, not the abdomen (Fig. 4.3a); put the bee at the spot on the intended area (Fig. 4.3b); remove the stinger within 1 min (Fig. 4.3c) and repeat steps a to c to treat different areas.

The indirect application – remove the sting using two tweezers or fine forceps (Figs. 4.3d, e); put the stinger on the target point for only few seconds (Fig. 4.3f); repeat step f, on up to ten different locations (using same stinger) and repeat steps d to f when it is necessary.

4.4.3   Inhalation

This method is used in clinics and hospitals in Russia. The remedy is composed of a unique blend of aromatized water and purified honeybee venom and it is applied by inhalation with the help of a porcelain tube. Nowadays, this treatment modality is being tested by physicians worldwide.

4.4.4   Iontophoresis

Iontophoresis (also called Electromotive Drug Administration, EMDA) is the introduction of drugs into the body by means of electrical current. This external application of bee venom is yet to be scientifically validated, however it might be useful for patients who cannot take injections.

4.4.5   Ointments

Ointments based on bee venom can easily be applied at home. However, this type of delivery can produce lesions and irritations on large areas of the skin and it is much less effective than intradermal injections.

4.5   Bee Venom Treatment (BVT) – Clinical Application

In case of BVT, the solution is administered intradermally using a 1.0 mL sterile syringe with 0.1 mL graduations, 25–27 gauge, with a 1/2 in. or 1/4 – 5/8 in. needle. A standard dose of 0.1 mL is given intradermally at each injection site. Prior to injection, the plunger should be retracted and to avoid intravascular injection. The injections are given in the painful or afflicted area first. It is considered more efficacious if injections are given to the tender or trigger points. Later, injections to the spinal area could be given according to the dermatome chart to achieve maximum effect.

4.5.1   Allergy Test

Like in bee sting therapy patients should always undergo an allergy test before initial treatments with honeybee venom (HBV). The test is conducted by injecting 1.0 mg/1.0 mL of bee venom on the flexor surface of the patient's forearm. The test area should be disinfected prior to injection and a sterile, disposable syringe should be used. The needle should be introduced into the superficial skin layer until the bevel of the needle is completely buried. 0.05 mL (one half of a standard single treatment dose) is slowly injected to create a small hemispherical bleb.

A pinpoint sized blood spot usually appears on the area of needle insertion and soon after a wheal with a diameter of 0.5–1.0 cm and a 2.5–4.0 cm large erythematous area appear in the injection site. The severity of the reaction is monitored according to the size of the wheal, size of erythema, and the appearance of any irregular spreading or pseudo-pod like projections on the test area. A patient is considered negative to the test if he/she does not develop any of the systemic reactions within 15–30 min after the intradermal injection of HBV.

4.5.2   Dosing Schedule

HBV is given in increments, depending on patient's ability to tolerate honeybee venom. Since sensitivity to venom differs in people, it is not possible to prescribe a dosage schedule that is universal to all patients. For overly sensitive patients, an individualized schedule should be employed.

In general, two sessions per week or treatments at 3 days interval are recommended. The doses of the venom should be increased with each succession, e.g. 3×, 5×, 7× and up to 20 injections, at one session. The total number of injections might be higher than 20 (2.0 mg/2.0 mL) depending physician's expertise and experience.

4.5.3 Treatment Duration

To achieve maximum results, it is necessary that patients receive an average of 12–20 sessions for most chronic inflammatory diseases. Chronic disabling conditions such as rheumatoid arthritis and multiple sclerosis may require longer treatment periods.

4.5.4 Instruction for Patients

For optimal treatment results patients should be instructed to adhere to the following recommendations during BVT:

1. Alcohol is strictly forbidden during treatment.
2. If a local reaction in form of swelling and/or itching appears, ice packs could be applied to the site.
3. It is recommended to take 2.000–3.000 mg of Vitamin C daily.
4. If fever and chills occur, acetaminophen could be taken (650 mg every 3–4 h as needed).
5. In case of severe itching an anti-histaminic drug might be prescribed.
6. If any serious reactions occur, the caring physician should be contacted immediately.

4.5.5 Cautions

Vyatchannikov and Sinka (1973) reported bee venom neurotoxicity is not severe or dangerous when administered correctly. Adolapin (a newly isolated analgetic and anti-inflammatory polypeptide from bee venom) has an analgesic effect in which central mechanisms may also be involved. This is suggested by the fact that Naloxone, a blocker of the opiate receptors eliminates this analgesic effect. Injection of Apamin (a neurotoxin found in bee venom) in the spinal cord causes an increase in monosysnaptic extensor reflex potentials and also in polysynaptic potentials due to flexor afferents. MCD-Peptide and Phospholipase A_2 may decrease blood pressure if administered incorrectly. Apamin may also induce an anti-arrhythmic or cardiac effect if administered improperly. Therefore it is important to note, that Honeybee Venom (HBV) must be administered by licensed physicians or by government approved practitioners.

All physicians and licensed practitioners must be thoroughly aware of potential adverse reactions before administering honeybee venom such as *Apitoxin/Apitox* to prevent over-dosage and adverse reactions during pregnancy. HBV must only be administered by physicians who are experienced with BVT to provide efficient and maximum tolerated dose levels. Due to the possibility of severe systemic reactions, the patient must always be fully informed with precautionary recommendations and must be under constant direct supervision by their physicians. In case of emergency the patient must be given an injection of subcutaneous epinephrine (adrenaline) immediately, and must be transferred to hospital.

4.6 Biochemical Properties

The honeybee contains 0.1–0.3 mg of venom in its poison glands. The venom concentration is transparent, aromatically pungent, and bitter in taste. Its specific gravity is 1.1313, and its pH 5.2–5.5. The venom is very resistant to heat and can withstand temperatures of up to 100 °C or 212 °F. Dehydration of the venom takes approximately 10-days and does not affect the potency of the venom. The venom is also very resistant to cold, moreover, the toxicity is not reduced in freezing conditions. Desiccated venom, if protected from moisture, can retain its potency for several years. Figure 4.4 shows samples of dry bee venom.

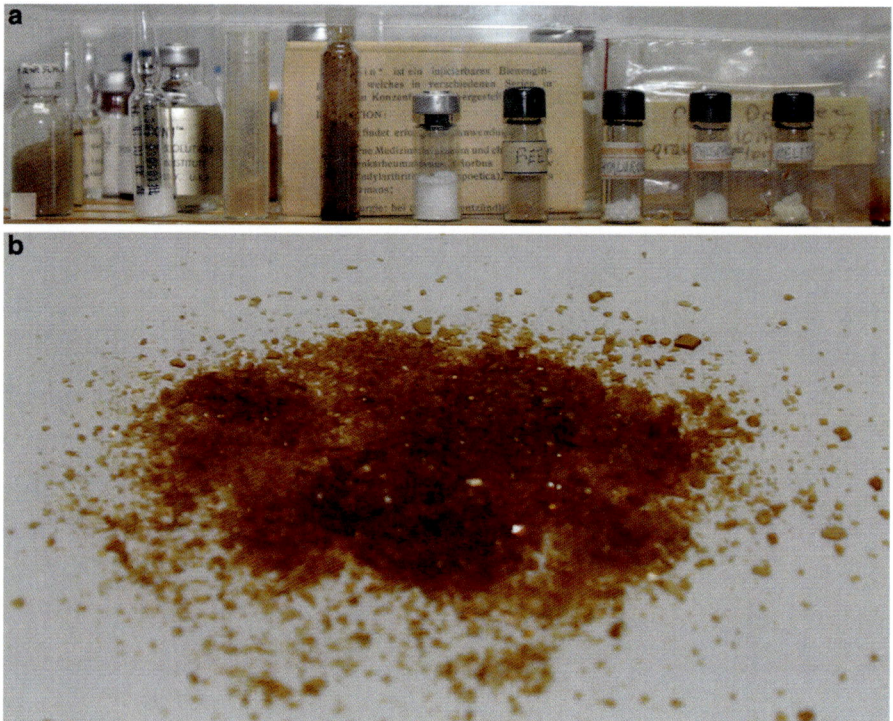

Fig. 4.4 (**a**) samples of dried bee venom (raw); (**b**) sample of dried Apitox

Table 4.3 General composition of honeybee venom

	Components	Mol. wt.	% (dry)
Peptides	Melittin	2,840	40–50
	Apamin	2,036	2–3
	MCD-Peptide (Peptide 401)	2,588	2–3
	Adolapin	11,500	1.0
	Protease inhibitor	9,000	<0.8
	Secarpin		0.5
	Tertiapin		0.1
	Melittin F		0.01
	Procamine A, B		1.4
	Minimine	6,000	2–3
Enzymes	Hyaluronidase	38,000	1.5–2
	Phospholipase A2	19,000	10–12
	-D-Glucosidase	170,000	0.6
	Acid phosphomonoesterase	55,000	1.0
	Lysophospholipase	22,000	1.0
Active amines	Histamine		0.6–1.6
	Dopamine		0.13–1.0
	Norepinephrine		0.1–0.7
	Glucose & fructose (*carbs*)		<2.0
	6 Phospholipids (*lipids*)		4–5
	r-aminobutyric acid (*amino acids*)		<0.5
	b-aminoisobutyric acid		<0.01

Adapted from Kim (1992)

There are numerous studies regarding the biochemical composition of honeybee venom (Tu 1977; O'Connor and Peck 1978; Shipolini 1984; Meier and White 1995). It is important to note that semi-purified fractions and purified concentrations of HBV are being produced, which should be even more effective in the treatment of patients (see Table 4.3).

4.7 Pharmacology

The pharmacologic profile of HBV has been determined with the aid of in-vitro analysis and research in animals and humans (Minton 1974; Tu 1977; O'Connor and Peck 1978; Habermehl 1981; Shipolini 1984; Meier and White 1995). The different ingredients of HBV are discussed below briefly.

4.7.1 Melittin

Melittin has anti-inflammatory properties that stimulate the pituitary-adrenal system which releases cortisol (Vick and Shipman 1972; Vick et al. 1972; Knepel and

Gerhards 1987). It is known to be 100 times more potent than hydrocortisone as evidenced in animal models (Vick et al. 1972). Melittin is also known to stabilize lysosome cell membranes and it is involved in the specific mechanisms that control against inflammation (Dufourcq 1986).

Melittin is known for its anti-bacterial, anti-fungal, and anti-viral properties (Dorman and Markey 1971; Fennell et al. 1968) as well as protection from X-irradiation by increasing resistance against irradiation (Shipman 1967; Ginsberg et al. 1968; Shipman and Cole 1968; Kanno et al. 1970). Furthermore, it has an anti-nociceptive effect (Son et al. 2007) and is known for its antitumor activities (Orsolic et al. 2003; Liu et al. 2008).

4.7.2 Apamin

Apamin stimulates the pituitary-adrenal system which releases cortisol (Vick and Shipman 1972). It reduces inflammation caused by dextran and serotonin induced inflammation. Additionally, it inhibits the complement system component C3, which is known to cause inflammation (Gencheva and Shkenderov 1986).

Habermann and Cheng-Raude (1975) reaffirmed this by stating that the neuro-toxic and postsynaptic effects (efferent system) of apamin blocks inhibition from α, not from β adreno-receptors. Furthermore, apamin is known to antagonize neurotensin-induced relaxation, while blocking most of the hyperpolarizing inhibi-tory effects (invertebrate smooth muscle) including: α-adrenergic, cholinergic, puri-nergic, and neurotensin-induced relaxations, and not β-adrenergic relaxation.

Apamin is a selective blocker of the Ca-dependent K^+ channels that are present in cell membranes (Hugues et al. 1982a). It also supports neural functional ability as concluded by with rat brain synaptosomes (Hugues et al. 1982b).

4.7.3 Mast-Cell Degranulating (MCD) Peptide (Peptide 401)

MCD peptide was originally named due to its biological action of causing release of histamine from mast cells (Jasani et al. 1979; Banks et al. 1990). MCD Peptide is known to block arachidonic acid and inhibits prostaglandin synthesis (Hanson et al. 1974). It has the property of inhibition of epilepsy and inhibiting a voltage-dependent potassium channel in brain membranes (Gandolfo et al. 1989).

4.7.4 Adolapin

Adolapin inhibits the microsomal cyclooxygenase and is 70 times stronger than indomethacin as shown in animal models. Moreover, adolapin inhibits platelet lipoxygenase, which includes hydroperoxy-eicosotetranonic acid (HPETE) and leukotriens, as well as thromboxane (TXA_2) and prostacycline (PGI_2), which are activated during inflammation. Additionally adolapin has anti-nociceptive and anti-inflammatory properties through inhibition of cyclooxygenase activity (Shkenderov and Koburova 1982). Adolapin has an analgesic effect (Koburova et al. 1985) and antipyretic properties (Koburova et al. 1984), in which central mechanism may also be involved.

4.7.5 Protease Inhibitor

The protease inhibitor is known to prevent infections and viruses by inhibiting prostaglandin E_1-, bradykinin- and histamine-induced inflammation. It's also known to inhibit chymotrypsin and leucine-aminopeptidase (Shkenderov 1986).

4.7.6 Phospholipase A₂ (PLA₂)

Phospholipase A_2 has inflammatory, nociceptive (Hartman et al. 1991; Landucci et al. 2000) and antigenic and allergenic effects (Minton 1974; Habermehl 1981; Shipolini 1984). PLA_2 is involved in nerve regeneration (Edstrom et al. 1996), facilitates neurotransmitter release (Yue et al. 2005), and delayed neurotoxic effects in vitro and in vivo (Clapp et al. 1995). PLA_2 shows complex interactions with melittin that can result in potentiation of secretory PLA_2 effects or in inhibition depending on the peptide/phospholipid ratio (Koumanov et al. 2003).

4.7.7 Hyaluronidase

Hyaluronidase break down hyaluronic acid in tissues such as in synovial bursa of rheumatoid arthritis patients (Barker et al. 1964). Hyaluronidase is very sensitive to heat and light so that it is a good indicator of the stability of bee venom (Kim 2012).

4.8 Clinical Studies

Studies over the past 25 years evaluating the safety and efficacy of Apitoxin/Apitox have been conducted in patients with chronic pain and inflammation such as rheumatoid arthritis, osteoarthritis, fibromyositis, peripheral neuritis, multiple sclerosis, and a multitude of other diseases. These studies are discussed below according to the disease or disorder type.

4.8.1 Arthritis, Chronic Pain, and Inflammation

Zaitsev and Poriadin (1961) reported that of 150 patients stricken with ankylosing spondylitis and polyarthritic deformity, symptomatic relief was obtained in 117 cases (78 %), satisfactory results in 30 cases (20 %), no change in two cases, and allergic reaction in one case.

A controlled clinical study was conducted by Steigerwaldt and Mathies (1966) comparing 50 cases treated with a standardized preparation of bee venom against 11 cases treated with a placebo (injection of physiologic sodium chloride solution). The results showed beneficial effects in 84 % of the cases and 55 % in those using a placebo. After elimination of the cases that showed only weak improvement, the proportion of patients showing beneficial effects was 66 % of those treated with venom and 27 % of those who had been given a placebo.

Hurkov (1971) studied the use of venom in 180 patients suffering from osteoarthritis. Ninety-four showed a significant improvement and no adverse reaction was observed. Nokolova (1973) reported a success rate of 94 % of his patients stricken with rheumatoid arthritis who had been treated without success by conventional treatments. Serban (1981) compared indomethacine (100 mg/day) with purified bee venom (Forapin) in two groups of 50 patients. The groups consisted of 20 cases of gonarthrosis, 10 cases of coxarthrosis, and 20 cases of spondylitis. After 24 days the bee venom treatment produced better responses than indomethacine. Feldsher et al. (1981) reported that bee venom therapy is highly effective for the treatment of chronic low back pain and lumbosacral radiculitis.

Mund-Hoym (1982) described the therapeutic results of 211 patients who suffered from mesenchymal diseases of the hip and knee joints. After 6 weeks of treatment, 70 % of patients showed marked improvement.

Forestier and Palmer (1983) reported that among 1.600 cases that were treated with bee venom, an 80 % success rate was obtained in the following cases: pain in the knee before the arthritis was advanced; chronic periarthritis of the shoulder and epicondylitis of the elbow that resisted the injections of cortisone; relief of pain at the base of the toes. In case of rheumatoid polyarthritis, positive results were noticed at the beginning, but effectiveness gradually diminished. There was a minimal effect observed in patients with coxarthrosis, ankylosing spondylitis, and vertebral osteoporosis of menopausal women. The authors finally conclude that an increasing dose of bee venom could end the intense pain of severe rheumatism, which has existed for months or even years.

Lonauer et al. (1985) presented results of 30 patients suffering from rheumatoid arthritis in stages I-III. The use of corticosteroid was excluded, and most of the patients did not use any non-steroidal anti-inflammatory drugs (NSAIDs). A marked improvement of articular joint pain was observed in 74 % of the cases, a moderate improvement in 9 % of the cases and no improvement in 17 %. Joint mobility was improved in 65 % of the cases and swelling diminished in 56.5 % of the cases. The use of NSAIDs could be reduced considerably and apitherapy constituted an efficient treatment and was well tolerated.

Kim (1989) conducted a study of bee venom therapy and a positive correlation of effectiveness between arthritic inflammation and bee venom was established. Kim (1991) conducted another study comprising with 180 randomized patients received injections of *Apitox* (purified *Apis mellifera* toxin) for over a period of 6 weeks (also known as *Apitoxin* in Korea). The injection dosage in this case was set at 1.0 mg/mL of honeybee venom, while the 50 % of each group (the control individuals) received 1.0 mL of histamine phosphate 0.275 mg/mL. The number of injections increased with each subsequent visit (e.g. 3, 6, 9, 12, 15, 18, and up to 20 injections per session); with a continuous dosage schedule of 20 injections/sessions for the remaining period of the study. Initially, the injections were administered directly to the area of pain. After the fifth session, when the number of injections increased gradually, injections were also administered to the corresponding dermatome area of the spine. A visual analogue scale (VAS) and a McGill Pain Questionnaire (MPQ) were used to assess the level of pain. Thermographic evaluations and physical examinations (to evaluate swelling, tenderness, and range of motion limitations) were also performed. Both groups experienced a reductions in pain (good or moderate scores) following treatment with *Apitox*. It was also shown that the *Apitox* treatment group demonstrated a greater improvement in their pain reduction scores compare to the control group. Scores after treatment with *Apitox* were established at 18 (pain level), while the control group scored at 57. After a period of 6 months, there was yet another significant difference in scores between the control and the *Apitox* treatment group. In this case, pain reduction scores were significantly higher, as the *Apitox* treatment group scored an average of 29, whereas the control group scored at 83. Overall, the *Apitox* treatment group demonstrated a remarkable improvement with their physical examinations as well as reduced inflammation in their computerized infrared thermographic readings.

Klinghardt (1990) reported anecdotally that among 128 patients with a wide spectrum of illnesses, all but 11 appeared to improve (90 % improvement). This report is typical of anecdotal apitherapy results that begin with stories of beekeepers recounting various health improvements after receiving accidental multiple stings from their bees. The patients had diagnoses of gout, rheumatoid arthritis, fibromyalgia, spinal strain (injury to either a muscle or tendon) or sprain (stretching or tearing of a ligament), spinal disc injuries, post-laminectomy pain, bunion (hallux valgus), postherpetic neuralgia (PHN), incomplete healing of fractured bone, intractable pain from large burn wounds, osteoarthritis, ankylosing spondylitis, vertigo, and multiple sclerosis.

4.8.2 Korean FDA, Phase I Clinical Study

Kang and Kim (1993) performed the toxicity study on human. The total 20 (10 male, 10 female) subjects between the ages of 23 and 45 years of age were enrolled and 15 completed the study. Five subjects dropped out of study because of failure to keep study appointments, not following study directions, or due to consumption of alcohol during the study. Each subject received an initial test dose of 0.05 mL and then 12 doses of Apitox, starting with 0.1 mL with the first intradermal injection and increasing to 0.2 mL (second injection), 0.25 mL (third injection), 0.3–0.7 mL (fourth through twelfth injections). Injections were administered 2–3 times per week over a period of 4–6 weeks. Physical examination, blood and urine laboratory evaluations, and vital sign measurements were performed.

There were no significant changes pre- and post-test of the hematology, blood chemistry, urinalysis, and vital signs after injection with Apitoxin. Also, there were no significant physiological changes in the clinical evaluations. Localized itching was the most common adverse experience (11/15). Edema (5/15), pain at injection site (2/15), and blister at injection site (1/15) were also reported, but no serious adverse experiences were reported. Thus, it was concluded that Apitox can be safely administered to humans when applied in therapeutic doses (Kim 1987).

4.8.3   Korean FDA, Phase II Clinical Study

This study was evaluated by the Korean FDA, labeled as Phase II Clinical Study, and was an active-controlled study. One hundred one randomized subjects with osteoarthritis of the knee or spine were given injections of *Apitoxin*. Maximum dose levels and schedules were set at: 0.7 mg twice a week for Group A; 1.5 mg for Group B; 2.0 mg for Group C, and 1,000 mg per os of Nabumetone (a non-steroidal anti-inflammatory drug) for Group D (control group) once daily for over a period of 6 weeks. A four-point Likert-like symptom severity rating scale was used to assess pain, disability, and physical mobility. In addition, a 5-point self-evaluation scale and assessment form was given to the research subjects. Overall, 81 out of the 101 subjects completed the study. Safety of the treatment was assessed through observations of adverse reactions or experience as well as through the assessment of blood and urine laboratory test.

There were no significant changes in vital signs or results of laboratory examinations of any patient in this clinical trial. Localized itching was experienced by all patients who received Apitox injections. Itching at the injection site generally lasted for 2–3 weeks; several patients had this reaction for a longer period. Chills and generalized body ache were reported by 81.7 % of Apitox-treated patients, and 60 % experienced pain at injection sites. One patient complained of nausea and abdominal pain after Apitox treatment, but it was not clearly due to the bee venom injection and the patient completed the trial. One patient dropped out the study because of blister formation at injection site; the blister was cured with local wound care. One patient developed urticaria following treatment but symptoms resolved after 1 week.

The results of this study demonstrate overall efficacy rates of 85 and 90 % in the moderate-dose and high-dose groups, respectively. The clinical study proved that the *Apitoxin* treatment group demonstrated significantly greater improvements than those in the Nabumetone group (p<0.01). Within the *Apitoxin* groups, Groups B and C demonstrated greater improvements than Group A (p<0.01) (Won et al. 1999).

4.8.4   Korean FDA, Phase III Clinical Study

The purpose of this clinical study was to compare efficacy and safety of Apitoxin and Nabumetone when the patients with osteoarthritis were given injection of Apitoxin and Nabumetone during 6 weeks. The clinical study was conducted with patients from four Korean University Medical Centers. The study was designed as randomized, active-controlled trial in which 405 subjects with osteoarthritis of the knee or spine were given intradermal injections of *Apitoxin* twice a week with maximum doses of 1.5 mg, while the control group received an oral dose of 1,000 mg Nabumetone once daily for a period of 8 weeks. Out of 405 participants 310 completed the study.

There were no significant changes in vital signs or results of laboratory examinations of any patient in this clinical trial. The adverse reactions were a little higher in the Apitoxin group, but there was no statistical significance. Adverse reactions caused by the medicine to study were similar in the two groups. No significant side effects were reported during this study.

Comparing the ratios of subjects who showed more than 20 % improvement in the total point of efficacy evaluation items during the 6th week of trial, the apitoxin group showed a better improvement compared to the Nabumetone group (69.69 vs. 46.15 %).

Furthermore, the improvement rate during the 2nd week after completion of the study of the Apitoxin group was 58.44 % while 42.95 % of the Nabumetone group improved. This tells us that the Apitoxin group showed a better sustained improvement rate in statistical significance than the Nabumetone group. The rate and severity of adverse reactions were similar in both groups. Muscle pain of the musculoskeletal system was more frequent in the Apitoxin group, while gastrointestinal pain was most common in the Nabumetone group. The authors concluded that *Apitoxin* was significantly more effective than the control drug Nabumetone in the treatment of pain and inflammation for osteoarthritis patients. Reduction of disability and physical progress was greater in the Apitoxin group than in the control group.

4.8.5   Korean FDA, Phase IV Clinical Study

In 2003 the Korean FDA issued a final report on the effects of bee venom injection treatment branded as *Apitoxin*. This drug became the first federally regulated biopharmaceutical medicine in the world (Guju and Apimeds 2003).

Following approval of *Apitoxin* in Korea a post-marketing survey (PMS) was conducted between the years 2003 and 2009 upon the conclusion of the 6-year study involving a total of 3,194 patients who voluntarily received *Apitoxin* treatment. The exit surveys concerned personal details, present illnesses, past history, present medication use, treatment dates for 12 or more sessions, dosage amounts, and any adverse reactions or experiences. A complete blood count was performed before the first and after the last treatment. The participants' physicians and Korean FDA were instructed to contact the acting pharmaceutical company involved to report any negative feedback or reactions. According to the survey no major adverse reactions were reported (Guju and Apimeds 2009).

The Pain Center at PC University Medical Center in Korea has documented the use of *Apitoxin* with over 6,132 patients with intractable medical conditions and autoimmune diseases between the years of 2003 and 2009. Minor adverse reactions included itching (injection site), swelling (injection site), slight pain, low-grade fever, flushing, headache, and diarrhea (Kim 2009).

4.8.6   Future Studies in the United States

Currently in the United States, *Apitox* is in Phase III for clinical study and is also in progress for United States FDA approval at the Center for Biologics Evaluation and Research (CBER). The study was labeled as a "Multi-Center, Randomized, Double-Blind, Active-Controlled, Phase III Parallel Group Clinical Study to Evaluate the Safety and Efficacy of Apitox vs. Histamine in Subjects with Refractory Osteoarthritic Pain and Inflammation". Twenty-six clinical centers in the US and India have been selected and the study might be completed by the end of 2013 (NIH 2010).

Kim (2012) reported the stability study of Apitox to the CBER, FDA. It is very safe and stable up to 36 months.

4.9   Case Studies and Research

There are hundreds of scientific case studies evaluating the effects of bee venom on both human and animal models. During a period of 50 years, scientists and physicians validated the restorative properties of honeybee venom as a treatment for many diseases and afflictions. The studies cited below are arranged according to disease type.

4.9.1   Arthritis and Rheumatoid Arthritis

Lorenzetti et al. (1972) conducted studies on the use of bee venom prophylactically and therapeutically to reverse adjuvant-induced arthritis in rats. Bee venom was injected three times a week for a period of 4 weeks subcutaneously. The researchers were able to prevent arthritic syndromes such as foot edema, secondary lesions, and reduction in inflammation. In this case, the experimenters proved that bee venom is effective as a therapeutic and as a prophylactic treatment producing immediate results.

Weissmann et al. (1973) found that the effects of daily injections of three individual bee venom components (melittin, apamin, and phospholipase A_2), along with the effects of daily injections of whole bee venom, prevented adjuvant arthritis from developing in rats.

Chang and Bliven (1979) administered bee venom for adjuvant-induced arthritis in rats. The sub-cutaneous injection dose was between 0.01 and 1.0 mL daily over a period of 17 days. It was shown that honeybee venom suppressed the development of adjuvant arthritis in a dose dependent manner. Single administration of bee venom also proved to suppress the development of carrageenan-induced paw edema. Moreover, bee venom was shown to effectively suppress the development of poly-arthritis. This suppressive effect would decrease progressively over time as dosing was delayed. Bee venom was found to be most effective when mixed and injected together with CFA (complete Freund's adjuvant), the disease-inducing agent. Similarly, bee venom mixed with egg albumin and CFA injected into the hind paw prevented the development of arthritis. These results suggest that at least two mech-anisms are involved in the anti-arthritic action of bee venom; one involving the alteration of immune response (most likely via antigen competition), and the second being the anti-inflammatory action and property of bee venom.

Eiseman et al. (1982) showed that the therapeutic effect of bee venom on arthritis was completely dependent on the site of administration. In particular, local injection of bee venom near the site of inflammation was more effective in inhibiting the development of adjuvant-induced arthritis.

Kim (1997) stated that according to TCM (traditional Chinese medicine) arthritis originates from a deficiency in the circulation of the blood and lymph inside tissue joints, which results in an accumulation of lactic acid where bacteria can multiply. He contended that the vasodilatatory effect of bee venom therapy can increase local circulation, which corrects deficient circulation and also works against the spread of irritation.

Kwon et al. (2001a) conducted a study to evaluate the anti-nociceptive effect of bee venom injections into a specific acupoint (Zusanli, Stomach point 36) as com-pared to a non-acupoint in a rat model with chronic arthritis. Subcutaneous injec-tions of bee venom (1.0 mg/kg per day) were found to dramatically inhibit paw edema caused by CFA injection. It was also shown that this treatment modality significantly reduced arthritis-induced nociceptive behaviors such as the nocicep-tive scores for mechanical hyperalgesia and thermal hyperalgesia. The anti-nociceptive and the anti-inflammatory effects of bee venom were observed for a period of 12–21 days post treatment. In addition, bee venom significantly sup-pressed the adjuvant-induced Fos-gene expression in the lumbar spinal cord at the 3rd week post-adjuvant injection. Finally, injection of bee venom into the Zusanli acupoint resulted in a significantly greater analgesic effect on arthritic pain as com-pared with bee venom injection into a more distant non-acupoint. The study has further demonstrated that bee venom injection into the Zusanli acupoint had both anti-inflammatory and anti-nociceptive effects on CFA-induced arthritis in rats. Altogether, these findings suggest that bee venom acupuncture is an effective ther-apy treatment for rheumatoid arthritis.

Kwon et al. (2001b) conducted a study with bee venom treatment for osteoarthritis. This was a 4-week comparison trial involving a total of 60 volunteers. The treatment group (n = 30) received honeybee venom acupuncture (bee venom injection at acu-puncture points), while the control group (n = 30) received traditional acupuncture treatment. It was observed that bee venom acupuncture produced maximum effective results for chronic pain relief rather than acupuncture treatment alone. Bee venom acupuncture showed a significant improvement (82.5 %) and almost all patients reported relief from pain and chronic ailments, while their computerized infrared thermograph readings were at normal levels.

Kang et al. (2002) effectively assessed the clinico-therapeutic effect of bee venom by administrating venom to 90 rats with adjuvant-induced arthritis rats. Clinical findings of lameness score, edema volume, hematological values, and histo-pathology (the interphalangeal joint of the right hind paw) were observed during the treatment period. In the treatment groups, the development of inflammatory edema and polyarthritis were suppressed effectively through bee venom treatment. In addition, the authors found that there were no significant changes regarding the hind paw edema volume and lameness score between the prednisolone and the venom treatment groups. Red blood cell count, hematocrit, and hemoglobin counts were not different between the groups; although significant leucocytosis was observed in the control group (p<0.01). Moreover, bee venom was found to effectively suppress erosions of articular cartilage and inflammatory cell infiltrations into the interphalangeal joints. The researchers concluded that honeybee venom successfully suppresses arthritic inflammation in rats.

Park et al. (2004) investigated the molecular mechanisms of the anti-inflammatory effects of bee venom using rat models. The rats suffered from carrageenan-induced acute edema in their paws and chronic adjuvant-induced arthritis. Bee venom was administered daily at 0.8 and 1.6 µg/kg into their hind paws. The results were consistent with the in vitro results, which showed an inhibitory effect of bee venom set between 0.5, 1.0, and 5.0 µg/mL per treatment. The study also showed that melittin (a major component in bee venom) applied at 5.0 and 10.0 µg/mL had an anti-inflammatory effect in rats with arthritis.

Yin et al. (2005) used HTB-94 human chondrosarcoma cells to establish the gene expression profiles of bee venom treatment. Out of the 344 genes profiled, 35 were down-regulated by bee venom; 16 were up regulated, whereas seven were down regulated by lipopolysaccharide (LPS). Furthermore, 32 were down regulated by LPS mixed with bee venom. The bee venom proved to reverse the upregulation caused by LPS for some genes, such as the IL-6 receptor, matrix metalloproteinase-15 (MMP-15), tumor necrosis factor, superfamily-10, caspase-6, and tissue inhibitor of metalloproteinase-1 (TIMP-1). The researchers effectively established the pharmacologic activity of bee venom in the treatment of arthritis.

4.9.2 Osteoarthritis and Chronic Pain

Vick et al. (1976) found a positive correlation between the effects of bee venom on cortisol levels related to increase of the activity of dogs that had hip dysplasia. Out of 24 dogs, 16 were normal and eight had severe hip dysplasia. The two groups were sub-divided into four treatment groups. Groups I and II included eight normal dogs each whereas Groups III and IV contained four dogs each with hip dysplasia.

Groups II and IV received 1.0 mg (about 0.067 mg/kg) of bee venom subcutaneously on days 30, 37, 50, and 60, and were crossed over to receive saline control treatment on days 90, 97, 110, and 120. Groups I and III first received a saline solution and later were crossed over to receive bee venom treatment. On day 90, treatment would be crossed-over again. The dogs receiving bee venom had an increase in plasma cortisol levels, and their physical activity increased inside the cages. The results showed that bee venom treatment stimulated the production of cortisol, which had anti-inflammatory properties, thus enhancing the mobility of dogs with hip dysplasia related arthritis.

Short et al. (1979) reported the results of treating 17 dogs that had been diagnosed as arthritic. Following bee venom therapy, 14 of 17 dogs improved significantly, returning to normal or near-normal movement. Four of five dogs treated for joint complications (hip displasia and arthritic joints) showed improved movement; four of six dogs treated for poor surgical recovery responded well and all dogs suffering from disc complications returned to normal or near-normal conditions after a series of bee venom injections administered at the sites of pain and stiffness. From the results of this study, the authors concluded that bee venom therapy may be highly beneficial in alleviating certain arthritic conditions in dogs.

Von Bredow et al. (1981) conducted a study in which he observed the effects of bee venom injections on eight arthritic horses, ranging in age from 8 to 17 years. Six of the eight horses showed significant improvement, with three of these six demonstrating a complete recovery.

Lee et al. (2001) examined the anti-nociceptive and anti-inflammatory effects of bee venom pretreatment on carrageenan-induced inflammation in rats. The experiments were designed to evaluate the effect of BV pretreatment on carrageenan (CR)-induced acute paw edema and thermal hyperalgesia. In addition, spinal cord Fos-gene expression induced by peripheral inflammation was quantitatively analyzed. In normal animals subcutaneous BV injection into the hindlimb was found to slightly increase Fos expression in the spinal cord without producing detectable nociceptive behaviors or hyperalgesia. In contrast pretreatment with BV (0.8 mg/kg) 30 min prior to CR injection suppressed both the paw edema and thermal hyperalgesia evoked by CR. In addition, there was a positive correlation between the percent change in paw volume and the expression of Fos positive neurons in the spinal cord. These results indicate that BV pretreatment has both antinociceptive and anti-inflammatory effects in CR-induced inflammatory pain. These data also suggest that BV administration may be useful in the treatment of the pain and edema associated with chronic inflammatory diseases.

4.9.3   Vascular System

Forster (1950) studied the pharmacological effects of bee venom based upon all its components, in particular the effect upon the blood circulation, the permeability of biological membranes, and the organic exchanges. Administration of venom results in vasodilation, accompanied by an increase of permeability of blood vessel caused by hyaluronidase. He also describes a hypophyseal effect demonstrated by the absence of an effect upon hypophysectomised animals. This effect upon the hypophysis brings about a discharge of cortisol from the adrenal glands.

Zaitsev and Poriadin (1973) studied 415 patients: 77 were suffering from endarteriitis obliterance (a form of vasculitis with unknown etiology), 138 had arteriosclerosis of the peripheral blood vessels, 65 Bechterew's disease (ankylosing spondylitis), 50 spondylarthritis deformans, 85 polyarthritis deformans. He reported a good result (80 %) of the arterial diseases of the extremities and satisfactory result (67 %) of the spine and the joint diseases.

Hanson et al. (1974) have shown that the intradermal injection of peptide 401 (mast cell degranulating peptide, a component of honeybee venom), substantially inhibited the edema provoked by subplantar injection of carrageenan or an intra-articular injection of turpentine in rats. Peptide 401 also suppressed the increased vascular permeability due to intradermal injection of various smooth muscle spasmogens (histamine, bradykinin, 5-hydroxytryptamine and prostaglandins). These results demonstrate that peptide 401 is a potent anti-inflammatory agent in the rat. Its effectiveness has proven superior to that of indomethacine, salicylate, and phenylbutazone.

Lee et al. (2010) studied the influence of bee venom on the expression of cellular adhesion molecules in the vascular endothelium. A great amount of information exists concerning the effects of an atherogenic diet on atherosclerotic changes in the aorta, but little is known about the molecular mechanisms and the levels of gene regulation involved in the anti-inflammatory process induced by BV. The experimental atherosclerosis was induced in mice by a lipopolysaccharide (LPS) injection and an atherogenic diet. The animals were divided into three groups, the NC groups of animals that were fed with a normal diet, the LPS/fat group was fed with the atherogenic diet and received intraperitoneal injections of LPS, and the LPS/fat+BV group was given LPS, an atherogenic diet and intraperitoneal BV injections. At the end of each treatment period, the LPS/fat+BV group had decreased levels of total cholesterol (TC) and tri-glyceride (TG) in their serum, compared to the LPS/fat group. The LPS/fat group had significant expression of tumor necrosis factor (TNF)-α and interleukin (IL)-1α in the serum, compared with the NC group ($p < 0.05$). The amount of cytokines was consistently reduced in the BV treatment groups compared with those in LPS/fat group. BV significantly reduced the amount of intercellular adhesion molecule-1 (ICAM-1), vascular cell adhesion molecule-1 (VCAM-1), transforming growth factor-α1 (TGF-α1) and fibronectin in the aorta, compared with the LPS/fat group ($p < 0.05$). A similar pattern was also observed in the heart. In conclusion, BV has anti-atherogenic properties via its lipid-lowering and anti-inflammatory mechanisms.

Kim et al. (2011) investigated the effects of melittin (a major component of bee venom) regulated athero-sclerotic changes in an animal model of atherosclerosis. The results showed that melittin decreased the total cholesterol and triglyceride levels in atherosclerotic mice. In addition, melittin decreased the expression levels of tumor necrosis factor (TNF)-β, interleukin (IL)-1β, vascular cell adhesion molecule (VCAM)-1, intercellular adhesion molecule (ICAM)-1, fibronectin and transforming growth factor (TGF)-β1 in atherosclerotic mice. In vitro, melittin decreased lipopolysaccharide (LPS)-induced THP-1 cells-derived macro-phases TNF-α and

IL-1β expression level and nuclear factor (NF)-κB signal pathway. The authors concluded that melittin has an anti-atherogenic effect by suppression of pro-inflammatory cytokines and adhesion molecules.

4.9.4 Immune System and Disease

Hyre and Smith (1986) conducted a study on the immunological effects of honeybee venom in mice. The authors state that mice injected with bee venom prior to, and following, an injection of sheep red blood cells produced significantly more direct IgM plaques than the sham-injected group. The results of this study indicate that bee venom can effects both T and B lymphocyte functions.

Hadjipetrou-Kourounakis and Yiangou (1988) conducted a study in order to determine whether the in-vivo effects of bee venom may be mediated by alterations in lymphokine production. With dosage schedules set at 0.5 mg/kg a day, bee venom was administered intramuscularly (IM) in rats for over a period of 17 days, which caused a reduction in interleukin (IL) production by splenocytes (white blood cell types purified from splenic tissue). Moreover, in vitro addition of IL-1 or IL-2 to the cultures resulted in an increase in normal response levels, suggesting that bee venom can affect the production of IL-1 by macrophages. These findings also indicate that bee venom can modify inflammatory responses by interfering with inflammatory cell functions.

Defendini et al. (1988) studies on adjuvant-induced symptoms in which a positive correlation between lymphokine production and the anti-inflammatory response of bee venom was demonstrated.

Rekka et al. (1990) showed that honeybee venom is able to inhibit significantly nonenzymatic lipid peroxidation. It also possesses a considerable hydroxyl radical scavenging activity, evaluated by its competition with dimethyl sulfoxide for HO (hydroxyl radicals). These results, in relation to the in vitro suppression mainly of interleukin-1 production offered by honey bee venom, may further support that antioxidant activity is involved in the anti-inflammatory activity of honey bee venom.

4.9.5 Anti-fungal, Anti-bacterial, and Anti-viral Properties

Schmidt-Lange (1941) wrote about the anti-bacterial properties of bee venom as an effective germicide. These findings were confirmed by Fennell et al. (1968) in their studies on the anti-bacterial action of melittin, which is an important component of honeybee venom. Furthermore, these researchers showed that honeybee venom and a polypeptide fraction of melittin, had an antibacterial property against a penicillin-resistant strain of Staphylococcus aureus (Strain 80). Both, whole bee venom and melittin inhibited the growth of 20 out of 30 different bacterial strains. Fennell and his team found that 86 % of the Gram-positive and 46 % of the Gram-negative bacteria were sensitive to bee venom and to melittin. Among the

Gram-positive micro-organisms, the antibacterial effect of 1.0 mg melittin was equal to that of 0.1–93 units of penicillin. Other studies by Dorman and Markey (1971) substantiated the anti-bacterial qualities of HBV.

Rauen et al. (1972) studied the anti-fungal properties of melittin and apamin on a variety fungi, while Hadjipetrou-Kourounakis and Yiangou (1984) showed that bee venom also had anti-viral properties.

Han et al. (2007) examined the antibacterial activity of bee venom against the most contagious agents of bovine mastitis. The results showed bee venom has significant antibacterial effects against seven major bacterial mastitis pathogens. The minimal inhibitory concentrations (MIC) against *Staphylococcus aureus*, methicillin resistant *Staphylococcus aureus* (MRSA) and *Escherichia coli* had a stronger effect as compared with standard drugs. The study indicates that bee venom has potential antibacterial effects, and provides justification for the evaluation of bee venom as an alternative to antibiotics for the treatment of bovine mastitis.

4.9.6   Anti-inflammatory Properties

There are numerous scientific articles that present sound evidence of the anti-inflammatory properties of bee venom.

Vick et al. (1972) studied the effect of whole bee venom on plasma cortisol levels in a monkey model. Subcutaneous injections of bee venom (1.0–100.0 mg), and melittin (1.0–10.0 mg) caused a marked and sustained elevation in plasma cortisol levels. This increase occurred approximately an hour after injection and lasted for 2–4 days. This effect appeared to be dose-related as higher doses of bee venom or melittin resulted in the production of the quickest and highest plasma cortisol levels. In an additional monkey subject, bee venom (1.0 mg) and melittin (0.1 mg) were administered each at 72–96 h respectively. This resulted in an immediate and sustained rise in cortisol which lasted for 20–30 days. Melittin alone appeared to be ten times more potent than whole bee venom. Necropsy conducted on the four monkeys that received the highest dosages of bee venom or melittin revealed no significant tissue damages. Overall, this study indicates that bee venom and its isolated component melittin stimulates the production of cortisol from the adrenal gland.

Banks et al. (1976) reported on the prostaglandins, which are implicated during the 'secondary inflammation' phase. The non-steroidal anti-inflammatory drugs (NSAIDs), such as aspirin and indomethacine, are inhibitors of the synthesis of the prostaglandins *in vivo* and *in vitro*. They have shown that peptide 401 is also an inhibitor in the conversion of arachidonic acid to prostaglandin E *in vitro*.

Menander-Huber (1980) demonstrated that melittin binds with calmodulin, and Jones et al. (1982) demonstrated that NADPH-oxidase contained in the membrane fraction of the activated leukocytes (which are responsible for the production of superoxide anions) depends on calmodulin. It could be that bee venom works via a similar mechanism. The principal fraction of venom, melittin, shows a great affinity to calmodulin, and is the only fraction of bee venom which has this characteristic.

Somerfield et al. (1986) explained the mechanism of bee venom's anti-inflammatory action by investigating its effect on neutrophil O_2-production. It is well established that oxygen radicals and their metabolites play a role in chronic inflammations and tissue destruction. Using human peripheral blood leukocytes, the polymorphonuclear fraction was isolated and used for in vitro assessments on the ability of melittin and other bee venom peptides on the production of reactive oxygen species (ROS). The results showed that melittin but not other bee venom fractions, inhibited ROS production both pre- and post-stimulation, suggesting that melittin may have a role in the in vivo regulation of radical production.

Dufourcq (1986) showed that melittin strongly disturbs the membrane structure of the liposomes into which it penetrates. It is not selective about the phospholipids, and it increases the permeability of all kinds of cells, like mastocytes and cytoplasmic bacterial membranes.

Gencheva and Shkenderov (1986) reported that the low molecular fraction of bee venom injected into rats in a daily dose of 100 μg/kg over a 3-week period reduced the activity of the complement system by 45 %.

4.9.7 Lyme's Disease

Tests measuring melittin's inhibitory actions against Lyme's Disease were carried out at the U.S. National Institute of Allergy and Infectious Diseases in Hamilton, Montana, Rocky Mountain Laboratories Microscopy Branch. Thorough research on *Borrelia burgdorferi* (the bacterial agent of Lyme disease) demonstrated that bee venom effectively resisted in vitro effects of powerful eukaryotic and prokaryotic metabolic inhibitors. Moreover, treatment of laboratory cultures of *B. burgdorferi* in Barbour-Stoenner-Kelly medium with melittin, showed immediate and profound inhibitory effects monitored with dark-field microscopy and optical density measurements. Furthermore, melittin concentrations as low as 100 mg/ml, ceased virtually all spirochete motility within seconds. The examination of the spirochetes under the electron microscope, which could reveal obvious alterations on the surface envelope of the spirochetes showed an extraordinary sensitivity of *B. burgdorferi* to melittin providing both a research tool in the study of selective permeability in microorganisms and also important clues for the development of new and effective drugs against Lyme disease with melittin and honeybee venom (Lubke and Garon 1997).

4.9.8 Radioprotective Effects

Shipman (1967) investigated the resistance of bee venom treated mice to irradiation with X-rays. The response of animals to whole-body irradiation with a lethal dose was modified by changes in their physiological state induced before exposure. The ability of bee venom in producing a degree of physiological stress in animals, eliciting a neuroendocrine response (pituitary-adrenal stimulation), increased the resistance of mice to radiation. In these experiments, pre-treatment with bee venom (IP dose range: 1.1–1.24 µg; SC dose range: 4.3–5.6 µg) resulted in greater survival rates as compared to saline control animals. Bee venom administered subcutaneously resulted in the greatest protection (70–80 % survival rate in a 30-day period) from irradiation. Only 7 % of the rats, which had 5.4 µg of melittin administered subcutaneously survived for a 30-day period. Based on these results, it was proposed that at least three mechanisms of action were attributed regarding the radio-protective effect of bee venom in mice. First, a stressor-like action that elicits an "adaptation syndrome"; second, the production of changes in the hematopoietic system; and third, the antibacterial properties of the venom.

Ginsberg et al. (1968) deduced that melittin provided a significant protection from x-irradiation in mice that received an SC injection dose of 60 mg/kg bee venom 24 h prior to irradiation.

Shipman and Cole (1968) came to the same conclusions, wherein mice were injected 1 day prior to radiation, as did Kanno et al. (1970) who reported similar results on the radio-protective action of bee venom.

4.9.9 Cancer

Bee sting therapy was used by Dr. Desjardins (1859) to treat skin tumors, and Yoannovoitch and Chahovitch (1932) to treat experimental cancerous tumors. They produced experimental cancers on the ears of rabbits and then treated them with bee venom. The tumors became soft, their bases diminished and parts fell off, followed by scar formation. The most noteworthy fact was that bee venom had an effect even on the tumors of the other ear, providing that it had not only local but also remote application and therefore systemic action.

Belliveau (1992) conducted research on colon cancer of rats induced by injection of adjuvant. He reported that bee venom is effective in treating an animal model of colon cancer.

Yun et al. (2000) reviewed the study on bee venom related to cancer in the PubMed database. He found 38 related articles published: leukemia (10), nonspecific tumor (5), neuroblastoma (4), lung cancer (3), pituitary tumor (3), pheochromocytoma (3), astrocytoma (2), glioma (2), lymphoma (2), bladder cancer (1), breast carcinoma (1), pancreatic carcinoma (1), and squamous cell carcinoma (1).

Jang et al. (2003) demonstrated that cells of the human lung cancer line NCI-H1299 treated with bee venom exhibit several features of apoptosis. In addition, reverse transcription-polymerase chain reaction and prostaglandin E_2 immunoassay were performed to verify whether BV possesses an inhibitory effect on the expression of cyclo-oxygenase (COX) and prostaglandin E_2 (PGE_2) synthesis. Expression of COX_2 mRNA and synthesis of PGE_2 were inhibited by BV. These results suggest the possibility that BV may exert an anti-tumor effect on human lung cancer cells in vitro.

Yin et al. (2005) also investigated cancer concentrating on gene expression and chondrosarcoma cells. The HTB-94 human chondrosarcoma cells were treated with BV, lipopolysaccharide (LPS), or both. Of the 344 genes profiled in this study, with a cut-off level of fourfold change in the expression, (1) 35 were down regulated following BV treatment, (2) 16 were up regulated and 7 down regulated following LPS treatment, and (3) 32 were down regulated following co-stimulation of BV and LPS. The results of this study shows that treatment with BV reversed the LPS-induced up regulation of such genes as interleukin-6 (IL-6) receptor, matrix metalloproteinase 15 (MMP-15), tumor necrosis factor (ligand) superfamily-10, caspase-6 and tissue inhibitor of metalloproteinase-1 (TIMP-1).

Kim et al. (2005) studied the effect of bee venom on the bone function in human osteoblastic cells. To provide insights into the effect of bee venom on aromatase activity in bone-derived cells, they examined the human leukemic cell line FLG29.1, which is induced to differentiate toward the osteoblastic phenotype by TPA and TGF-beta 1, and the primary first-passage osteoblastic cells (hOB). The authors demonstrated that cells of the osteoblastic lineage synthesize aromatase in vitro by the local cytokine of TGF-beta 1 and bee venom.

More recently, BVT was also considered as a potential cancer treatment (Hu et al. 2006a, b; Liu et al. 2008; Orsolic et al. 2003; Putz et al. 2006; Ip et al. 2012; Jo et al. 2012; Orsolic 2012).

4.9.10   Peripheral Nerve Paralysis

Kim et al. (2007a, b) administered Apitoxin/Apitox to dogs with hind limb paralysis, which resulted in favorable therapeutic responses. However, more studies are needed to prove the therapeutic effect of bee venom in cases of canine hind limb paralysis as well as of paralysis of the face.

4.9.11   Multiple Sclerosis

Hauser et al. (2001) proposed that BVT as an alternative therapy for the treatment of multiple sclerosis (MS). A study was made to evaluate the efficacy of bee venom injections to halt or reverse the course of MS. Fifty-one patients with clinically documented MS were first tested to ensure safety of participation before they receive higher doses of venom. The venom was administered one to three times per week, consisting of an average of 11 intradermal injections (0.1 mL) per session. The patients' clinical responses were evaluated every 3 months for a year. A positive correlation between BVT and the improvement of MS symptoms was shown. Fifty-eight percent of the participants experienced positive results; 29.8 % experienced no benefits, and only one patient reported a worsening of his condition. BVT was effective against fatigue and showed a 42–44 % improvement in the overall condition.

Significant improvements were also seen in bowel control (32.2 %) and body coordination (31.4 %). After only 12 months of treatment, average scores for the *Related Observable Symptoms Scale* (ROSS) survey went up from 36.1 to 48.6. Overall, 68 % of the patients enrolled in the study experienced positive and long-lasting effects from bee venom treatment.

Wesselius et al. (2005) performed a randomized crossover study of bee sting therapy for multiple sclerosis (MS). A total of 26 patients with relapsing-remitting or relapsing secondary progressive MS was enrolled, and assigned to 24 weeks of bee sting therapy or 24 weeks of no treatment. Live bees (up to a maximum of 20) were used three times a week. The results showed that there was no significant reduction in the cumulative numbers of new gadolinium-enhancing lesions. The T2-weighted lesion load further progressed, and there was no significant reduction in relapse rate. There was no improvement of symptoms of MS. Bee sting therapy was well tolerated, and there were no serious adverse events. The authors concluded that the treatment with live bee sting in patients with relapsing MS did not reduce disease activity, disability, or fatigue and did not improve the quality of life.

Castro et al. (2005) evaluated the safety of bee venom extract as a possible treatment for patients with progressive forms of multiple sclerosis (MS). A total of nine patients (with no history of bee venom allergy) with progressive forms of MS, who were 21–55 years of age with no other illnesses, were entered into four groups on a structured 1-year immunization schedule. Although no serious adverse allergic reactions were observed in any of nine subjects, four experienced worsening of symptoms, necessitating their termination of the study; this could not be ascribed to side effects of the therapy. Of remaining five subjects, three felt that the therapy had subjective amelioration of symptoms and two showed objective improvement. Although this study suggests safety, because of the small numbers studied, there was no definite conclusion regarding efficacy. Larger and more carefully conducted multicenter studies will be required to establish efficacy.

4.10   Case Reports

4.10.1   Multiple Sclerosis (MS)

SR is a 47-year old female suffering from MS for the last 15 years. Despite multiple conventional treatments, her quality of life was worsening progressively. She continuously felt tingling and numbness in her legs, lost 40 % of bladder control, lost strength in her extremities, and she was only walking with the help of a cane or walker. After a total 28 treatment sessions, a marked improvement was observed in her condition (Table 4.4). Accordingly, she regained 100 % control over her bladder and she did no longer need a cane or walker. Today, she only comes for follow-up visits every 3–4 weeks to monitor progress.

Table 4.4 Clinical evaluation of patient SR before and after treatment of MS

	Before treatment	After treatment
Muscle weakness	++++	+
Spasticity	+++++	++
Bladder control	+++	+++++
Balance	Poor	Good
Using a cane or walker	+	–

Legend: (+) 20 %, (++) 40 %, (+++) 60 %, (++++) 80 %, (+++++) 100 %

Table 4.5 Clinical evaluation before and after treatment, RA

	Before treatment	After treatment
RA titer (<20=negative)	268	18
ESR (normal:0–10)	49	7
Range of motion (Flexion: 0–70)	Wrists – right:15, left:10	Wrists – right:55, left:60
Tenderness and swelling	+++++	+
Visual analog scale	98	15

Legend: (+) 20 %, (++) 40 %, (+++) 60 %, (++++) 80 %, (+++++) 100 %

Table 4.6 Clinical evaluation before and after treatment, Ankylosing Spondylitis

	Before treatment	After treatment
Tenderness and swelling	+++++	+
Visual analog scale	95	20

Legend: (+) 20 %, (++) 40 %, (+++) 60 %, (++++) 80 %, (+++++) 100 %

4.10.2   Rheumatoid Arthritis (RA)

NG, a 39-year old female was suffering from RA for the last 19 years. Previously, she was treated with a large variety of prescriptions, including a new cancer drug to help her condition. The pain and swelling of her hands and wrists were unbearable and she became disabled. She underwent 24 sessions of BVT with a 95 % improvement rating (Table 4.5). Her condition stabilized and she reported that she no longer suffered from pain and swelling.

4.10.3   Ankylosing Spondylitis

MO, a 53-year old male had long-standing lower back problems, along with multiple operations with fusions. He underwent 18-months of rehabilitation therapy without relief. After 26 treatments with BVT, he showed excellent improvement of symptoms (Table 4.6). Medication was no longer needed, thus, only monthly follow-up visits were required.

Table 4.7 Clinical evaluation before and after treatment, Ankylosing Spondylitis

	Before treatment	After treatment
Tenderness and swelling	+++++	+
Visual analog scale	92	14

Legend: (+) 20 %, (++) 40 %, (+++) 60 %, (++++) 80 %, (+++++) 100 %

Table 4.8 The treatment sessions, concentration of each bee venom injections, and dosing schedule

Treatment	Concentration per injection		Total dose (mg)
1st	1×0.07		0.07
2nd	1×0.07	+ 0.05	0.12
3rd	2×0.07		0.14
4th	2×0.07	+ 0.05	0.19
5th	3×0.07		0.21
6th	3×0.07	+ 0.05	0.26
7th	4×0.07		0.28
8th	4×0.07	+ 0.05	0.33
9th	5×0.07		0.35
10th	5×0.07		0.35
11th	5×0.07		0.35
12th	5×0.07		0.35
13th	5×0.07		0.35

4.10.4   Chronic Surgical Inflammation and Pain

BC, a 49-year old male had more than ten abdominal surgeries due to bleeding ulcers, alcoholic liver cirrhosis, and intestinal obstructions. Due to a phantom gallbladder pain and general surgical pain, he was heavily addicted to multiple medications. After 28 sessions with BVT concentrating at his surgical scars, relief was rated at nearly 100 % (Table 4.7).

4.10.5   Report Index for Healthy Normal Human Volunteers

Fourteen healthy volunteers (10 males and 4 females) between 26 and 52 years were enrolled in this study. They were given injections twice a week subcutaneously with each dose being dependent upon the tolerance level of the previous dose. All subjects were administered a maximum dose of 0.07 mg of bee venom which were continued for a week, before changing to an arthritic injection schedule. Each subject received 13 doses of bee venom according to the following schedule (Table 4.8)

All were administered skin tests, vital sign measurements, blood and urine laboratory analyses. One subject who became ill was subsequently hospitalized and was diagnosed with an adenovirus infection. Another subject had a delayed local reaction (the subject developed a 4+ wheal with a very large flare, 18 h after the injection). All other subjects who received the same dosage schedule did not show any delayed local reaction or systemic reaction. The volunteers reported that they did not notice any changes in their vital signs. None of the subjects reported any significant pain or adverse reaction.

In a similar study, 20 subjects (10 males and 10 females) between the ages of 23 and 45 years were enrolled. Five subjects dropped out due to failure of keeping appointments, failure to adhere to research instructions, and/or to consumption of alcohol during the study. Each subject received an initial test dose of 0.05 mL, as well as 12 doses of Apitoxin throughout the course of the study. The injections began at 0.1 mL for the first intradermal injection; later increasing to 0.2 mL for the second injection; 0.25 mL for the third injection, and last, anywhere from 0.3 to 0.7 mL (through the fourth up to the twelfth injection). These injections were administered 2–3 times per week over a period of 4–6 weeks. A physical examination along with blood and urinary laboratory analyses were conducted. No significant changes or major adverse reactions were noted. Minor reactions included localized itching as the most common adverse effect for 11 out of the 15 participants, followed by edema (5/15), pain felt at injection site (2/15), and blister formation at injection site (1/15). To conclude, there were no serious adverse experiences or complications reported. The results above confirm that bee venom can be safely administered to humans when applied in therapeutic doses (Kim 1987).

4.11   Conclusions

In some ways, apitherapy is a classic alternative therapy. It has ancient roots, widespread worldwide use and, although discarded by mainstream medicine, has survived in folk medicine. Bee sting therapy, consisting of a series of honeybee stings at regular intervals, is often administered by private beekeepers and other non-medically qualified practitioners and entails a risk of fatal allergic reactions.

The first injectable bee venom to treat human patients with bee venom therapy was approved by the Korean Food and Drug Administration in 2003 (Guju Pharma, Apimeds 2003), but its efficacy and safety have not yet been approved by the U.S. FDA (NIH 2010). Thus, it still remains a challenge for physician to accept bee venom therapy as mainstream therapy.

References

Banks BE, Rumjanek FD, Sinclair NM, Vernon CA (1976) Possible therapeutic use of a peptide from bee venom. Bull Pasteur Inst 74:137–144

Banks BE, Dempsey CE, Vernon CA, Warner JA, Yamey J (1990) Anti-inflammatory activity of bee venom peptide 401 (mast cell degranulating peptide) and compound 48/80 results from mast cell degranulation in vivo. Br J Pharmacol 99:350–354

Barker SA, Bayyuk SH, Brimacombe JS, Hawkins CF, Stacey M (1964) The structure of the hyaluronic acid compound of synovial fluid in rheumatoid arthritis. Clin Chim Acta 9:339–343

Belliveau J (1992) The effectiveness of bee venom on adjuvant induced colon cancer of the rats. Second American Apitherapy Society conference, Boston

Castro HJ, Mendez-Inocencio JI, Omidvar B, Omidvar J, Santilli J, Nielsen HS, Pavot AP, Richert JR, Bellanti JA (2005) A phase I study of the safety of honeybee venom extract as a possible treatment for patients with progressive forms of multiple sclerosis. Allergy Asthma Proc 26(6):470–476

Chang YH, Bliven ML (1979) Anti-arthritic effect of bee venom. Agents Actions 9:205–211

Clapp LE, Klette KL, Ma DC, Bernton E, Petras JM, Dave JR, Laskosky MS, Smallridge RC, Tortella FC (1995) Phospholipid A2-induced neurotoxicity in vitro and in vivo in rats. Brain Res 693:101–111

Defendini M, Ayeb M, Regnier VA, Pierres M (1988) H-2A-linked control of T cell and antibody response to bee venom. Immunogenetics 28(2):139–141

Dorman LC, Markey LD (1971) Solid phase synthesis and antibacterial activity of N-terminal sequences of melittin. J Med Chem 14:5–9

Dufourcq J (1986) Molecular details of melittin-induced lysis of phospholipid membranes as revealed by deuterium and phosphorus NMR. Biochim Biophys Acta 859(1):33–48

Edstrom A, Briggman M, Ekstrom PA (1996) Phospholipase A2 activity is required for regeneration of sensory axons in cultured adult sciatic nerves. J Neurosci Res 43:183–189

Eiseman JL, von Bredow J, Alvares AP (1982) Effect of honeybee (Apis mellifera) venom on the course of adjuvant-induced arthritis and depression of drug metabolism in the rat. Biochem Pharmacol 31(6):1139–1146

Feldsher AS, Solodovnikox GI, Gorobets GN (1981) Bee venom treatment of lumbosacral radiculitis. Feldsher Akush (USSR) 46(4):55–57

Fennell JF, Shipman WH, Cole LJ (1968) Antibacterial action of melittin, a polypeptide from bee venom. Proc Soc Exp Biol Med 127:707–710

Forestier F, Palmer M (1983) Apitherapy; rheumatology: 1600 cases investigated thoroughly. Fr Rev Apic 421:1–10

Forster KA (1950) Forty years of experience with bee venom therapy. Che Med

Gandolfo G, Gottesmann C, Binnard JN, Lazdunski M (1989) K+ channels openers prevent epilepsy induced by the bee venom peptide MCD. Eur J Pharmacol 159:329–330

Gencheva G, Shkenderov S (1986) Inhibition of complement activity by certain bee venom components. Acad Bulg Sci 39(9):137–139

Ginsberg NJ, Dauer M, Slotta KH (1968) Melittin used as a protective agent against X-irradiation. Nature 220:1334

Guju Pharma, Apimeds (2003) Final report filed to KFDA

Guju Pharma, Apimeds (2009) PMS report filed to KFDA

Habermann E, Cheng-Raude D (1975) Central neurotoxicity of apamin, crotamin, phospholipase A2 and alpha-amanitin. Toxicon 13:465–467

Habermehl GG (1981) Venomous animals and their toxins. Springer, New York

Hadjipetrou-Kourounakis L, Yiangou M (1984) Bee venom and adjuvant induced disease. J Rheumatol 1(5):720

Hadjipetrou-Kourounakis L, Yiangou M (1988) Bee venom, adjuvant induced disease and interleukin production. J Rheumatol 15:1126–1128

Han SM, Lee KG, Yeo JH, Jweon HY, Kim BS, Kim JM, Baek HJ, Kim ST (2007) Antibacterial activity of the honeybee venom against bacterial mastitis pathogens infecting daily cows. Int J Ind Entomol 14(2):137–142

Hanson JM, Morley J, Soria-Herrera C (1974) Anti-inflammatory property of 401 (MCD-peptide), a peptide from the venom of the bee Apis mellifera (L). Br J Pharmacol 50:383–392

Hartman DA, Tomchek LA, Lugay JR, Lewin AC, Chau TT, Carlson RP (1991) Comparison of anti-inflammatory and anti-allergenic drugs in the melittin- and D49 PLA$_2$-induced mouse paw edema models. Agents Actions 34:84–88

Hauser RA, Daguio M, Wester DE, Hauser M, Kirchman A, Skinkis C (2001) Bee venom therapy for treating multiple sclerosis: a clinical trial. Altern Complement Ther 7(1):37–45

Hu H, Chen D, Zhang X (2006a) Effect of polypeptides in bee venom on growth inhibition and apoptosis induction of the human hepatoma cell line SMMC-7721 in vitro and Balb/c nude mice in vivo. J Pharm Pharmacol 58:83–89

Hu H, Chen D, Liu Y, Yang S, Qiao M, Zhao J, Zhao X (2006b) Target ability and therapy efficacy of immune liposomes using a humanized antihepatoma disulfide-stabilized Fv fragment on tumor cells. J Pharm Sci 95:192–199

Hugues M, Romey G, Duval D, Vincent JP, Lazdunski M (1982a) Apamin as a selective blocker of calcium dependent potassium channel in neuroblastoma cells: voltage-clamp and biochemical characterization of the toxin receptor. Proc Natl Acad Sci 79:1308–1312

Hugues M, Duval D, Kitabgi P, Lazdunski M, Vincent JP (1982b) Preparation of pure monoiodo derivative of bee venom neurotoxin apamin and its binding properties to rat brain synaptosomes. J Biol Chem 257:2762–2769

Hurkov S (1971) Electrophoresis of the bee venom preparation Melivenon in the treatment of osteoarthritis. Kurort Fizioter 8(3):128–131

Hyre HM, Smith RA (1986) Immunological effects of honeybee venom using balb/c mice. Toxicon 24(5):435–440

Ip SW, Chu YL, Chen PY, Ho HC, Yang JS, Huang HY, Chueh FS, Lai TY, Chung JG (2012) Bee venom induces apoptosis through intracellular Ca++– modulated intrinsic death pathway in human bladder cancer cells. Int J Urol 19(1):61–70

Jang MH, Shin MC, Lim S, Han SM, Park HJ, Shin I, Lee JS, Kim KA, Kim EH, Kim CJ (2003) Bee venom induces apoptosis and inhibits expression of cyclooxygenase-2 mRNA in human lung cancer line NCI-H1299. J Pharmacol Sci 91(2):95–104

Jasani B, Kreil G, Mackler BF, Stanworth DR (1979) Further studies on the structural requirements for polypeptide mediated histamine release from rat mast cells. Biochem J 181:623–632

Jo M, Park MH, Kollipara PS, An BJ, Song HS, Han SB, Kim HJ, Song MJ, Hong JT (2012) Anticancer effect of bee venom toxin and melittin in ovarian cancer cells through induction of death receptors and inhibition of JAK/2/STAT3 pathway. Toxicol Appl Pharmacol 258(1):72–81

Jones HP, Chai G, Petrone WF (1982) Calmodulin dependent stimulation of the NADPH oxidase of human neutrophils. Biochim Biophys Acta 714:152–156

Kang JK, Kim CMH (1993) Toxicity test of apitoxin (14 p). Phase I study, Final Report to KFDA

Kang SS, Pak SC, Choi SH (2002) The effect of whole bee venom on arthritis. Am J Chin Med 30(1):73–80

Kanno I, Ito Y, Okuyama S (1970) Radioprotection by bee venom. J Jpn Med Radiat 29:30

Kim CMH (1986) Bee venom therapy. Manag Stress Pain 1(4):1–6

Kim CMH (1987) The final report of the safety and toxicity of Apitox. Phase I clinical trial, FDA

Kim CMH (1989) Bee venom therapy for arthritis. Rheumatologie 41:67–72

Kim CMH (1991) Honey bee venom therapy for arthritis (RA, OA), fibromyositis (FM) and pheripheral neuritis (PN). J Korean Pain Res 1(1):55–65

Kim CMH (1992) Bee venom therapy and bee acupuncture therapy. Korean Ed Publishing, Seoul, 515 pp

Kim CMH (1997) Potentiating health and the crisis of the immune system. Chapter 24; apitherapy (Bee Venom Therapy) literature review. Plenum Press, New York, pp 243–270

Kim CMH (2009) Report to FDA, and update of the DMF BB13, 130

Kim CMH (2011) Personal communication with American Apitherapy Society

Kim CMH (2012) Stability test: 3 years follow up (92 p). Report to FDA – Phase III clinical trial

Kim KS, Choi US, Lee SD, Kim KH, Chung KH, Chang YC, Park KK, Lee YC, Kim CH (2005) Effects of bee venom on aromatase expression and activity in leukemic FLG 29.1 and primary osteoblastic cells. J Ethanol pharmacol 99(2):245–252

Kim DH, Kim CMH, Jun HK, Park SK, Hsu CY, Hsu CL, Liao JC, Chueh HJ, Cheng HW (2007a) Treatment by injection-acupuncture with Apitox combined by Chinese herbal medicine in patients with canine hind limb paralysis. J Vet Clin 24(2):225–228

Kim DH, Kim CMH, Oh JW, Lee HH, Jeong SM, Choi SH (2007b) Therapeutic effect of bee venom and dexamethasone in dogs with facial nerve paralysis. J Vet Clin 24(4):503–508

Kim SJ, Park JH, Kim KH, Lee WR, Kim KS, Park KK (2011) Melittin inhibits atherosclerosis in LPS/high-fat treated mice through atheroprotective actions. J Atheroscler Thromb 18(12): 1117–1126

Klinghardt D (1990) Bee venom therapy for chronic pain. J Neuro Ortho Med Surg 11(3):195–197

Knepel W, Gerhards C (1987) Stimulation by melittin of adrenocorticotrophin and beta-endophin release from rat adenohypophysis *in vitro*. Prostaglandins 33(3):479–490

Koburova KL, Michailova SG, Shkenderov SV (1984) Antipyretic effect of polypeptide from bee venom – adolapin. Eksp Med Morfol 23:143–148

Koburova KL, Michailova SG, Shkenderov SV (1985) Further investigation on the antiinflammatory properties of adolapin – bee venom polypeptide. Acta Physiol Pharmacol Bulg 2(2):50–55

Koumanov K, Momchilova A, Wolf C (2003) Bimodal regulatory effect of melittin and phospholipase A2 activating protein on human type II secretory phospholipase A2. Cell Biol Int 27:871–877

Kwon YB, Lee JD, Lee HJ, Mar WC, Kang SK, Beitz AJ, Lee JH (2001a) Bee venom injection into an acupuncture point reduces arthritis associated edema and nociceptive responses. Pain 90(3):271–280

Kwon YB, Kim JH, Yoon JH, Lee JD, Han HJ, Mar WC, Beitz AJ, Lee JH (2001b) The analgesic efficacy of bee venom acupuncture for knee osteoarthritis: a comparison study with needle acupuncture. Am J Chin Med 29(2):187–199

Landucci EC, Toyama M, Marangoni S, Oliveira B, Cirino G, Antunes E, de Nucci G (2000) Effect of crotapotin and heparin on the rat raw edema induced by different secretory phospholipase A2. Toxicon 38:199–208

Langer J (1897) Uber das Gift Unserer Honigbiene. ALeipz 38:381–396

Lee JH, Kwon YB, Han HJ, Mar WC, Lee HJ, Yang IS, Beitz AJ, Kang SK (2001) Bee venom pretreatment has both an antinociceptive and antieffect on Carrageenan inflammation. J Vet Med Sci 63(3):251–259

Lee WR, Kim SJ, Park JH, Kim KH, Chang YC, Park YY, Lee KG, Han SM, Yeo JH, Park KK (2010) Bee venom reduces atherosclerotic lesion formation via anti-inflammatory mechanism. Am J Chin Med 38(6):1077–1092

Liu S, Yu M, Xiao L, Wang F, Song C, Sun S, Ling C, Xu Z (2008) Melittin prevents liver cancer cell metastasis through inhibition of the Rac1-dependent pathway. Hepatology 47:1964–1973

Lonauer G, Meyers A, Kastner D, Kalveram K, Forck G, Gerlach U (1985) Treatment of rheumatoid arthritis with a new purified bee venom. Abstract, XXX Apomondia

Lorenzetti OJ, Fortenberry B, Busby E (1972) The influence of bee venom in the adjuvant induced arthritic rat model. Res Commun Chem Pathol Pharmacol 4(2):339–352

Lubke LL, Garon CF (1997) The antimicrobial agent melittin exhibits powerful in vitro inhibitory effects on the Lyme disease spirochete. Clin Infect Dis 25(Suppl 1):S48–S51

Meier J, White J (1995) Handbook of clinical toxicology of animal venoms. CRC Press, New York

Menander-Huber J (1980) Melittin bound to calmodulin. NMR assignments and global conformation features. Exp Biochem 112:236

Minton SA (1974) Venom disease. Charles C Thomas, Springfield

Mund-Hoym WD (1982) A report of the results of treating a total of 211 patients with bee venom. Med World 33(34):1174–1177

Neumann W, Habermann E, Amend G (1952) Zur Papierelektrophoretischen Fraktionierung Tierischer Gifte. Naturwissenschaften 39:286–287

NIH (1995) Apitherapy, alternative medicine: expanding medical horizons, NIH Pub., Bethesda, pp 172–175

NIH (2010) ClinicalTrials.gov Identifier, NCT01112722

Nokolova V (1973) A study of the therapeutic value of electrophoresis with bee venom in children with rheumatoid arthritis. Probl Pediatr 16:101–106

O'Connor R, Peck ML (1978) Venoms of apidae. Arthropod venoms. Springer, New York, pp 613–659

Orsolic N (2012) Bee venom in cancer therapy. Cancer Metastasis Rev 31(1–2):173–194

Orsolic N, Sver L, Versovsek S, Terzic S (2003) Inhibition of mammary carcinoma cell proliferation in vitro and tumor growth in vivo by bee venom. Toxicon 41:861–870

Park HJ, Lee SH, Son DJ, Oh KW, Kim KH, Song SH, Kim GJ, Oh GT, Yoon DY, Hong JT (2004) Antiarthritic effect of bee venom: inhibition of inflammation mediator generation by suppression of NF-kappaB through interaction with the p50 subunit. Arthritis Rheum 50(11):3504–3515

Putz T, Ramoner R, Gander H, Rham A, Bartsch G, Thurnher M (2006) Antitumor action and immune activation through cooperation of bee venom secretory phospholipase A2 and phosphotidylinositol-(3, 4)-bis-phosphate. Cancer Immunol Immunother 55:1374–1383

Rauen HM, Schriewer H, Ferie F (1972) Alkylans alkylandum reactons. 10. Antialkylating activity of bee venom, melittin, and apamin. Arzneim-Forsch 22:1921

Rekka E, Kourounakis L, Kourounakis P (1990) Antioxidant activity of and interleukin production affected by honey bee venom. Arzneimittel Forschung – Drug Res 40:912–913

Schmidt-Lange W (1941) The germicidal effect of bee venom. Muench Med Wochenschr 83:935

Serban E (1981) Bee venom and rheumatism. Fr Rev Apitherapy, p 399

Shipman WH (1967) Increased resistance of mice to X-irradiation after the injection of bee venom. Nature 215:311–312

Shipman WH, Cole LJ (1968) Increased radiation resistance of mice injected with bee venom one day prior to exposure. Report USNRDL-TR-67-4, US Naval Radiological Defense Lab, San Francisco, pp 1–10

Shipolini RA (1984) Biochemistry of bee venom. Arthropod venoms, vol 48, Handbook of experimental pharmacology. Springer, New York, pp 49–85

Shkenderov S (1986) Anti-inflammatory effect of bee venom protease inhibitor on a model system of acute inflammatory edema. Comptes rendus de l'Academie bulgare des Sciences 39:151–154

Shkenderov S, Koburova K (1982) Adolapin – a newly isolated analgesic and anti-inflammatory polypeptide from bee venom. Toxicon 20:317–321

Short T, Jackson R, Beard G (1979) Usefulness of bee venom therapy in canine arthritis. NAAS Proc 2:13–17

Somerfield SD, Stach JL, Mraz C, Gervais F, Skamene E (1986) Bee venom melittin blocks neutrophil O_2-production. Inflammation 10:175–182

Son DJ, Lee JW, Lee HY, Song HS, Lee CK, Hong JT (2007) Therapeutic application of antiarthritis, pain releasing, and anti-cancer effects of bee venom and its constituent compounds. Pharmacol Ther 115:246–270

Steigerwaldt F, Mathies DF (1966) Standardized bee venom (SBV) therapy of arthritis. Controlled study of 50 cases with 84 % benefit. Ind Med Surg 35:1045–1050

Terc P (1904) Lecture from the monthly assembly of beekeepers, 11 February 1904. Bee Venom: the natural curative for arthritis and rheumatism, appendix H, G.P. Putnam's Sons, New York, pp 183–197

Tu AT (1977) Bee venom 501–515. Venoms: chemistry and molecular biology. Wiley, New York

Vick JA, Shipman WH (1972) Effects of whole bee venom and its fractions (apamin and melittin) on plasma cortisol levels in the dog. Toxicon 10:377–380

Vick JA, Mehlman B, Brooks R, Shipman WH (1972) Effect of bee venom and melittin on plasma cortisol in the unanesthetized monkey. Toxicon 10:581–586

Vick JA, Warren GB, Brooks RB (1976) The effects of treatment with whole bee venom on cage activity and plasma cortisol levels in the arthritic dog. Inflammation 1:167–174

Von Bredow J, Short T, Beard G, Reid K (1981) Effectiveness of bee venom therapy in the treatment of canine arthritis. NAAS Proc 4:45–48

Vyatchannikov NK, Sinka AY (1973) Effect of melittin, the major constituent of bee venom, on the central nervous system. Farmakol Toksikol 36:625

Weissmann G, Zurier RB, Mitnick D, Bloomgarden D (1973) Effects of bee venom of experimental arthritis. Ann Rheum Dis 32:466–470

Wesselius T, Jeersema DJ, Mostert JP, Heerings NP, Admiraal-Behloul F, Talebian A, van Buchem MA, De Keyser J (2005) A randomized crossover study of bee sting therapy for multiple sclerosis. Neurology 65:1764–1768

Won JH, Choi ES, Kim CMH, Hong SS (1999) The effectiveness of bee venom on osteoarthritis patients. K Rheumatol 6(3):218–226

Yin CS, Lee HJ, Hong SJ, Chung JH, Koh HG (2005) Microarray analysis of gene expression in Chondro- sarcoma cells treated with bee venom. Toxicon 45:81–91

Yoannovotich G, Chahovitch X (1932) Le traitement des tumeurs par le venin des abeilles. In Achard C, Renault J (eds), Bulletin de l'Académie nationale de médecine, 3e série, tome 107. Masson et Cie, Paris, pp. 892–893

Yue HY, Fujita T, Kumamoto E (2005) Phospholipase A2 activation by melittin enhances spontaneous glutamatergic excitatory transmission in rat substantia gelatinosa neurons. Neuroscience 135:485–495

Yun HS, Lee JD, Lee YH (2000) Systemic review: the study on bee venom related to cancer in PubMed. KJAMS 17(4):69–78

Zaitsev GP, Poriadin VT (1961) Bee venom in the treatment of ankylosing spondylitis and polyarthritis, Moscow National Institute of Medicine, Moscow

Zaitsev GP, Poriadin VT (1973) Bee venom in the treatment of the arterial vessels of the extremities and of the diseases of the spine and joints. XVIII Apimondia Congress Press, pp 1–9

Chapter 5
Apitherapy – The Use of Honeybee Products

Theodore Cherbuliez

5.1 Introduction

In the bee world, the unit of life is the colony, not the bee. None of the three casts, i.e. queen, worker or drone, can survive alone nor reproduce itself. Swarming is a collective behavior by which the colony reproduces itself. For this purpose, the workers build special cells in which the queen deposits eggs. The hatching larvae are fed with a secretion of the worker bees, the royal jelly, which is necessary for the growth of the future queens. When laying eggs the queen is too heavy to fly; the bees then starve her to prepare her for swarming. When the queen cells are ready and the queen able to fly, about one-half of the bee population, together with the old queen, leaves the original nest to find another home.

By looking at the products of the hive one can see that here too, a product is not created by any single bee. In fact, honey, venom, pollen, propolis, royal jelly and wax each owe its existence to a succession of bee activities, each pooling their individual contribution to the "pot." Accordingly, for any use, only the amount contributed by many bees is large enough to be significant. An exception would be when the venom of a single bee is sufficient to achieve a desired result, e.g., applied on the big toe for gout would lead to a significant and prompt reduction in pain.

5.2 Apitherapy

The name Apitherapy comes from Latin "Apis", meaning bee, and Greek: θεραπεία, meaning serving and caring for. Interestingly, even the name Apitherapy is the result of a synergy.

T. Cherbuliez, M.D. (✉)
Consulting Medical Staff at Spring Harbor Hospital, Maine Medical Center,
13 Main Street, PO Box 155, South Freeport, ME 04078-0155, USA
e-mail: tcherbuliez@gmail.com

M. Grassberger et al. (eds.), *Biotherapy - History, Principles and Practice:*
A Practical Guide to the Diagnosis and Treatment of Disease using Living Organisms,
DOI 10.1007/978-94-007-6585-6_5, © Springer Science+Business Media Dordrecht 2013

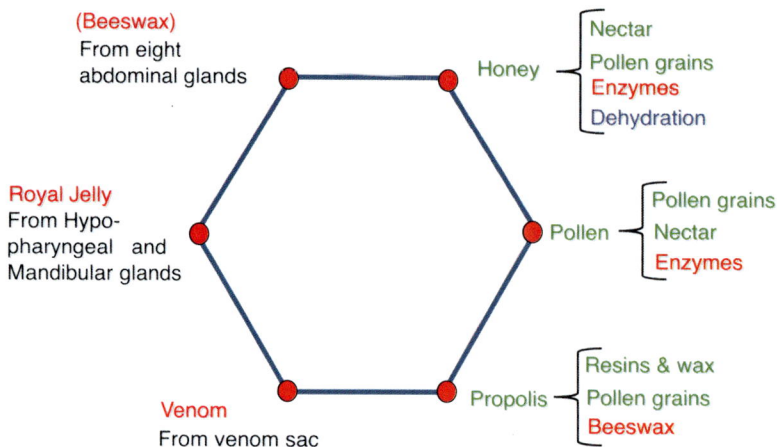

Fig. 5.1 Bee products: secretion (*red*), collection (*green*), modification (*blue*)

Apitherapy is the science (and art) of the use of honeybee products, to maintain health and assist the individual in regaining health when sickness or accident interferes.

In the past, the products of the hive, i.e. honey, pollen and propolis were frequently used as natural remedies for health maintenance and bee venom for treatment of ailments. More recently, the products of the hive have been incorporated into Western medical practice, where the focus of attention is mainly the illness and its prevention. This practice tends to use these products after they have undergone some processing, e.g. the irradiation of honey to insure sterility, or the use of bee venom extracts for ease of utilization. It is interesting to note that the concepts underlying the thinking in Apitherapy as a natural medicine differs from the corresponding ones used in Western medical culture. These concepts will be addressed below.

The main products of the hive used in Apitherapy and their origin are shown in Fig. 5.1.

5.2.1   Bee Products

Honey, pollen and propolis have in common that they begin as botanical material, respectively nectar, male gametes of flowers, and resins. These are collected by the bees who then add their secretions. The contribution of each bee is diverse, due to the variation of the botanical sources. The story of venom is different: its composition depends on the pollen the bee consumed and on the age of the bee as it matures with time. Royal jelly and wax are secretions of the bees, and they remain constant in their composition.

Table 5.1 Comparison of "Traditional" with "Western medicine" concepts

Traditional concepts	Western concepts
Uncertainty	Certainty
Variability	Constancy
Open system	Closed system
Synergy: multi-molecular systems	Specific molecule for identified action

5.3 Concepts of Apitherapy

As shown in Table 5.1, the concepts used in Apitherapy can be those of "Traditional Medicine" or those used in "Western Medicine."

I do not suggest that one set of concepts is superior to the other. Rather, that they are complementary and that Apitherapy can be viewed from either lens.

5.3.1 Uncertainty – Certainty

"Traditional concept" – **Uncertainty:** determining precisely the chemical composition of honey, pollen, propolis and venom is limited to establishing the formula of the sample tested, with the awareness that all other samples may be somewhat different. If we take the example of bee venom, we can see that if we take the venom of a given bee, we can either examine it to quantify its contents or effects, after which it is no longer available to use, or we can use it and we will not know exactly its composition.

"Western concept" – **Certainty.** The composition of a product can be reliably known and the product can be reproduced. One sample is identical to any other sample of the same product. It is then possible to know exactly the composition of a product one uses. It is also possible to create any amounts of the same product. When dealing with the products of the hive, working with one component, for instance melittin (out of the peptides of bee venom), respects the laws of certainty.

5.3.2 Variability – Constancy

"Traditional concept" – Products from the hive are **variable,** in function of both time and place, coming from many sources, such as pollen from different plants, and venom, which varies in its composition based on the age of the bee. Thus the products used in Apitherapy are never exactly the same, from one application to the next, and are not known nor are they reproducible or precisely measurable. The clinical consequences of this dimension are of great importance. A number of bee products have antibiotic properties. The products vary in their composition; these variations cannot be predicted or known by germs. They, therefore, cannot get accustomed (become resistant) to a presence that varies constantly. Variability may have another value, that is the possibility, empirically often observed with natural products, that administration of several doses of a product showing small variations in their composition over time may be more effective than products that are rigorously homogeneous.

"Western concept" – A given product has the same known composition regardless of where and when it was made. **Constancy** in the composition of products allows the running of clinical protocols in very different circumstances, places, and times and pooling the results. Further consequence of this approach is the creation of multidrug resistant, MDR or "superbugs". This resistance turns hospitals into zones of risks, and communities into sources of epidemics of Methicillin-resistant Staphylococcus aureus (MRSA) and vancomycin resistant enterococci (VRE). In the United States, the Center for Disease Control and Prevention (CDC) estimated (2006) that roughly two million hospital-associated infections, from all types of bacteria combined, cause or contribute to 99,000 deaths each year, more than doubling the mortality and morbidity risks of any admitted patient.

5.3.3 Open and Closed Systems

"Traditional concept": In an **open system** not all relevant variables are fully known, nor measureable nor controllable. (It is also the system in which life can be sustained). An Open system allows flexibility in the means used to achieve results. The participation and contribution of the individuals involved becomes one of the most important variables taken into account, a variable that increases in importance as the patient has more experience. In a therapeutic context, the lack of control over variables prevents proof of treatment effectiveness.

"Western concept": A **closed system** is necessary to reach a control, and a measure, of the relevant variables; those variables that cannot be controlled are to be discarded. The variables retained are described in detail, which allow proofs, valid within the limits of the observed situation. When the process observed takes place in people, the discipline leading to retain only controllable variables to define a process represents a "simplification" of the person(s) referred to. The progresses that such simplification of the person has allowed, in the development of medicine and of the knowledge of illnesses and their treatment, are immeasurable. The drawback of this simplification is the elimination of individual's humanity, initiative, adaptation, other than compliance with, and tolerance of, the instructions received. It should be emphasized that this only concerns the concepts of Western medicine and does not refer to the attitude of physicians treating patients. Western medicine, in its thinking, does not engage the individual to take responsibility nor does it acknowledge his power to do so.

5.3.4   Synergy

"Traditional concept" – An action is typically caused by a **synergy** of molecules, of components, or effects. The variations within time and space of many of these bee products, add to the fact that they are often used together.

"Western concept" – The emphasis is on reproducibility and precision. The goal is to find a **specific componen**t, for instance, a molecule, for each action of a product. Further, the search is for the reproducibility of such a component or molecule and for the predictability of its actions.

5.4   Bee Venom

The distinction between bee venom, defined by what a bee delivers when it stings, and Apitoxin, a patented name reserved to derivatives or extracts of bee venom, merits attention as there is major confusion in the literature regarding these two entities. The differences between bee venom and Apitoxin were evaluated by the Apitherapy Commission of Apimondia (Domerego 2012), which was carried out a study in the Chemistry Department of the Catholic University of Louvain in Brussels, Belgium. The study showed that fresh bee venom has about 3 % essential oils, from which 60 % are represented by esters. The latter are practically absent in Apitoxin and seem to be replaced mainly by acids and alcohols (Fig. 5.2). Observations further suggested that there are differences between the clinical effects of the two products, an issue deserving further research.

5.4.1   Treatment with Bee Venom

The treatment with natural remedies often includes the active participation of the patient and Apitherapy is particularly well suited to serve this way. Treatment for any condition can become an opportunity to involve patients intensely at each step, while their collaboration and decision making should utilized to the point that they should be able to continue their care on their own, assuming responsibility for their health.

A particularity of bee venom therapy (BVT) is that with each sting, there is an opportunity to observe several effects. A major part of the treatment is teaching patients to observe what is happening to their bodies with each sting, and progressively to determine themselves the indication to add another sting and if yes, where it should be applied. At first, before having any experience with this approach, people think mainly about the associated pain. As the treatment proceeds, they become more interested in the responses communicated by their body, such as differences in balance, a general feeling of relaxation and with some experience can even define when they have received an optimal amount of stings.

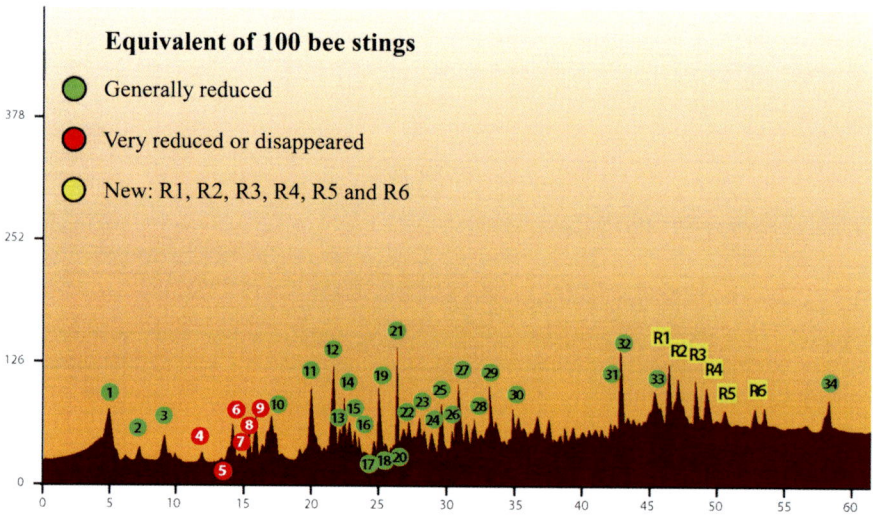

Fig. 5.2 Chromatographic analysis of essential oils in fresh honeybee venom (*above*) compared to the chromatographic analysis of Apitoxin (*below*)

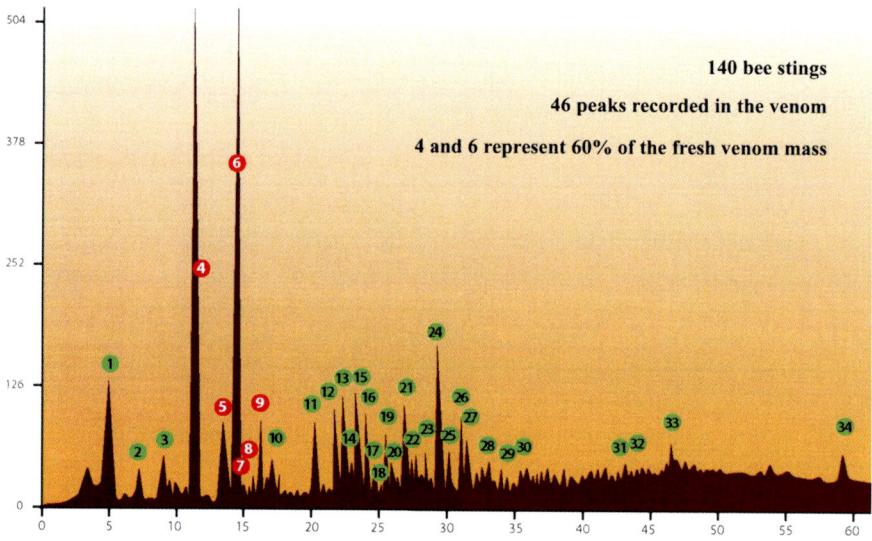

Fig. 5.3 Separating the
stinger from the bee

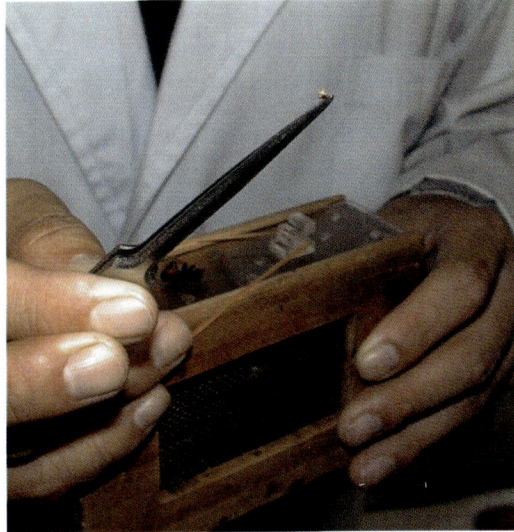

5.4.2 Practical Considerations When Administering BVT

In the practice, of BVT, because there is no possibility to combine venom with an anesthetic, as in the case with Apitoxin, one has to address the element of pain and apprehension from the very beginning. There are several techniques to decrease the effect of pain, one being the dosing of the venom determined by the type of sting.

There are three kinds of stings:

1. The regular or full sting, when the bee's stinger is kept in place for >5 min and the quantity of venom administered is approximately 150 µg.
2. The mini-sting, when the time before removing the stinger is shortened. In this case, the patient can also decide when to remove the stinger, which represents an important step in putting him in charge of his treatment. Here, the quantity varies usually from 20 to 30 µg to a full dose, depending on how long the stinger is left in place. Either the practitioner or the patient decides when to remove the stinger.
3. The micro-sting, when the therapist removes the stinger from the bee and applies it to the patient a number of times (Figs. 5.3 and 5.4). Experts in this technique can apply more than 100 micro-stings with one stinger. It is estimated that an averages of 1–2 µg venom is delivered with each sting. Another approach uses warm compresses just before application of the micro-stings. The patient feels only the light touch of the practitioner's fingers, not the stings themselves. This technique is totally painless and gives the clinician control over the stinging that even babies can be treated without discomfort.

Fig. 5.4 The technique of
micro-sting. It can be noted
that on this case the skin had
been prepared with cupping

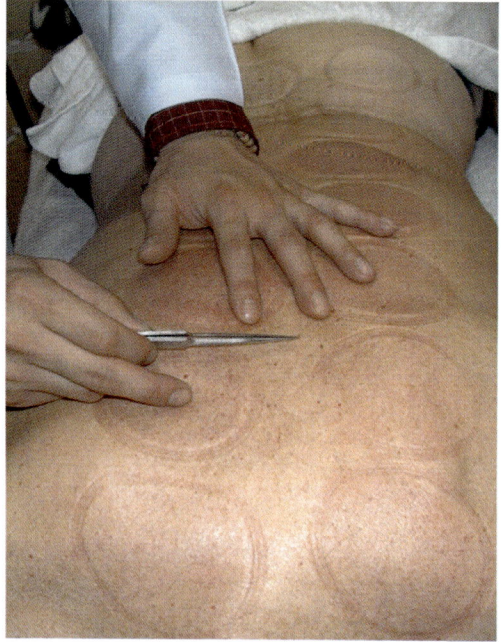

Fig. 5.5 Intensity of pain
related to bee sting in *cooled*
and *un-cooled* skin

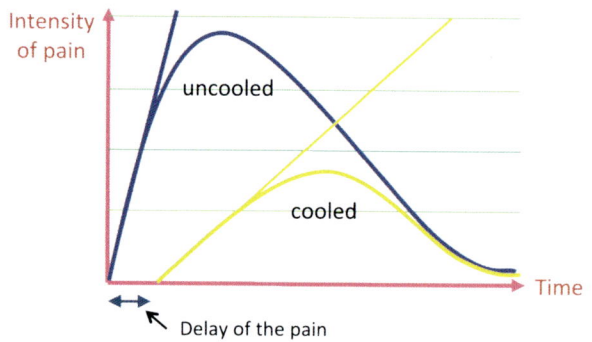

Another way to decrease the pain experience is the cooling of the site before
stinging, e.g. with a frozen object. The main effect of the cooling is the decrease of
the gradient of pain (Fig. 5.5). For many people the pain of the bee sting has a
marked component of fright due to the steepness of the gradient, which adds to the
perceived pain. Cooling is contra-indicated while the stinger is on the skin, because
this would stop the pumping action of the venom sack and impede the administration of the venom.

5.4.3   Protocol of Treatment

The treatment protocol of the present author includes having a trusting, personal relationship with the patient at the very beginning of the therapy. Both, the therapist and the patient, have to be clear about what is expected from the treatment, and, perhaps more important, how the therapy will engage the therapist and the patient. The patient is considered as the expert in his knowledge of himself, his abilities and goals. Accordingly, therapist and patient form a team and the team treats the patient. Anticipated reactions to venom are described in advance and the treatment is not initiated unless the potential reactions are acceptable to the patient. Discussing reactions and responding to patients concerns about reactions, minimizes fear, enhances trust and supports the relationship between the therapist and the patient.

In this context, one has to clarify the distinction between a tissue reaction to venom, which proceeds by extension from the site of the sting and is not, regardless of its dimension, an indication of allergy. An example is if a person is stung on the hand and has swelling from this area up to the elbow. As this swelling extends from the sting site, it is only an inflammation and not an allergic reaction. An allergic reaction is mediated systemically and accordingly its manifestations take place also on other sites of the body, e.g., if a person develops hives on the stomach after receiving a sting on the hand.

The American Apitherapy Society (AAS) has elaborated a desensitization program that proposes two starting points, which are shown in Table 5.2. The patient makes always the choice between these two.

The values presented in this table are presented as maxima; these numbers should be respected and only exceeded by very experienced practitioners who have achieved a trusting relationship with their patient. An important principle applicable to desensitization is the continuity of these administrations. Any interruption of the process exceeding a few days requires starting the process over from the beginning.

For this purpose, AAS gives yearly courses on BVT and on the testing for sensitivities. For example, AAS recommends that those practicing BVT have epinephrine available in case of an anaphylactic reaction. The use of anti-histamine tablets is also taught, with less emphasis on their use.

5.4.4   Side-Effects of BVT

Instances of allergy to venom are not rare; they practically never take place at the beginning of therapy and can be easily handled: the therapist faces the patient who lays down, relaxes and describes his experience in detail. The therapy is resumed as soon as the patient is ready, using small amounts of venom and increasing the quantity gradually. An injectable dose of epinephrine and anti-histamine tablets should always be available in the office of practitioner.

Table 5.2 Protocol for desensitization to bee venom with micro- and mini stings

Time	Starting with micro-stings Program 1	Time	Starting with mini-stings Program 2
Day 1	5 tests with micro-stings		
Day 2	Day of rest		
Day 3	5 tests with micro-stings		
Day 4	Day of rest		
Day 5	5 tests with mini-stings	Day 1	5 tests with mini-stings
Day 6	Day of rest	Day 2	Day of rest
Day 7	5 tests with mini-stings	Day 3	5 tests with mini-stings
Day 8	Day of rest	Day 4	Day of rest
Session 5	1 test with a 30" mini	Session 3	1 test with a 30" mini
Session 6	1 test with a 60" mini	Session 4	1 test with a 60" mini
Session 7	1 test with a 90" mini	Session 5	1 test with a 90" mini
Session 8	1 test with a 2 min sting	Session 6	1 test with a 2 min sting
Session 9	1 test with a 5 min full sting	Session 7	1 test with a 5 min full sting
Session 10	2 tests with full stings	Session 8	2 tests with full stings
Session 11	1 more sting than the previous time	Session 9	1 more sting than the previous time

There are numerous observations suggesting that a trusting relationship between the two people involved in BVT, where the apprehension about venom has been reduced to a low, tolerable level, might lead to a decrease of the strength of allergic reaction. Some observations speak strongly in favor of the importance of the relationship in the protection against massive allergic reaction. In nearly 70 years of stinging experience, Charles Mraz, called the grandfather of BVT in the USA, never had to use epinephrine, the treatment of choice in bee venom related anaphylaxis. In 45 years of Apitherapy the present author observed only twice an anaphylactic reaction in a person stung. Both times were at congresses, where a speaker asked the audience for a volunteer for a demonstration of bee venom administration. In both instances a young man volunteered. The volunteer and the therapist did not know each other and they had no pre-established relation. The allergic reaction appeared very soon after the sting but was treated successfully thereafter.

5.5 Pollen

5.5.1 A Historical Note

Back as far back as 2735 B.C., Shen Nung, a Chinese emperor, had medical texts discussing the merits of pollen. There are also Egyptian papyri in which pollen is referred to as life-giving dust. In the 400s B.C., Hippocrates recommended pollen

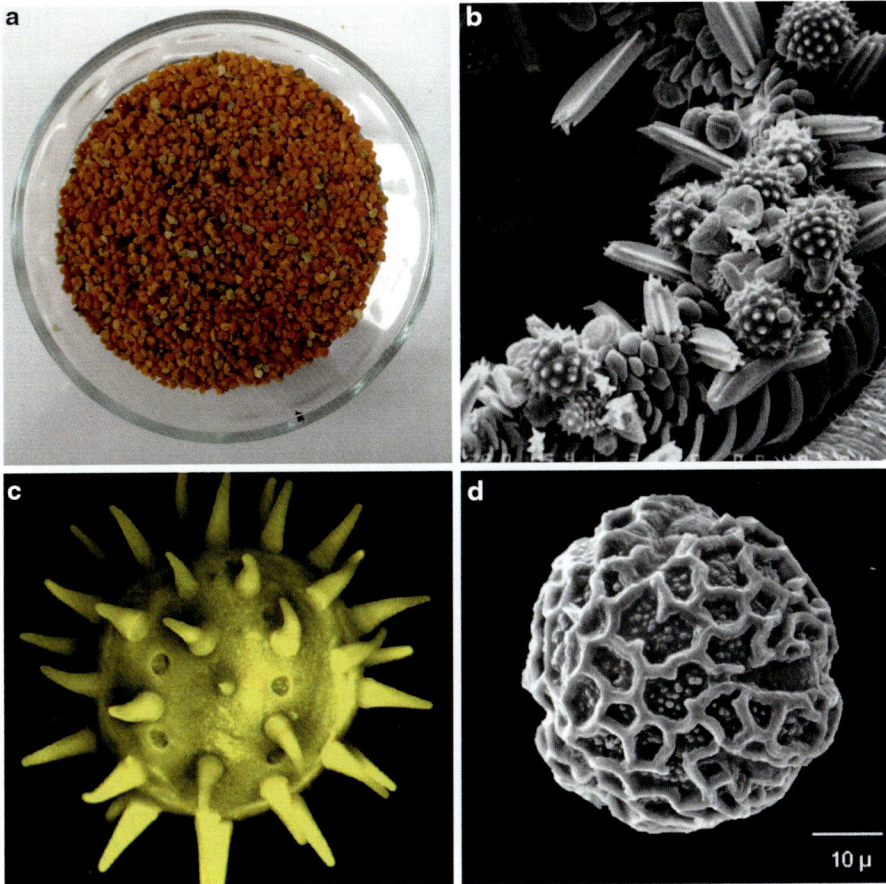

Fig. 5.6 Pollen grains collected by bees

as a remedy for multiple conditions and the Hindus taught that eating honey and pollen could produce health, vigor, happiness and wisdom.

5.5.2   What is Pollen?

Pollen grains are the male reproduction units (gametophytes) formed in the anthers of the higher flowering plants. The pollen is transferred onto the stigma of a flower (a process called pollination), by either wind, water or various animals (mostly insects), among which bees are the most important ones. Pollen from each species of flower is different and no one pollen type can contain all the characteristics ascribed to pollen in general (Fig. 5.6). The chemical composition of pollen is shown in Table 5.3.

Table 5.3 Chemical composition of pollen

Compound	Percentage and additional explanations
Water	5 % (dry) – 15 % (fresh)
Peptides	7.3–35 %
Sugars	15–55 %: Composed of fructose (60 %), glucose (40 %) and sucrose
Lipids	2–14 %: Primarily free unsaturated fatty acids, lecithin and phospholipids
Esters	Such as auxins, brassins, gibberellines and kinins
Carotenoids	Pro vitamin A
Antioxidants	Food with the highest value as measured by oxygen radical absorbance capacity test
Pigments	Yellow and orange
Gonadotropins	Increased levels of testosterone-like hormone
Estrogenic compounds	Antioxidant and chemo-protective activity
Minerals	>3 %: Bo, Ca, Cr, Cu, F, Fe, I, K, Mg, Mn, Mo, Na, Ni, P, S, Se, Si, Zn
Vitamins	A, B, C, D, E, K
Enzymes	>200
Hydroxycinnamic acid	Strong antioxidant activity (from a Brazilian pollen)
Undetermined	3 %

5.5.3 Foraging for Pollen

A foraging honeybee rarely collects both pollen and nectar from more than one species of flowers during one trip. Thus, the resulting pollen grains on her hind legs contain usually only one pollen species. The different colors are due to different flavonoids and carotenoids, which besides being anti-oxidants are also pigments.

During their collecting trips, bees add their saliva and nectar as they collect pollen. The collection process has two steps:

1. The bee, foraging for nectar, puts her head deep into the flower and accumulates pollen all over her body (Fig. 5.7a).
2. She then "cleans" herself by putting the pollen in the baskets of her third pair of legs, adding nectar and saliva to create a ball. The vibration of the bee's body during flight causes the pollen basket to become compact (Fig. 5.7b).

The nectar contains between five and eight strains of lactobacteria and three yeasts (Olofssos and Vásquez 2009). Honeybees therefore possess an abundant, diverse and ancient lactic acid bacteria microbiota in their honey crop with beneficial effects for bee health, defending them against microbial threats (Vásquez et al. 2012). They prevent the growth of other bacteria that may cause the pollen to spoil. This bacteriological "microflora" is perfectly preserved when the pollen is frozen, and varies in terms of quantities from between one and ten million bacteria per gram of pollen. However, when the pollen is dried, this microflora is destroyed. In itself, the pollen does not contain antibiotics; however, its microflora has a "barrier effect" that keeps the harmful bacteria such as *Proteus vulgaris* and *Proteus mirabilis* at bay (Percie du Sert 2006).

Fig. 5.7 Bees with pollen: (**a**) first step of collection; (**b**) pollen attached to the third leg of the bee

5.5.4 Bee Bread

Bees build three kinds of cells in the hive. By far the largest number of cells is for the workers, which are also the smallest. The cells destined for drones are larger in which bees can also store pollen. Both types are set horizontally. Cells that are built for queen larvae hang vertically. These queen cells are the only ones allowed to encroach into the space between combs and are never used to store pollen.

When a bee returns to the hive with a load of pollen grains, she transfers them to another bee, who puts them in a cell and compresses each grain. When the cell is filled, they cover it with a thin layer of propolis; they have created bee bread (Fig. 5.8). Each cell contains the nutrition needed for one larva, in order to develop to an adult bee.

5.5.5 Trapping and Conservation

Pollen traps are devices that force bees to go through narrow holes as they enter the hive. The holes are calibrated so that the bees, passing through, drop approximately 70 % of their loads. These traps should be installed only on strong colonies, and for relatively short times (2–3 weeks), separated by similar intervals when bees may bypass them. Bees respond to pollen trapping by increasing their pollen trips. However, traps retaining more than 70 % of the load discourage bees from foraging further. A careful handling of traps allows for collecting about 6 kg of pollen a year from each hive, which is 10 % of the colony's production.

Several techniques are used to conserve the collected pollen, the easiest being the immediate freezing after removal from the hive. A good trap will yield pollen that is without debris, such as bee parts, and does not need to be cleaned. Special equipment is needed to freeze dry pollen. It can be dried in the dark at 30 °C for 20 h. Dried pollen kept at room temperature looses about 50 % of its antioxidant components, but none of its minerals, essential amino-acids or vitamins.

Fig. 5.8 Bee bread of *Apis mellifera*

5.5.6 Medicinal Properties of Pollen

Pollens from specific plants are known for their health effects, e.g. thyme: stimulating and antiseptic; sage: diuretic, regulates GI function, menstrual effects; sunflower: diuretic and laxative; canola: varicose veins; apple: myocardial effects; false acacia: sedative; chestnut tree: venous and arterial circulation, liver and prostate decongestant effects.

5.5.7 Commercial Preparations of Pollen

Pollen is sold as grains (fresh-frozen or dried), powder, tablets, capsules, coated pills, or as mixtures with honey or other foods. Formulations of pollen are prepared with water and alcoholic solutions in form of ointments, creams, lotions, capsules, suppositories, ovules often mixed with other bee products and/or essential oils.

5.5.8 Tolerance and Side-Effects

No contra-indications are known with the consumption of pollen, even in pregnancy. Allergy to pollen can be handled generally easily. (See below) There are no known incompatibilities with other therapeutics. Pollen is well tolerated even in long lasting administration and has no toxic effects even in high dosage. Recently, Hurren and Lewis (2010) reported a case, which suggests a probable drug interaction between bee pollen and warfarin.

The side-effects of pollen consumption or treatments with pollen formulations include unpleasant feeling of taste and flavor, when pollen becomes moldy, mild intestinal disorder such as diarrhea during the first days of use, gastric pain (when pellets are not well dissolved) and very rarely allergic reactions such as anaphylaxis.

5.5.9   Indications

5.5.9.1   Nutritional

Although pollen is a perfect food for the bee, this is not the case for humans as the nutrients are not always in the exact ratio needed for the human body. Particularly, pollen lacks the right amount of vitamin A, however it has a maximum percentage of essential amino-acids. Pollen is rich in proteins and 100 g of pollen can provide as much protein as seven eggs or 400 g of beef. Pollen contains all 22 essential amino-acids, as well as high amounts of proline and hydroxyproline, which serve as building blocks for collagen. Clinical tests have shown that orally ingested pollen is rapidly and easily absorbed, while most of its components pass directly from intestines into the blood stream. It should be also noted that pollens supply, in average, about 250 kcal per 100 g.

5.5.9.2   Treatment of Allergies to Pollen

The treatment of allergies should be started with one granule of pollen and the dose should be doubled each day as tolerated. Local unfiltered honey taken regularly over long periods can also keep the patient free of allergies. Whether taken directly or in honey, maintenance treatment, after cessation of symptoms, requires one dose of pollen monthly.

5.5.9.3   Asthma

In vivo activity assessment of a honey and pollen mixture formulation is claimed to be effective for the treatment of asthma, bronchitis, cancers, peptic ulcers, colitis, various types of infections including hepatitis B, and rheumatism by the herb dealers in northeast Turkey. Küpeli Akkol et al. (2010) concluded that their studies have clearly proved that mixing pure honey with bee pollen significantly increased the healing potential of honey and provided additional support for its traditional use.

5.5.9.4   Rheumatoid Arthritis and Digestive Track Disorders

The positive effect of flower pollen in patients with rheumatoid arthritis and concomitant disorders of the gastro-duodenal and hepatobiliary systems has been shown by clinical and biochemical, endoscopic and ultrasonic investigations (Voloshyn et al. 1998).

5.5.9.5   Ophthalmology

In Lithuania bee products such as honey, propolis, bee pollen and royal jelly are widely used in ophthalmology (Jankauskiene et al. 2006). The Mayas used to treat eye diseases with Melipona honey (Vit 2002).

5.5.9.6   Urinary Tract

In a double-blinded placebo controlled study on 60 men with benign prostate hypertrophy (BPH), pollen and placebo was administrated for a period of 6 months. At that time, 60 % of the treated and 30 % of the control men had improved or became symptom free (Buck et al. 1990). In another study with patients having BPH and prostatitis, alpha-blockers were recommended, but also included saw palmetto and bee pollen (Nickel 2006).

5.5.9.7   Metabolism

In 157 overweight and obese patients, who received honey and pollen, 18.3 % showed a decrease in the total cholesterol and 23.9 % in the LDL cholesterol levels. However, the improvement of blood lipid composition in those overweight (body mass index 25–30) and obese (BMI over 30) patients occurred only after loss in body mass (Kas'ianenko et al. 2011).

Chinese studies on humans and animals have demonstrated that consuming bee pollen for several days prior to moving to high altitude reduces the incidence of altitude sickness, and apparently improves the ability to adapt to lower levels of oxygen in the air (Peng 1990).

5.5.9.8   Radiation Protection

Mice bred to develop and die from tumors, were fed bee pollen at ratio of 1:10.000 in mice chow. In untreated mice, tumors developed in 100 % at an average of 31.3 weeks, while in pollen fed mice the average onset was at 41.1 weeks and some were still tumor free at about 60 weeks, when the test terminated (Robinson 1948).

5.5.9.9   Inhibition of Growth

A number of prostate inhibitory components were identified in pollen extract. A fraction designated as FV-7 maintained a strong time- and dose-dependent inhibitory effect on the growth of a prostate cancer cell line (Habib et al. 1995), while a sample of synthesized hydroxamic acid, structurally indistinguishable from FV-7, showed the same in vitro effects (Zhang et al. 1995).

It should be noted that there is a substantial number of warnings in publications, without confirming evidence, of the danger of bee pollen from anaphylactic reactions to a vast array of symptoms that cleared with the discontinuation of pollen.

5.6   Honey

It has been estimated, that bees must carry 120,000–150,000 loads of nectar and together have to fly about 50,000 km or 1.25 times around the equator to produce 1 kg of honey. They go through 120–150 regurgitation and swallowing cycles for each load. A strong bee colony is capable of producing up to 460 kg of honey per year.

5.6.1   Standards for Honey

The US Federal Drug Administration (FDA) does not have an official definition for honey. For internal operations, the Department of Agriculture established voluntary grade standards for comb honey in 1933 and for extracted honey in 1985. For exported honey, the standards included in The Codex Alimentarius Commission of the United Nation's Food and Agriculture Organization (FAO), are used. In the US, there are 48 floral varieties of honey, the essential oils of which, account, among other constituents, for its therapeutic properties.

5.6.2   Toxic Honeys

Some honeys can be toxic to humans, but are innocuous to bees and their larvae. An example is honey from *Rhododendron ponticum,* which contains alkaloids. The symptoms of honey intoxication vary from case to case and may include weakness, sweating, hypotension bradycardia, Wolff-Parkinson-White syndrome, gastritis, peptic ulcer, nausea, vomiting, faintness, leukocytosis, mild paralysis, dizziness, vertigo, blurred vision, convulsions and respiratory rate depression (Mayor 1995). Outside of Historical references, there are very few reports of accidental poisoning with honey, none fatal.

5.6.3   Composition of Honey

Carbohydrates comprise the major component of honey, which include monosaccharides, disaccharides and oligosaccharides. The monosaccharides in honey are fructose and glucose; the disaccharides include sucrose, maltose, isomaltose, maltulose, turanose and kojibiose. The oligosaccharides present in honey, formed from incomplete breakdown of the higher saccharides present in nectar and honeydew, include erlose, theanderose and panose (Bogdanov et al. 2008; Erejuwa et al. 2012).

Various proteins and amino acids are also ingredients of honey. Enzymes include invertase, which converts sucrose to glucose and fructose; amylase, which breaks starch down into smaller units; glucose oxidase, which converts glucose to gluconolactone; which in turn yields gluconic acid and hydrogen peroxide; catalase which breaks down the peroxide formed by glucose oxidase to water and oxygen; and acid phosphorylase, which removes inorganic phosphate from organic phosphates. There are 18 free amino acids, the most abundant of which is proline.

Vitamins found in honey include trace amounts of the B vitamins riboflavin, niacin, folic acid, pantothenic acid and vitamin B6 and ascorbic acid (vitamin C) (Bogdanov et al. 2008). Additionally there are minerals, including calcium, iron, zinc, potassium, phosphorous, magnesium, selenium, chromium and manganese. Antioxidants include flavonoids, of which one, pinocembrin, is unique to honey and bee propolis. Ascorbic acid, catalase and selenium are other antioxidants in honey.

Organic acids such as acetic acid, butanoic acid, formic acid, citric acid, succinic acid, lactic acid, malic acid, pyroglutamic acid and gluconic acid, as well as a number of aromatic acids are found in honey. The main acid present is gluconic acid, formed in the breakdown of glucose by glucose oxidase (Bogdanov et al. 2008).

5.6.4 Indications for Honey

5.6.4.1 Health Maintenance

The earliest information on using honey in children's nutrition dates to about the ninth century BC. Ancient Germanic tribes and Greeks handled milk combined with honey and melted butter. In numerous historic texts, information about the therapeutic and prophylactic use of honey can be found. In Egypt, 3,500 years ago in the "Book of the Preparation of Medicines for all Parts of the (Human) Body", there are many recipes, which include honey. In an ancient Chinese script, honey was given this description: *"Honey heals internal organs, gives strength, reduces fever; long-term use increases its power, gives ease to the body, retains his youth, extending the years of life."* The old Indian "book of life" says that extending human life can only be achieved with elixirs and diet, including honey and milk. In India, honey was used as an antidote to intoxication of any kind. The ancient philosophers and physicians of Rome and Greece including Pythagoras, Aristotle, Hippocrates, Dioscorides, Homer, and Galen wrote on the high healing properties of honey. The philosopher Democritus believed that honey had anti-aging properties. In the ancient script "The canon of the Avesta" from Iran, the use of honey and wax among other animal products are recommended. Avicenna, the tenth century Persian physician and philosopher, in his book "The Canon of Medicine" refers to dozens of recipes, which included honey and wax as ingredients. For people older than 45 years, he recommended the systematic use of honey and walnuts. In the Middle Ages, honey was also widely used in medical therapy to treat wounds.

Honey as a healing product remained important in later centuries, and even in our days. It is widely used in folk medicine and is recognized as an official medicine throughout the world. Honey has earned fame as a universal energy product and for its remarkable pharmacological properties. It has tonic, antimicrobial, antitoxic, anti-inflammatory and sedative properties. Experience and tradition show that honey

- does not irritate the digestive system,
- does not need energy or enzymes to be digested,
- offers a fast availability of energy for mental and physical activity,
- does not generate byproducts to be metabolized by either kidney or liver,
- normalizes the intestinal microflora,
- shows a very light, natural, general soothing effect on the whole organism,
- has a very gentle diuretic effect,
- is safe to eat honey during pregnancy,
- calms the mood and is a most respected sleeping remedy and
- is a universal antitoxic remedy.

5.6.4.2   General Mechanisms for Wound Healing

Honey has many ways to assist in wound repair (Molan and Betts 2008). It is an effective thermal insulator and nurtures repair with a protective biofilm (Black and Costerton 2010). It reduces wound pain and thereby enhances patients' cooperation. The unlinked dehydrated gel of fructose and glucose in honey has twice the osmotic strength of an equivalent sucrose solution. Honey combats bacteria through osmotic pressure and draws fluid from wounds, decreasing tissue edema (Molan 2001). With its enzyme glucose oxidase, it reduces atmospheric oxygen to hydrogen peroxide (H_2O_2), which acts as a bactericide (Dustmann 1979). Hydrogen peroxide is toxic to bacteria but not to the fibroblasts and induces angiogenesis in the wound, particularly so in granulation tissue (Bang et al. 2003). The strength of this hydrogen peroxide is much lower than pharmacologic hydrogen peroxide, which is damaging to the healing environment of a wound. Few microbes grow in an acid milieu and chronic wounds have a high pH. Honey lowers the wound pH, and every 1 % reduction in pH is associated with a 1 % reduction in wound size (Gethin et al. 2008). Blaser et al. (2007) found a positive effect of medical honey on MRSA-colonized wounds and achieved healing where antiseptics and antibiotics had previously failed to eradicate the clinical signs of infection. Honey's glucose is a source of energy for cells involved in wound healing like fibroblasts, myofibroblasts and macrophages. Chronic wounds tend to dry, which promotes scab formation, associated with delay in wound closure. Honey inhibits drying and scab formation.

5.6.4.3 Infected Wounds

Between 1984 and 2009, at the Centre Hospitalier Universitaire de Limoges, France, 3,012 lesions, both infected and uninfected mainly on the abdominal wall were treated with honey (Descottes 2009). For 33 sacral cysts and 102 stoma closures, honey was systematically applied immediately after the end of the operation. Wound healing varied from 21 days for a lesion 10 cm squared, to 75 days for necrotizing abdominal wall lesions of over 30 cm squared. Wound healing was always aesthetically satisfactory except in cases where skin had been previously affected by radiotherapy.

A randomized controlled study of 108 patients with venous leg ulcers comparing Medihoney™ (standardized antibacterial *Leptospermum* [Manuka] honey from New Zealand) and hydrocolloid gel showed that honey eradicated MRSA (methicillin-resistant *Staphylococcus aureus*) in 70 % of patients, while hydrocolloid gel in 16 % only. For wounds contaminated by MRSA, honey supported better wound healing (Gethin and Cowman 2008).

As is often the case, Apitherapy takes more time than the standard medical approach, but also gets good results. Figure 5.9 illustrates the effects of treatment with honey of an infected wound caused by a fish bone puncture wound infected with *Staphylococcus aureus*. Healing was facilitated by 10 days of maggot therapy and 25 days of honey dressing (Dr. F. Feraboli 2006, personal communication).

5.6.4.4 Burns

Burn wounds are particularly at risk for infection and fluid losses of the patient if large areas are affected. Therefore, healing requires prevention or treatment of bacterial infections and prevention of excess fluid losses. There are several studies that present the effectiveness of honey in treating burn wounds (Molan 2001; Subrahmanyam 1991; Subrahmanyam et al. 2001).

Sukur et al. (2011) evaluated the effect of Tualang honey on wound healing in bacterial contaminated full-thickness burn wounds in rats. They found that topical application of honey on burn wounds contaminated with *Pseudomonas aeruginosa* and *Acinetobacter baumannii* gave the fastest rate of healing compared with other treatments.

The patient depicted in Fig. 5.10 was treated with local application of honey with added propolis extract for severe gasoline burns (Dr. A. Piñiero-Perez 2005, personal communication).

5.6.4.5 Prevention of Radiation Mucositis

In a recent meta-analysis Song et al. (2012) found promising results for the prevention of radiation induced mucositis with honey, but contended that further studies are needed to strengthen the current evidence prior to a firm clinical recommendation being given.

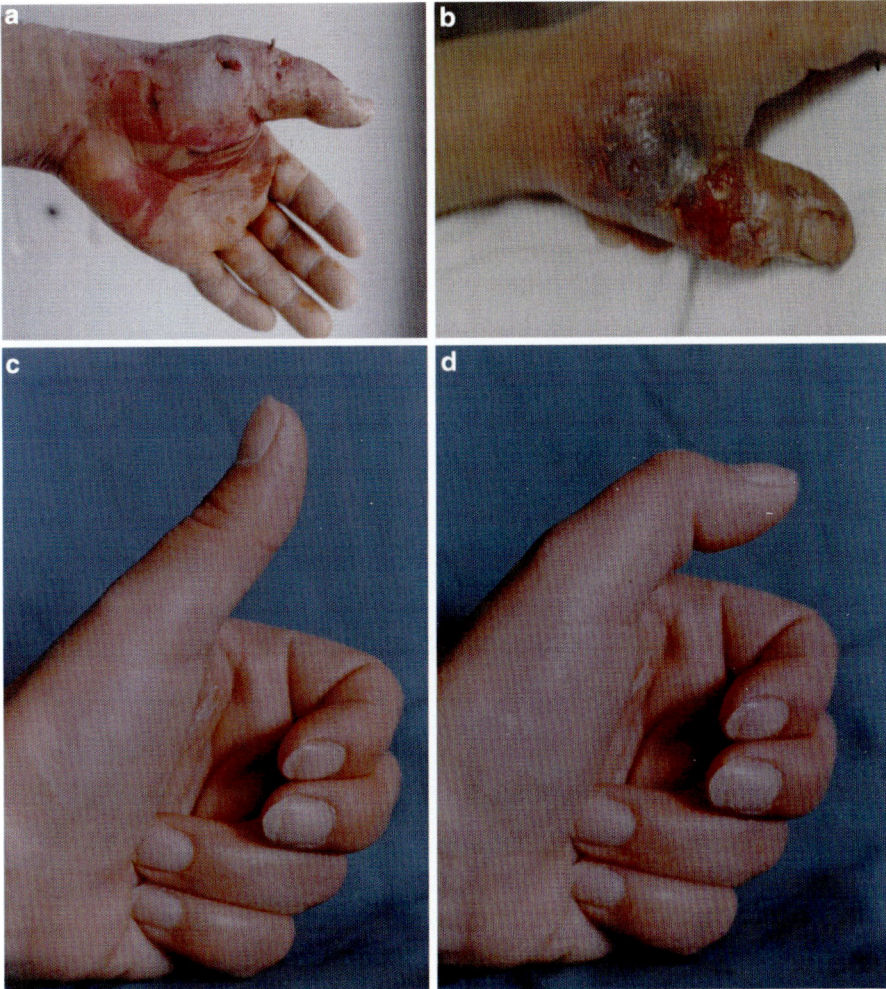

Fig. 5.9 (**a**) and (**b**) *Staphylococcus aureus* infected puncture wound after 7 days of unsuccessful antibiotic treatment; (**c**) and (**d**) The situation 1 year later after treatment with maggot therapy and honey (photo courtesy of Dr. F. Feraboli, Cremona, Italy)

5.6.4.6 Respiratory Tract

In a systematic review of honey for acute cough in children, Oduwole et al. (2012) found that honey is better than 'no treatment' and may be better than diphenhydramine (a first-generation antihistamine) in the symptomatic relief of cough.

Cohen et al. (2012) compared the effects of a single nocturnal dose of honey to placebo on nocturnal cough and difficulty sleeping associated with childhood upper respiratory tract infections (URI). Although there was a significant improvement in both

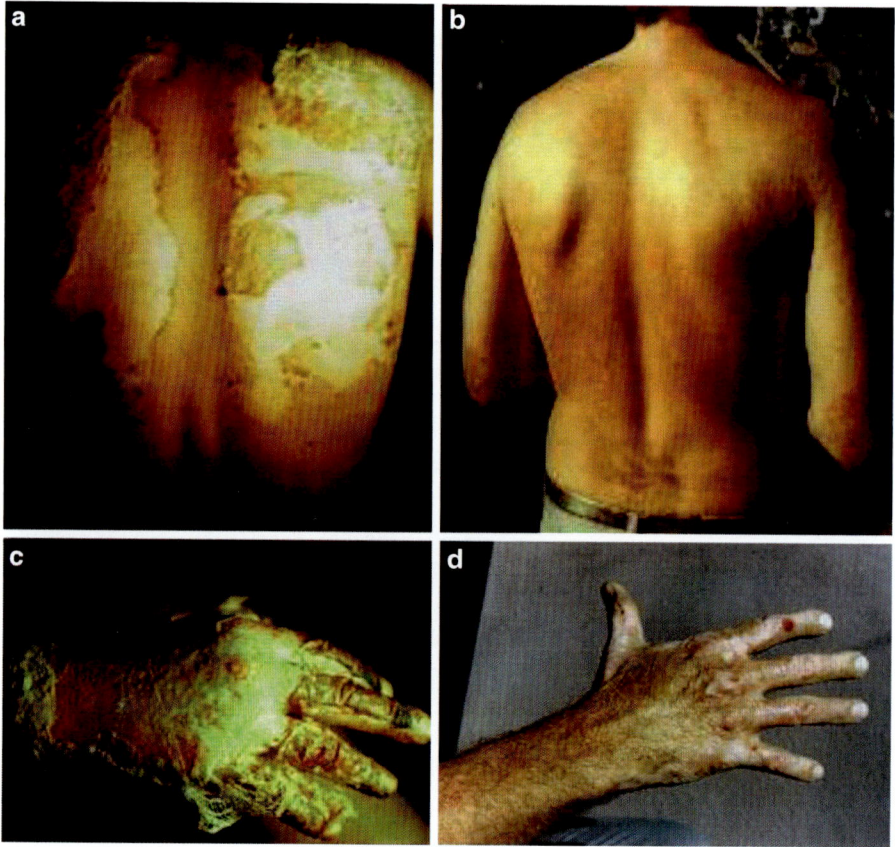

Fig. 5.10 A patient with gasoline burns soon after the accident (*panels* **a** and **b**) and healed 1 year later (*panels* **c** and **d**) (photo courtesy of Dr. J. Ramos, La Havana, Cuba)

groups, the improvement was greater in the honey group. They conclude that honey may be a preferable treatment for cough and sleep difficulty associated with childhood URI.

5.6.4.7 Gastro-Intestinal Tract

The finding that the bacterium, *Helicobacter pylori*, is a cause of stomach ulcers and the causative agent in many cases of dyspepsia has raised the possibility that the therapeutic action of honey for symptoms of dyspepsia may be due to manuka-honey's antibacterial properties. Somal et al. (1994) demonstrated that visible growth of *H. pylori* over the incubation period of 72 h was completely prevented by the presence of 5 % honey. A similar in-vitro antimicrobial effect of manuka-honey was reported against *Campylobacter spp.* (Lin et al. 2009).

5.6.5 Honey and Botulism

Honey sometimes contains dormant endospores of the bacterium *Clostridium botulinum*. Botulism causing bacteria thrive in areas with low acidity. In infants, the spores can survive and cause disease because the acidity of a baby's digestive tract is not strong enough to destroy the spores, leading to illness and even death. Therefore, the longstanding recommendation has existed to avoid giving honey to children under 1 year of age (Koepke et al. 2008). This position towards honey for the infant is not shared by all experts. The Apitherapy Commission of Apimondia is currently reviewing its scientific basis as well as its clinical history. However, there are no reports of botulism cases with honey applied on wounds.

5.7 Propolis

5.7.1 Description and Function in the Hive

Propolis is a complex compound of resins collected by bees on young buds, to which they add secreted enzymes and beeswax. The main function of propolis is to protect the colony, as it is used by bees as "architectural" and anti-infectious agent, which

- acts as space filler and "glue" to immobilize all structural elements,
- shapes and restricts entrances, making them easier to defend,
- excludes external moisture and light,
- smoothest the walls, therefore protecting the wings of the bees from fraying,
- inhibits fungal and bacterial growth in the hive and
- is used to mummify killed invading animals that are too heavy to be thrown out.

Basically, the warm (35 °C), humid (70–100 % humidity) and sweet (nectar and honey) hive is an ideal milieu for growth of germs, which the bees bring back from their trips. Bacterial overgrowth is inhibited by propolis. Additionally, propolis has been shown to have probiotic properties, supporting the presence of beneficial microorganisms in the bee. The best-known varieties are the red propolis from European poplar (*Populus spp.*) and the green propolis from *Baccharis dracunculifolia*, a medical plant from Brazil.

Although more than 300 constituents have been identified in propolis samples, biological activity is mainly due to few substances, such as flavonoids, terpenes, caffeic, ferulic and cumaric acids and esters. Propolis from different areas varies considerably in biochemistry. However, the medically active properties of propolis vary much less than its constituents. The principal composition of propolis is listed in Table 5.4.

Table 5.4 Composition of
propolis

Component	Percentage
Resins/Balsams	45–55
Waxes and fatty acids	25–35
Essential oils	10
Pollen	5
Other organics & minerals	5

5.7.2 Major Properties of Propolis

5.7.2.1 Natural Antioxidants

Propolis is rich in phenolic compounds, which act as natural antioxidants, and are becoming increasingly popular because of their potential role in contributing to human health. These compounds can also be used as indicators in studies into the floral and geographical origin of the honey and propolis themselves (Russo et al. 2002).

5.7.2.2 Protection from Radiation

Propolis is effective in reducing and delaying radiation-induced mucositis in the animal model (Ghassemi et al. 2010). Free radical scavenging and antioxidant activities are probably the mechanisms that protect cells from ionizing radiation (Montoro et al. 2011).

5.7.2.3 Propolis and Cancer

In a recent in-vitro study, Kamiya et al. (2012) found that an ethanol extract of Brasilian red propolis induces apoptosis in human breast cancer cells. Búfalo et al. (2009), evaluating the effect of Brazilian green propolis on human laryngeal epidermoid carcinoma (HEp-2) cells demonstrated that propolis exhibited a cytotoxic effect in-vitro against HEp-2 cells, in a dose- and time-dependent way. Shimizu et al. (2005) investigated the effect of Artepillin C in Brazilian propolis on colon carcinogenesis and found that it dose-dependently inhibited cancer growth through the induction of a cell-cycle arrest. Additionally, inhibition of angiogenesis by propolis has been reported by several authors (Song et al. 2002; Dornelas et al. 2012; Kunimasa et al. 2011).

Abdulrahman et al. (2012) after a randomized controlled pilot study on honey and a mixture of honey, beeswax, and olive oil-propolis extract in treatment of chemotherapy-induced oral mucositis, recommend using honey and possibly other bee products and olive oil in future therapeutic trials targeting chemotherapy-induced mucositis. A detailed review on the anticancer activity of propolis can be found in Sawicka et al. (2012).

5.7.2.4   Tissue Regeneration and Wound Treatment

The effect of propolis on wound healing, especially chronically infected wounds has been studied extensively. Barroso et al. (2012) studying the effect of propolis on mast cells in wound healing, found that the anti-inflammatory action of propolis mediated by mast cells was more effective than dexamethasone in the inflammatory phase of healing. Additionally, propolis exhibited a significantly favorable effect on healing in experimental colon anastomosis in rats (Temiz et al. 2008).

Diabetic ulcers occur in 15 % of all patients with diabetes and precede 84 % of all lower leg amputations. Studying wound healing in a rodent model of experimental diabetes McLennan et al. (2008) found that propolis could possibly accelerate wound healing in diabetes. They conclude that their results and the established safety profile of propolis provide a rationale for studying topical application of this agent in a clinical setting.

One study has shown propolis to be effective in decreasing the number of recurrences and improve the quality of life in patients who suffer from recurrent aphthous stomatitis, a common, painful, and ulcerative disorder of the oral cavity of unknown etiology (Samet et al. 2007). Guney et al. (2011) found that propolis has some time-dependent beneficial effects on fracture healing.

5.7.2.5   Antibacterial Properties

Propolis was found to have antibacterial activity against a range of commonly encountered cocci and Gram-positive rods, including the human tubercle bacillus, but only limited activity against Gram-negative bacilli (Grange and Davey 1990).

Uzel et al. (2005) studied the antimicrobial activity of four different Anatolian propolis samples. Although propolis samples were collected from different regions of Anatolia, all showed significant antimicrobial activity against the Gram-positive bacteria and yeasts. They suggest that propolis can prevent dental caries and oral disease since it demonstrated significant antimicrobial activity against microorganisms such as *Streptococcus mutans*, *Streptococcus sobrinus* and *Candida albicans*.

5.7.2.6   Antiviral Properties

Schnitzler et al. (2010) analyzed the antiviral effect of propolis extracts and selected constituents (e.g. caffeic acid, p-coumaric acid, benzoic acid, galangin, pinocembrin and chrysin) against herpes simplex virus type 1 (HSV-1) in cell culture. Since propolis extracts exhibited high levels of antiviral activity against HSV-1 in viral suspension tests and plaque formation was significantly reduced by >98 %, propolis extracts might be suitable for topical application against herpes infection.

Sartori et al. (2011) investigated whether brown Brazilian hydroalcoholic propolis extract (HPE) protects against vaginal lesions caused by herpes simplex virus type 2 (HSV-2) in female mice. HPE promoted protective effect on HSV-2 infected animals by acting on inflammatory and oxidative processes, probably due to its antioxidant and anti-inflammatory properties.

Studying the anti-HIV-1 activity of propolis in CD4+ lymphocyte and microglial cell cultures Gekker et al. (2005) showed that propolis inhibited viral expression in a concentration-dependent manner. Similar anti-HIV-1 activity was observed with propolis samples from several geographic regions.

Shimizu et al. (2008) reported an anti-influenza virus activity of Brazilian propolis along with a reduction of influenza symptoms in mice. The authors conclude that the Brazilian propolis studied may be a possible candidate for an anti-influenza dietary supplement for humans.

5.7.3   Clinical Effectiveness and Fields of Application

The Natural Medicines Comprehensive Database of the U.S. National Library of Medicine rates effectiveness based on scientific evidence and classified propolis as "possibly effective" for:

- Cold sores: Applying a specific 3 % propolis ointment might help improve healing time and reduce pain from cold sores;
- Genital herpes: Applying a 3 % propolis ointment might improve healing of recurrent genital lesions caused by herpes simplex virus type 2 (HSV-2). Some research suggests that it might heal lesions faster and more completely than the conventional treatment with 5 % acyclovir ointment;
- Improving healing and reducing pain and inflammation after mouth surgery.

According to the database more evidence is needed to rate propolis for indications like cancer sores, tuberculosis, infections, nose and throat cancer, improving immune response, ulcers, stomach and intestinal disorders, common cold, wounds, inflammation and minor burns. Therefore, additional clinical research is needed to gain more evidence of the effectiveness of propolis for these conditions.

5.7.4   Contraindications and Side Effects

Generally, propolis is well tolerated and exhibits a favorable safety profile. Not enough is known about its use during pregnancy and breast-feeding.

Propolis should be used with caution in individuals allergic to conifers, poplars, Peru balsam, and salicylates and it should be avoided in patients with asthma. Allergic reactions to propolis are generally limited to cutaneous manifestation. These usually resolve after discontinuation.

However, Li et al. (2005) report a patient with propolis-induced acute renal failure who required subsequent hemodialysis. The patient had cholangiocarcinoma and had ingested propolis for 2 weeks before presentation. Renal function improved after propolis withdrawal, deteriorated again after re-exposure, and then returned to a normal level after the second propolis withdrawal.

5.8 Royal Jelly

5.8.1 Description and Function in the Hive

Royal Jelly (RJ) is a remarkable substance, both for its role within the colony but also for its composition and for the biological role it can play for our human cells.

It is secreted by the hypo-pharyngeal glands (clear fluid) and mandibular glands (white secretion) that are located in the front part of the head of the nurse bee, who is then 5–15 days old. Each egg laid in the hive is bathed during the first 3 days of its existence in Royal Jelly. Thanks to its presence, the egg's weight increases by 1,500 times in the first 6 days.

Different compositions of Royal Jelly are given to the bees in the hive:

- RJ given for 3 days to larvae of workers;
- RJ given for 3 days to larvae of drones;
- RJ given to the queen larva for the whole larval period;
- RJ given to queen during her lifetime as exclusive food.

Observing the life of the queen reveals Royal Jelly's full power. Every day of the summer the queen lays between 1,500 and 2,000 eggs per day, equaling almost her own weight.

Table 5.5 compares the queen with the worker bee demonstrating some of Royal Jelly's potential.

Kucharski et al. (2008) demonstrated that silencing the expression of DNA methyltransferase Dnmt3, allows for the development of queens with fully developed ovaries.

Royal Jelly has been known about for ages but has only been available in the last 50–60 years in quantities large enough for human consumption. China is acknowledged as the largest producer (over 2,000 tons yearly) and consumer (about 1,000 tons) of Royal Jelly. Table 5.6 lists the average composition of Royal Jelly. The pH of Royal Jelly is 3.6.

5.8.2 Pharmaceutical Preparations

Several pharmaceutical preparations of Royal Jelly are available: Fresh frozen, lyophilized (1 g fresh frozen RJ equals to 200–300 mg lyophilized RJ), tablets, granules, coated pills, lotions, creams, ointments, shampoos, emulsions, salves, suppositories and ovules.

Table 5.5 Parameters of queen bees compared to worker bees

	Queen bee	Worker bee
Food	Royal Jelly	Honey and pollen
Development	15.5 days	21 days
Size	17 mm	12 mm
Weight	200 mg	125 mg
Lifespan	3–5 years	4–6 weeks

Table 5.6 Average composition of Royal Jelly

Compound	Percentage and additional information
Water	67 %
Sugars	11 % (fructose 6 %, glucose 4 %, sucrose 1 %)
Proteins	13 % (fur)
Fat	5.6 %
Minerals	1 % (includes Ca, Cu, Fe, Mg, Mn, Na, K, Zn, Si)
Sterols	7–9 (sterols including sitosterol, cortisol, cholesterol)
Lipids	four phospholipids (from which cell walls are made); five glycolipids (which provide energy)
Vitamins	B1, B2, B3, B4, B6, B7, B9, B12 (Royal Jelly does not contain vitamins A, C; contains vitamin K in traces)
Gamma globulin	Mostly immunoglobulins which strengthen the immune system
10-Hydroxydecanoic acid	20–60 mcg/g; (its anti-bacterial and anti-fungal quality keeps Royal Jelly sterile)
Gelatin	Precursor of collagen for skin, tendon, ligaments
Acetylcholine	Up to 1 mg/g of Royal Jelly (important in nerve transmission and production and release of glandular secretions. Provokes adrenaline secretion)
Nucleic acids	DNA and RNA (the building blocks of genetic material)

Stocker et al. (2005) showed that in the RJ samples collected the concentrations of trace and mineral elements were highly constant, independently of the proportion present in the environment.

The best way to maintain the quality of RJ is by freezing it. Refrigerated and combined with honey it keeps for 3 weeks. However, lyophilized RJ will keep at room temperature indefinitely. Li et al. (2008) identified a protein (MRJP5) as a reliable freshness marker.

5.8.3 Mechanisms of Action and Possible Indications

5.8.3.1 Metabolic Activity

Narita et al. (2009) demonstrated on rats that ingested RJ tended to compensate for age-associated decline in pituitary functions. Inoue et al. (2003) showed on mice that RJ extended the life span by 25 %, possibly through the mechanism of reduced oxidative damage.

5.8.3.2 Hormonal Activity and Osteoporosis

Kafadar et al. (2012) investigated whether RJ and bee pollen reduce the bone loss due to osteoporosis in an oophorectomized rat model and found that bone tissue calcium and phosphate levels were higher in RJ and bee pollen groups compared to controls.

5.8.3.3 Wound Healing

Siavash et al. (2011) evaluated the efficacy of topical Royal Jelly on healing diabetic foot ulcers and concluded that Royal Jelly dressings may be an effective method for treating diabetic foot ulcers besides standard treatments. Koya-Miyata et al. (2004) showed that RJ promoted collagen production in skin fibroblasts by inducing TGF-beta 1 production. Park et al. (2012) studied RJ's protection against skin aging in rats with ovariectomy-induced estrogen deficiency and found that RJ may protect against skin aging by enhancing collagen production.

5.8.3.4 Irradiation Protection

Azab et al. (2011) investigated the possible protective effects of RJ against radiation induced oxidative stress, hematological, biochemical and histological alterations in male Wister albino rats. The authors suggested that the biochemical, hematological and histological amelioration observed in RJ treated irradiated rats might be due to the antioxidant capacity of RJ active constituents.

5.8.3.5 Anti-cancer Properties

Nakaya et al. (2007) investigated the effect of RJ on an environmental estrogen (Bisphenol A) that stimulates proliferation of human breast cancer cells. Royal jelly inhibited the growth-promoting effect of Bisphenol A (BPA) on human breast cancer MCF-7 cells, even though it did not affect the proliferation of cells in the absence of BPA. In addition, the observed inhibiting effect of RJ was heat-stable.

5.8.3.6 Anti-infectious Properties

Fujiwara et al. (1990) showed that Royalisin (a new potent antibacterial protein found in royal jelly of the honeybee with extensive sequence homology to two other insect derived antibacterial proteins) exhibited potent antibacterial activity at low concentrations against Gram-positive bacteria, but not against Gram-negative bacteria. They speculated that Royalisin may be involved in a defense system active against bacterial invasion of the honeybee.

5.8.3.7 Increase of Vigor and Physical Strength

Kamakura et al. (2001) investigated the antifatigue effect of RJ in mice in a swimming experiment. Their findings suggest that RJ can ameliorate physical fatigue after exercise, and that this effect seems to be associated with the freshness of RJ.

5.8.3.8 Biologic Activity

Nomura et al. (2007) investigated the effects of RJ on insulin resistance in Otsuka Long-Evans Tokushima Fatty (OLETF) rats, a type 2 diabetic model. RJ treatment tended to decrease systolic blood pressure and significantly decreased serum levels of insulin and the Homeostasis Model Assessment ratio, an index of insulin resistance. These results suggest that RJ could be an effective and functional food to prevent the development of insulin resistance.

Morita et al. (2012) conducted a randomized placebo-controlled, double-blind trial to investigate the effect of RJ in healthy volunteers. Six-month ingestion of RJ in humans improved erythropoiesis, glucose tolerance and mental health.

5.8.3.9 Immune Disorders

Sugiyama et al. (2012) reviewed the molecular mechanisms underpinning the biological activities of 10-Hydroxy-trans-2-decenoic acid or "royal jelly acid", which could lead to new therapeutic targets for the treatment of immune disorders.

5.8.3.10 Adverse Effects

Thien et al. (1996) observed symptoms of asthma and in some cases anaphylaxis following ingestion of RJ. These symptoms were true IgE-mediated hypersensitivity reactions. Katayama et al. (2008) reported the case of a Japanese woman who developed anaphylaxis after drinking a beverage of crude Royal Jelly including honey. They contended that Royal Jelly should be considered as a causative allergen in food-induced anaphylaxis.

References

Abdulrahman M, El Barbary NS, Ahmed Amin D, Saeid Ebrahim R (2012) Honey and a mixture of honey, beeswax, and olive oil-propolis extract in treatment of chemotherapy-induced oral mucositis: a randomized controlled pilot study. Pediatr Hematol Oncol 29(3):285–292

Azab KS, Bashandy M, Salem M, Ahmed O, Tawfik Z, Helal H (2011) Royal jelly modulates oxidative stress and tissue injury in gamma irradiated male Wister Albino rats. N Am J Med Sci 3(6):268–276

Bang LM, Bunttig C, Molan P (2003) The effect of dilution on the rate of hydrogen peroxide production in honey and its implication for wound healing. J Altern Complement Med 9: 267–273

Barroso PR, Lopes-Rocha R, Pereira EM, Marinho SA, de Miranda JL, Lima NL, Verli FD (2012) Effect of propolis on mast cells in wound healing. Inflammopharmacology 20(5):289–294

Black CE, Costerton W (2010) Current concepts regarding the effect of wound microbial ecology and biofilms on wound healing. Surg Clin N Am 90(6):1147–1160

Blaser G, Santos K, Bode U, Vetter H, Simon A (2007) Effect of medical honey on wounds colonised or infected with MRSA. J Wound Care 16(8):325–328

Bogdanov S, Jurendic T, Sieber R, Gallmann P (2008) Honey for nutrition and health: a review. J Am Coll Nutr 27(6):677–689

Buck AC, Cox R, Rees RW, Ebeling L, John A (1990) Treatment of outflow tract obstruction due to benign prostatic hyperplasia with the pollen extract, cernilton. A double-blind, placebo-controlled study. Br J Urol 66(4):398–404

Búfalo MC, Candeias JM, Sforcin JM (2009) In vitro cytotoxic effect of Brazilian green propolis on human laryngeal epidermoid carcinoma (HEp-2) cells. Evid Based Complement Alternat Med 6(4):483–487

Cohen HA, Rozen J, Kristal H, Laks Y, Berkovitch M, Uziel Y, Kozer E, Pomeranz A, Efrat H (2012) Effect of honey on nocturnal cough and sleep quality: a double-blind, randomized, placebo-controlled study. Pediatrics 130(3):465–471

Descottes B (2009) Cicatrisation par le miel, l'expérience de 25 années. Phytothérapie 7(2): 112–116

Domerego R (2012) La thérapie au venin d'abeille, Baroch editions, pp 30–33

Dornelas CA, Fechine-Jamacaru FV, Albuquerque IL, Magalhães HI, Dias TA, Faria MH, Alves MK, Rabenhorst SH, de Almeida PR, de Lemos TL, de Castro JD, Moraes ME, Moraes MO (2012) Angiogenesis inhibition by green propolis and the angiogenic effect of L-lysine on bladder cancer in rats. Acta Cir Bras 27(8):529–536

Dustmann JH (1979) Antibacterial effect of honey. Apiacta 14:7–11

Erejuwa OO, Sulaiman SA, Wahab MS (2012) Honey–a novel antidiabetic agent. Int J Biol Sci 8(6):913–934

Fujiwara S, Imai J, Fujiwara M, Yaeshima T, Kawashima T, Kobayashi K (1990) A potent antibacterial protein in royal jelly. Purification and determination of the primary structure of royalisin. J Biol Chem 265(19):11333–11337

Gekker G, Hu S, Spivak M, Lokensgard JR, Peterson PK (2005) Anti-HIV-1 activity of propolis in CD4(+) lymphocyte and microglial cell cultures. J Ethnopharmacol 102(2):158–163

Gethin G, Cowman S (2008) Bacteriological changes in sloughy venous leg ulcers treated with manuka honey or hydrogel: an RCT. J Wound Care 17(6):241–244, 246–247

Gethin GT, Cowman S, Conroy RM (2008) The impact of manuka honey dressings on the surface pH of chronic wounds. Int Wound J 5(2):185–194

Ghassemi L, Zabihi E, Mahdavi R, Seyedmajidi M, Akram S, Motallebnejad M (2010) The effect of ethanolic extract of propolis on radiation-induced mucositis in rats. Saudi Med J 31(6): 622–626

Grange JM, Davey RW (1990) Antibacterial properties of propolis (bee glue). J R Soc Med 83(3):159–160

Guney A, Karaman I, Oner M, Yerer MB (2011) Effects of propolis on fracture healing: an experimental study. Phytother Res 25(11):1648–1652

Habib FK, Ross M, Lewenstein A, Zhang X, Jaton JC (1995) Identification of a prostate inhibitory substance in a pollen extract. Prostate 26(3):133–139

Hurren KM, Lewis CL (2010) Probable interaction between warfarin and bee pollen. Am J Health Syst Pharm 67(23):2034–2037

Inoue S, Koya-Miyata S, Ushio S, Iwaki K, Ikeda M, Kurimoto M (2003) Royal jelly prolongs the life span of C3H/HeJ mice: correlation with reduced DNA damage. Exp Gerontol 38(9):965–969

Jankauskiene J, Vasiliauskas S, Jankauskaite D (2006) Bee products in ophthalmology in Lithuania. Apimedica, Athens, 15 Oct 2006

Kafadar İH, Guney A, Türk CY, Öner M, Silici S (2012) Royal jelly and bee pollen decrease bone loss due to osteoporosis in an oophorectomized rat model. Agricultural Faculty of Erciyes University, Kayseri

Kamakura M, Mitani N, Fukuda T, Fukushima M (2001) Antifatigue effect of fresh royal jelly in mice. J Nutr Sci Vitaminol (Tokyo) 47(6):394–401

Kamiya T, Nishihara H, Hara H, Adachi T (2012) Ethanol extract of Brazilian red propolis induces apoptosis in human breast cancer MCF-7 cells through endoplasmic reticulum stress. J Agric Food Chem 60(44):11065–11070

Kas'ianenko VI, Komisarenko IA, Dubtsova EA (2011) Correction of atherogenic dyslipidemia with honey, pollen and bee bread in patients with different body mass. Ter Arkh 83(8):58–62

Katayama M, Aoki M, Kawana S (2008) Case of anaphylaxis caused by ingestion of royal jelly. J Dermatol 35(4):222–224

Koepke R, Sobel J, Arnon SS (2008) Global occurrence of infant botulism, 1976–2006. Pediatrics 122(1):e73–e82

Koya-Miyata S, Okamoto I, Ushio S, Iwaki K, Ikeda M, Kurimoto M (2004) Identification of a collagen production-promoting factor from an extract of royal jelly and its possible mechanism. Biosci Biotechnol Biochem 68(4):767–773

Kucharski R, Maleszka J, Foret S, Maleszka R (2008) Nutritional control of reproductive status in honeybees via DNA methylation. Science 319(5871):1827–1833

Kunimasa K, Ahn MR, Kobayashi T, Eguchi R, Kumazawa S, Fujimori Y, Nakano T, Nakayama T, Kaji K, Ohta T (2011) Brazilian propolis suppresses angiogenesis by inducing apoptosis in tube-forming endothelial cells through inactivation of survival signal ERK1/2. Evid Based Complement Alternat Med. doi:10.1093/ecam/nep024

Küpeli Akkol E, Orhan DD, Gürbüz I, Yesilada E (2010) In vivo activity assessment of a "honeybee pollen mix" formulation. Pharm Biol 48(3):253–259

Li YJ, Lin JL, Yang CW, Yu CC (2005) Acute renal failure induced by a Brazilian variety of propolis. Am J Kidney Dis 46(6):e125–e129

Li JK, Feng M, Zhang L, Zhang ZH, Pan YH (2008) Proteomics analysis of major royal jelly protein changes under different storage conditions. J Proteome Res 7(8):3339–3353

Lin SM, Molan PC, Cursons RT (2009) The in vitro susceptibility of Campylobacter spp. to the antibacterial effect of manuka honey. Eur J Clin Microbiol Infect Dis 28(4):339–344

Mayor A (1995) Mad honey [toxic honey in history]. Archaeology 48(6):32–40

McLennan SV, Bonner J, Milne S, Lo L, Charlton A, Kurup S, Jia J, Yue DK, Twigg SM (2008) The anti-inflammatory agent Propolis improves wound healing in a rodent model of experimental diabetes. Wound Repair Regen 16(5):706–713

Molan PC (2001) Potential of honey in the treatment of wounds and burns. Am J Clin Dermatol 2(1):13–19

Molan PC, Betts JA (2008) Using honey to heal diabetic foot ulcers. Adv Skin Wound Care 21:313–316

Montoro A, Barquinero JF, Almonacid M, Montoro A, Sebastià N, Verdú G, Sahuquillo V, Serrano J, Saiz M, Villaescusa JI, Soriano JM (2011) Concentration-dependent protection by ethanol extract of propolis against γ-ray-induced chromosome damage in human blood lymphocytes. Evid Based Complement Alternat Med. doi:10.1155/2011/174853

Morita H, Ikeda T, Kajita K, Fujioka K, Mori I, Okada H, Uno Y, Ishizuka T (2012) Effect of royal jelly ingestion for six months on healthy volunteers. Nutr J 11:77

Nakaya M, Onda H, Sasaki K, Yukiyoshi A, Tachibana H, Yamada K (2007) Effect of royal jelly on bisphenol A-induced proliferation of human breast cancer cells. Biosci Biotechnol Biochem 71(1):253–255

Narita Y, Ohta S, Suzuki KM, Nemoto T, Abe K, Mishima S (2009) Effects of long-term administration of royal jelly on pituitary weight and gene expression in middle-aged female rats. Biosci Biotechnol Biochem 73(2):431–433

Nickel JC (2006) The overlapping lower urinary tract symptoms of benign prostatic hyperplasia and prostatitis. Curr Opin Urol 16(1):5–10

Nomura M, Maruo N, Zamami Y, Takatori S, Doi S, Kawasaki H (2007) Effect of long-term treatment with royal jelly on insulin resistance in Otsuka Long-Evans Tokushima Fatty (OLETF) rats. Yakugaku Zasshi 127(11):1877–1882

Oduwole O, Meremikwu MM, Oyo-Ita A, Udoh EE (2012) Honey for acute cough in children. Cochrane Database Syst Rev 3:CD007094. doi:10.1002/14651858.CD007094.pub3

Olofssos T, Vásquez A (2009) Lactic acid bacteria – can honeybees survive without them? Apimondia presentation, Montpellier, 18 Sept 2009

Park HM, Cho MH, Cho Y, Kim SY (2012) Royal jelly increases collagen production in rat skin after ovariectomy. J Med Food 15(6):568–575

Peng H (1990) The effect of pollen in enhancing tolerance to hypoxia and promoting adaptation to highlands. Zhonghua Yi Xue Za Zhi 70(2):77–81

Percie du Sert P (2006) The healing power of pollen. Editions Guy Tredaniel, Paris, p 25

Robinson W (1948) Delay in the appearance of palpable mammary tumors in C3H mice following the ingestion of pollenized food. J Natl Cancer Inst 9(2):119–123

Russo A, Longo R, Vanella A (2002) Antioxidant activity of propolis: role of caffeic acid phenethyl ester and galangin. Fitoterapia 73(Suppl 1):S21–S29

Samet N, Laurent C, Susarla SM, Samet-Rubinsteen N (2007) The effect of bee propolis on recurrent aphthous stomatitis: a pilot study. Clin Oral Investig 11(2):143–147

Sartori G, Pesarico AP, Pinton S, Dobrachinski F, Roman SS, Pauletto F, Junior LC, Prigol M (2011) Protective effect of brown Brazilian propolis against acute vaginal lesions caused by herpes simplex virus type 2 in mice: involvement of antioxidant and anti-inflammatory mechanisms. Cell Biochem Funct 30(1):1–10

Sawicka D, Car H, Borawska MH, Nikliński J (2012) The anticancer activity of propolis. Folia Histochem Cytobiol 50(1):25–37

Schnitzler P, Neuner A, Nolkemper S, Zundel C, Nowack H, Sensch KH, Reichling J (2010) Antiviral activity and mode of action of propolis extracts and selected compounds. Phytother Res 24(Suppl 1):S20–S28

Shimizu K, Das SK, Hashimoto T, Sowa Y, Yoshida T, Sakai T, Matsuura Y, Kanazawa K (2005) Artepillin C in Brazilian propolis induces G(0)/G(1) arrest via stimulation of Cip1/p21 expression in human colon cancer cells. Mol Carcinog 44(4):293–299

Shimizu T, Hino A, Tsutsumi A, Park YK, Watanabe W, Kurokawa M (2008) Anti-influenza virus activity of propolis in vitro and its efficacy against influenza infection in mice. Antivir Chem Chemother 19(1):7–13

Siavash M, Shokri S, Haghighi S, Mohammadi M, Shahtalebi MA, Farajzadehgan Z (2011) The efficacy of topical Royal Jelly on diabetic foot ulcers healing: a case series. J Res Med Sci 16(7):904–909

Somal N, Coley KE, Molan PC, Hancock BM (1994) Susceptibility of helicobacter pylori to the antibacterial activity of manuka honey. J R Soc Med 87(1):9–12

Song YS, Park EH, Jung KJ, Jin C (2002) Inhibition of angiogenesis by propolis. Arch Pharm Res 25(4):500–504

Song JJ, Twumasi-Ankrah P, Salcido R (2012) Systematic review and meta-analysis on the use of honey to protect from the effects of radiation-induced oral mucositis. Adv Skin Wound Care 25(1):23–28

Stocker A, Schramel P, Kettrup A, Bengsch E (2005) Trace and mineral elements in royal jelly and homeostatic effects. J Trace Elem Med Biol 19(2–3):183–189

Subrahmanyam M (1991) Topical application of honey in treatment of burns. Br J Surg 78:497–498

Subrahmanyam M, Sahapure AG, Nagane NS, Bhagwat VR, Ganu JV (2001) Effects of topical application of honey on burn wound healing. Ann Burns Fire Disaster 14:143–145

Sugiyama T, Takahashi K, Mori H (2012) Royal jelly acid, 10-hydroxy-trans-2-decenoic acid, as a modulator of the innate immune responses. Endocr Metab Immune Disord Drug Targets 12(4):368–376

Sukur SM, Halim AS, Singh KK (2011) Evaluations of bacterial contaminated full thickness burn wound healing in Sprague Dawley rats treated with Tualang honey. Indian J Plast Surg 44(1):112–117

Temiz M, Aslan A, Canbolant E, Hakverdi S, Polat G, Uzun S, Temiz A, Gonenci R (2008) Effect of propolis on healing in experimental colon anastomosis in rats. Adv Ther 25(2):159–167

Thien FC, Leung R, Baldo BA, Weiner JA, Plomley R, Czarny D (1996) Asthma and anaphylaxis induced by royal jelly. Clin Exp Allergy 26(2):216–222

Uzel A, Sorkun K, Onçağ O, Cogŭlu D, Gençay O, Salih B (2005) Chemical compositions and antimicrobial activities of four different Anatolian propolis samples. Microbiol Res 160(2):189–195

Vásquez A, Forsgren E, Fries I, Paxton RJ, Flaberg E, Szekely L, Olofsson TC (2012) Symbionts as major modulators of insect health: lactic acid bacteria and honeybees. PLoS One 7(3):e33188

Vit P (2002) Effect of stinglessbee honey in selenite induced cataracts. Apiacta 3

Voloshyn OI, Pishak OV, Seniuk BP, Cherniavs'ka NB (1998) The efficacy of flower pollen in patients with rheumatoid arthritis and concomitant diseases of the gastroduodenal and hepato-biliary systems. Lik Sprava 4:151–154

Zhang X, Habib FK, Ross M, Burger U, Lewenstein A, Rose K, Jaton JC (1995) Isolation and characterization of a cyclic hydroxamic acid from a pollen extract, which inhibits cancerous cell growth in vitro. J Med Chem 38(4):735–738

Chapter 6
Ichthyotherapy

Martin Grassberger and Ronald A. Sherman

6.1 Introduction

Ichthyotherapy is defined as the treatment of skin disease (so far mainly cornification disorders like psoriasis and ichthyosis) with the so-called "doctor fish of Kangal", *Garra rufa* (Heckel 1843). In accordance with other biotherapy terminologies, such as maggot therapy (use of sterile fly larvae, or "maggots"), hirudotherapy (use of the medicinal leech, *Hirudo medicinalis*) and apitherapy (use of the honeybee, *Apis mellifera*), the term "Ichthyotherapy" was proposed in 2006 (Grassberger and Hoch 2006) and readily adopted. The name derives from the Greek word for fish ("*Ichthys*"). Other descriptive terms frequently used are: "Dr. Fish treatment", "Kangal Fish therapy" or "Nibblefish therapy".

In recent years, many reports in the media featured this kind of treatment due to its apparent peculiarity. Wellness spas opened all over the world offering predominantly foot care for callus removal. "Fish foot therapy" was offered in recreational "fish spas", shopping malls, department stores and tea saloons in Japan, South Korea and Singapore, and later spread throughout Southeast Asia and all over the world. In 2008, the fish spa hype swept through the United States, with countless fish foot spas opening in many states. Due to the lack of regulation and industry standards regarding the treatment itself, the fish species used, humane treatment protections for the fish, and the absence of hygienic precautions for the clients, it did not take long until the first negative reports appeared in the media. Authorities soon

M. Grassberger, M.D., Ph.D. (✉)
Institute of Pathology and Microbiology, Rudolfstiftung Hospital
and Semmelweis Clinic, Juchgasse 25, 1030 Vienna, Austria
e-mail: martin.grassberger@mac.com

R.A. Sherman, M.D., M.Sc.
BTER Foundation, Urey Court 36, Irvine, CA 92617, USA
e-mail: rsherman@uci.edu

M. Grassberger et al. (eds.), *Biotherapy - History, Principles and Practice:*
A Practical Guide to the Diagnosis and Treatment of Disease using Living Organisms,
DOI 10.1007/978-94-007-6585-6_6, © Springer Science+Business Media Dordrecht 2013

took notice of this, sometimes fishy business. Consequently, numerous spas were shut down by the health authorities due to concerns over the potential for spreading communicable diseases as a result of customers sharing the same fish.

Notwithstanding, the publication of pilot clinical studies (Özcelik et al. 2000; Grassberger and Hoch 2006) and numerous anecdotal reports indicate that ichthyotherapy is a promising treatment for psoriasis, and deserves further study.

6.2   Historical Aspects

When news of the "Doctor Fish of Kangal" first came out of the central Anatolia region of Turkey, this alternative therapy must have seemed quite odd. The treatment was first mentioned in *The Lancet* in 1989 by Warwick and Warwick (1989) but details were not published until more than 10 years later by Özcelik et al. (2000) in the *Journal of Dermatology*. This Turkish study reported considerable benefits for patients suffering from psoriasis.

6.2.1   The Kangal Hot Springs

In the hot pools of Kangal, where food plankton is reportedly scarce, two fish species from the carp and minnow family feed on the skin scales of patients with illnesses such as psoriasis and atopic dermatitis (Timur et al. 1983; Özcelik et al. 2000). The two species are *Cyprinion macrostomus* and *Garra rufa*. The activity of the fish (especially *G. rufa*) is associated with reducing the scales and symptoms of these skin diseases (Fig. 6.1).

The Kangal Spa was built in 1900, but did not open to the public until 1963. It is located 98 km from Sivas and 13 km north of Kangal, Turkey, at an altitude of approximately 1,660 m above sea level. There are several pools with a mean temperature of 37 °C. According to the Kangal Spa authorities, it attracted the attention of the public in 1917, when a shepherd hurt his foot, "only to see it healed by the water of the spring".

6.2.2   Clinical Studies

Thus far, the medical and scientific communities have given ichthyotherapy very little attention. Treatment efficacy has been evaluated in only two published studies.

The first clinical study, published in the *Journal of Dermatology* by Özcelik et al. (2000), involved 87 patients from the Kangal Hot Spring Spa. The second study, published in 2006 by Grassberger and Hoch, was a retrospective analysis of 67 psoriasis patients who underwent 3 weeks of ichthyotherapy in an outpatient treatment facility in Austria.

Fig. 6.1 Patient seeking relief of his psoriasis symptoms in one of the pools at the Kangal hot spring (Photo courtesy of Dr. Wim Fleischmann)

At the *8th International Conference on Biotherapy*, Grassberger (2010) presented initial data from a questionnaire-based health related quality of life study in a cohort of 82 patients. To the best of our knowledge no additional clinical research on this promising treatment has been published.

6.3 The Reddish Suction Barbel – *Garra rufa*

Two different types of fish live in the pools of the Kangal hotspring: *Cyprinion macrostomus* and *Garra rufa*. Both fish are members of the carp and minnow family (Cyprinidae). *Garra rufa*, also called the "reddish log sucker" or "reddish suction barbel" is regarded as the primary therapeutic species. Because of its history of use in the Kangal hot springs and most of the ichthyotherapy facilities around the world, it is now regarded as the "gold standard" species for ichthyotherapy.

Garra rufa is a non-migratory freshwater fish found in rivers throughout much of Iraq, Israel, Jordan, Turkey, Syria, and possibly also Oman and Saudi Arabia (Berg 1949; Menon 1964; Abdoli 2000; Teimori 2006; Coad 2012). *Garra rufa* is found in a variety of habitats: rivers, lakes, ponds and muddy streams. This fish preferentially lives at the bottom of flowing rivers, where it adheres to rocks by suction using

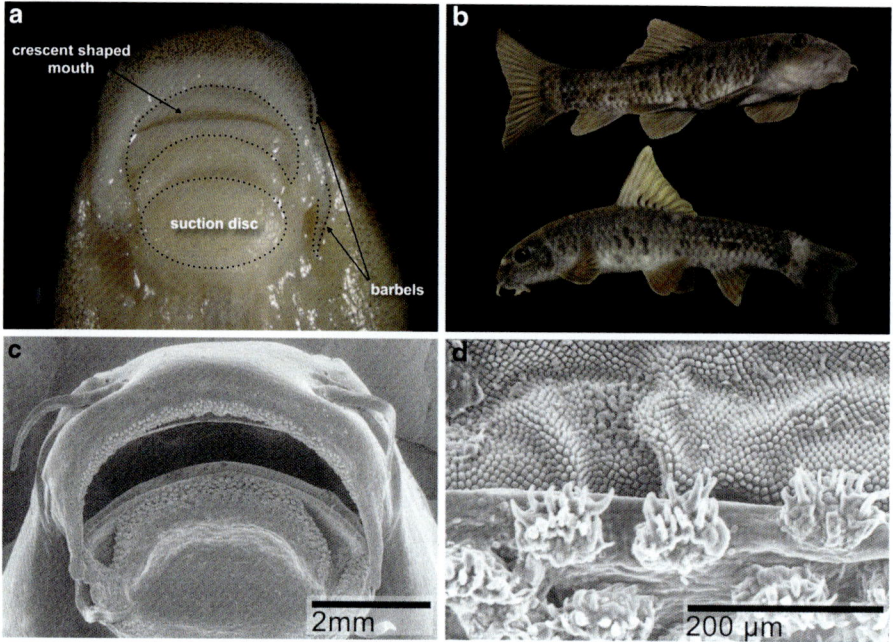

Fig. 6.2 Anatomy of *Garra rufa*. (**a**) Gross view of ventral crescent shaped mouth with suction disc and lateral barbels; (**b**) lateral and ventro-lateral view; (**c**, **d**) scanning electron microphotographs of the mouth with its surrounding tiny spines

its ventral crescent-shaped mouth (organ of attachment; Fig. 6.2a). Accordingly the head is depressed and bears two dorsolaterally placed eyes, resulting in an overall very distinct shape (Fig. 6.2b). It feeds on phyto- and possibly zooplankton.

As in other Cyprinidae, the barbels are slender, fleshy structures on the snout and chin, used for touch and taste. The crescent-shaped mouth of *G. rufa* has no external teeth (Fig. 6.2c). However, scanning electron microphotographs demonstrate a fringe of grouped minute spines (Fig. 6.2d), which appears to serve the fish as a scraper while feeding on outgrowths or human epidermis. The adhesive organ in this species is a very complex combination of integumentary modifications, as described in detail by Teimori et al. (2011).

There are currently about 180 species described in the Genus *Garra* that can be distinguished by morphological characteristics (www.fishbase.org; Kullander and Fang 2004). Whether any other species in this genus exhibit the same therapeutic potential as *G. rufa* is unknown, and could be an area for future research.

6.3.1 "Chin Chin Fish"

When photographs from various "fish foot therapy" spas were published, it became apparent that many of these new spas used fish species quite different from *G. rufa.*

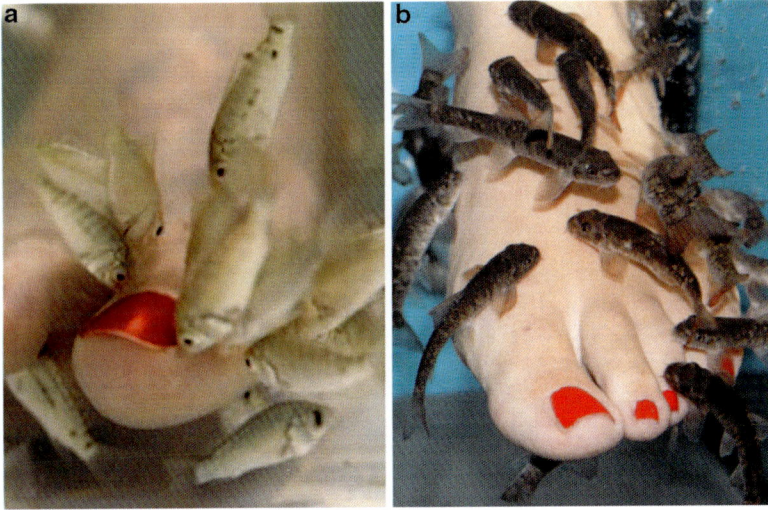

Fig. 6.3 Young Tilapia species, the so-called Chin Chin fish (**a**) in a fish foot spa compared to *Garra rufa* (**b**)

Most of these fish have been identified as juvenile *Tilapia* species (e.g. *Oreochromis niloticus*). When kept hungry, these fish will aggressively nibble on anything. However, no therapeutic properties have ever been associated with this species. As long as they are small (approx. 2–3 cm), they may look quite similar to *G. rufa* to the untrained eye. Nevertheless, closer observation reveals a strikingly different anatomy. Its mouth is not crescent-shaped and ventral as in *Garra* spp., but rather positioned apically, resulting in a completely different nibbling pattern (Fig. 6.3). As these fish grow larger, they feed more aggressively on the skin, and can be painful. They can cause skin wounds that act as portals for water- or fish-borne pathogens. Since they sometimes cause bleeding, they could cause human pathogens from one client to contaminate the water and infect another client. Without an effective water sanitation system, clients could develop severe soft tissue infections or could even acquire blood-born infectious.

6.4 Clinical Efficacy of Ichthyotherapy

In 2000, Özcelik et al. published their study of "Kangal Hot Spring with Fish" as a treatment for psoriasis. The study followed 87 patients (49.4 % male, 50.6 % female, mean age mean 32.5 ± 14.3 years) with psoriasis vulgaris. The mean length of stay in the spa was 11.5 ± 6.6 days. The patients used the pools twice daily and the mean length of stay of the patients in the pools was 7.4 ± 1.1 h in a day. Sixty-five patients (74.7 %) had plaque type psoriasis vulgaris and the mean duration of their symptoms was 10.6 ± 5 years. Fifty-two patients (59.7 %) in the study group came to the hot springs of Kangal for the first time, while 35 (40.3 %) had visited the hot springs previously.

The patients were evaluated by a dermatologist throughout the study, using the standardized Psoriasis Area and Severity Index (PASI score). The PASI is a physician-assessed score, recognized by the US Food and Drug Administration to assess efficacy of psoriasis therapies in clinical trials. The PASI score takes into account the extent of involved skin surface and the severity of erythema, desquamation, and plaque induration. The composite score ranges from 0 to 72, with higher numbers indicating more severe disease and a reduction in score representing improvement.

The researchers demonstrated significant improvement in psoriasis symptoms and PASI score within just 3 days ($P<0.01$). Follow-up at 21 days was available only for 14 patients, but was associated with complete and lasting benefits in 8 (57 %) of them. The remaining six patients had improvement in their symptoms, but not complete resolution. The 35 patients previously treated at the hot springs reported longer remission following their last spa treatment than they normally have following corticosteroid treatment.

According to the authors, many different factors may have contributed to the observed beneficial effects of spa therapy. Apart from the strikingly rapid clearing of the skin scales by the fish and the better penetration of ultraviolet light after removal of the scales, a positive psychological effect because of the initially rapid treatment progress was advocated. Additionally, the authors attributed the observed treatment effect in part to the high selenium level (1.3 mg/L) in the spring water. The only side effect reported was first-degree sunburn in two patients at the beginning of the therapy.

Interestingly, the authors described the fish as "attacking" areas with lesions, and also pointed out that the fish may "invade normal skin". These observations may be due to the fact that there are two fish species living in the Kangal Hot Springs: *Garra rufa* (referred to as "licker") and *Cyprinion macrostomus* (also called "striker"). *C. macrostomus*, being the more aggressive fish, may have been the cause of the small skin wounds. In the pools of the remote Kangal hot spring patients are required to take their bath with up to 20 patients simultaneously, which might be unacceptable for some patients and could be of concern for public health issues regarding the possible transmission of infectious disease.

In 2006, Grassberger and Hoch published the results of a retrospective analysis of 67 patients (39 male and 28 female) diagnosed with moderate to severe chronic plaque psoriasis who underwent 3 weeks of ichthyotherapy at an outpatient treatment facility in Austria. Patient's ages ranged between 10 and 75 years [mean 41 years, 95 % confidence interval (CI) 37.5, 44.5]. The mean duration of psoriasis at baseline was 13.9 years (range 1–35 years; 95 % CI 11.6, 16.2). All patients referred themselves to the treatment facility.

In contrast to the Kangal Hot Springs study (Özcelik et al. 2000), each patient completed a full 3-week treatment course. In addition, the daily "fish bath" lasted only 2 h, and each patient was allocated to a single bathing tub for the entire study period (no two patients shared the same tub at any time during the study). According to the authors this shortened daily treatment time probably made ichthyotherapy more acceptable for the patients and considerably improved compliance.

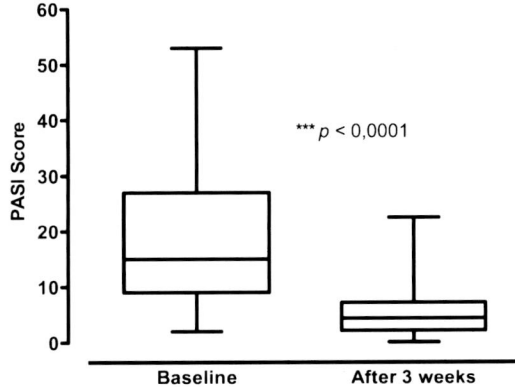

Fig. 6.4 Box plot of PASI scores before and after treatment. *Horizontal lines* Indicate the 75th percentile, median, and 25th percentile; *whiskers* indicate the range. From Grassberger and Hoch (2006)

Patients with no contraindication to UV exposure used a commercially available stand-up rapid-tan facility (UVA) for 3–5 min after each bath session. After UV exposure, the patients applied a generic skin lotion containing glycerine, Shea butter and *Aloe vera* extract. If the psoriasis significantly involved the scalp, then the patient's head was shaved before treatment.

As in the study by Özcelik et al. (2000), the primary efficacy outcome measure was the overall total reduction in PASI score and the proportion of patients with 50 % and 75 % improvement in PASI score (denoted as "PASI-50" and "PASI-75," respectively) at week 3, relative to baseline. The PASI-75 is the currently recognized benchmark of end-points used in psoriasis clinical trials. PASI-50 is also regarded as a clinically significant end-point (Carlin et al. 2004).

PASI scores were assessed using high-resolution digital color photographs taken at baseline and at the end of the 3-week treatment period. Baseline measurements were those made just prior to the beginning of treatment. Additionally, response to treatment was defined according to the rate of improvement in PASI score. Patient-reported outcomes were evaluated by a short questionnaire administered immediately after the 3-week course of treatment, and by a follow-up questionnaire sent to all patients 3–36 months after the study treatment. The follow-up questionnaire included questions concerning the duration of remission, the number of different treatment regimens prior to ichthyotherapy, the severity of a possible relapse, and the personal satisfaction with ichthyotherapy when compared to other treatments.

At the end of the 3-week treatment course, 31 of the 67 patients (46 %) achieved PASI-75 and an additional 30 patients (45 %) achieved PASI-50. For all patients, the average reduction in PASI score compared to baseline was 71.7 % (P<0.0001) (Figs. 6.4, 6.5, and 6.6). All of the patients experienced clinical improvement: complete resolution in three patients (4.5 %), marked improvement in 29 (43.3 %), moderate improvement in 29 (43.3 %) and slight improvement in 6 (8.9 %).

Assessment of patient-reported outcomes demonstrated substantial satisfaction with the treatment. The reported mean remission period was 8.58 months (95 % CI, 6.05–11.11). Overall, 87.5 % of patients reported a more favorable outcome with ichthyotherapy, compared to their previous therapies. This might be due, in part, to the

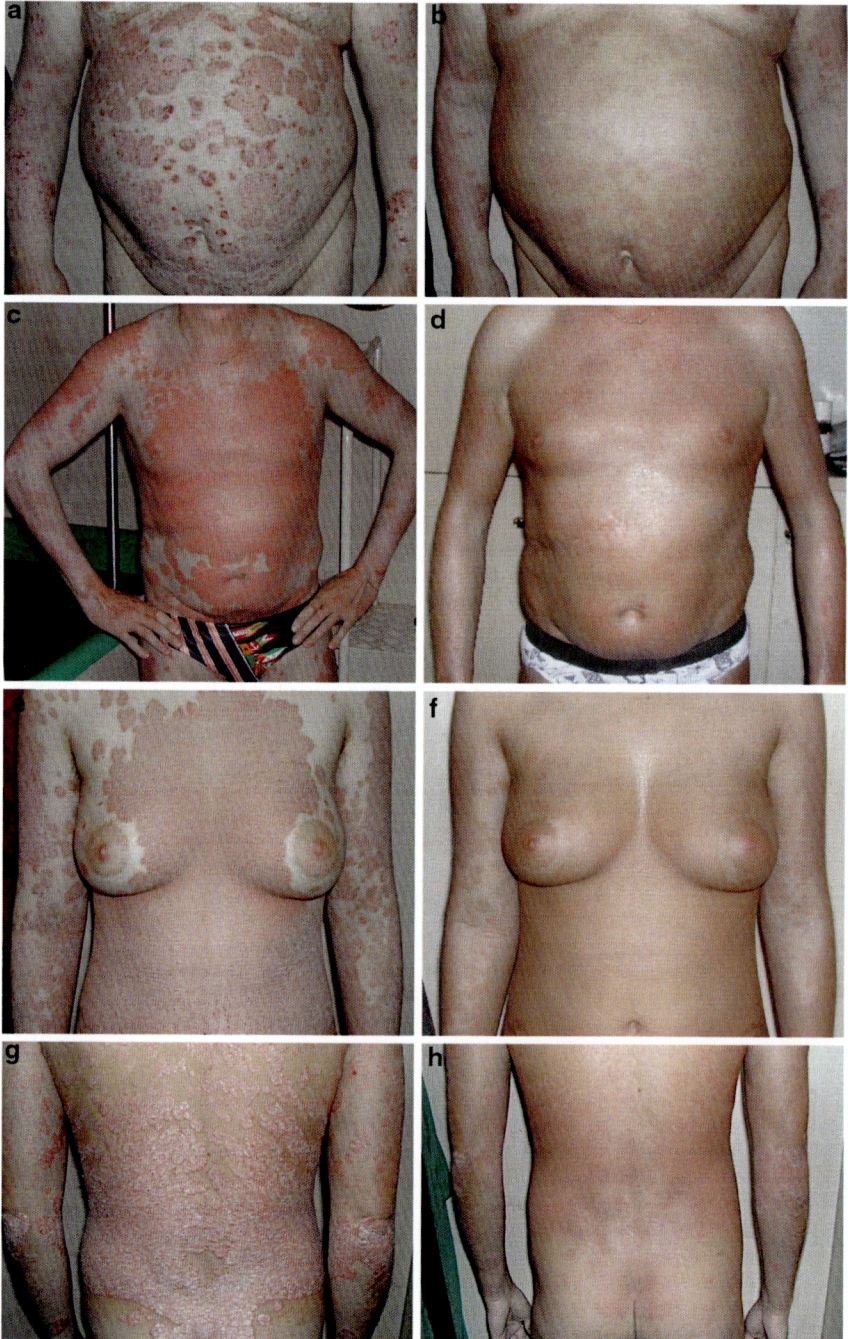

Fig. 6.5 Four patients suffering from psoriasis before (*left*) and after a 3-week course of treatment with ichthyotherapy (2 h daily fish bath) combined with UVA radiation (*right*)

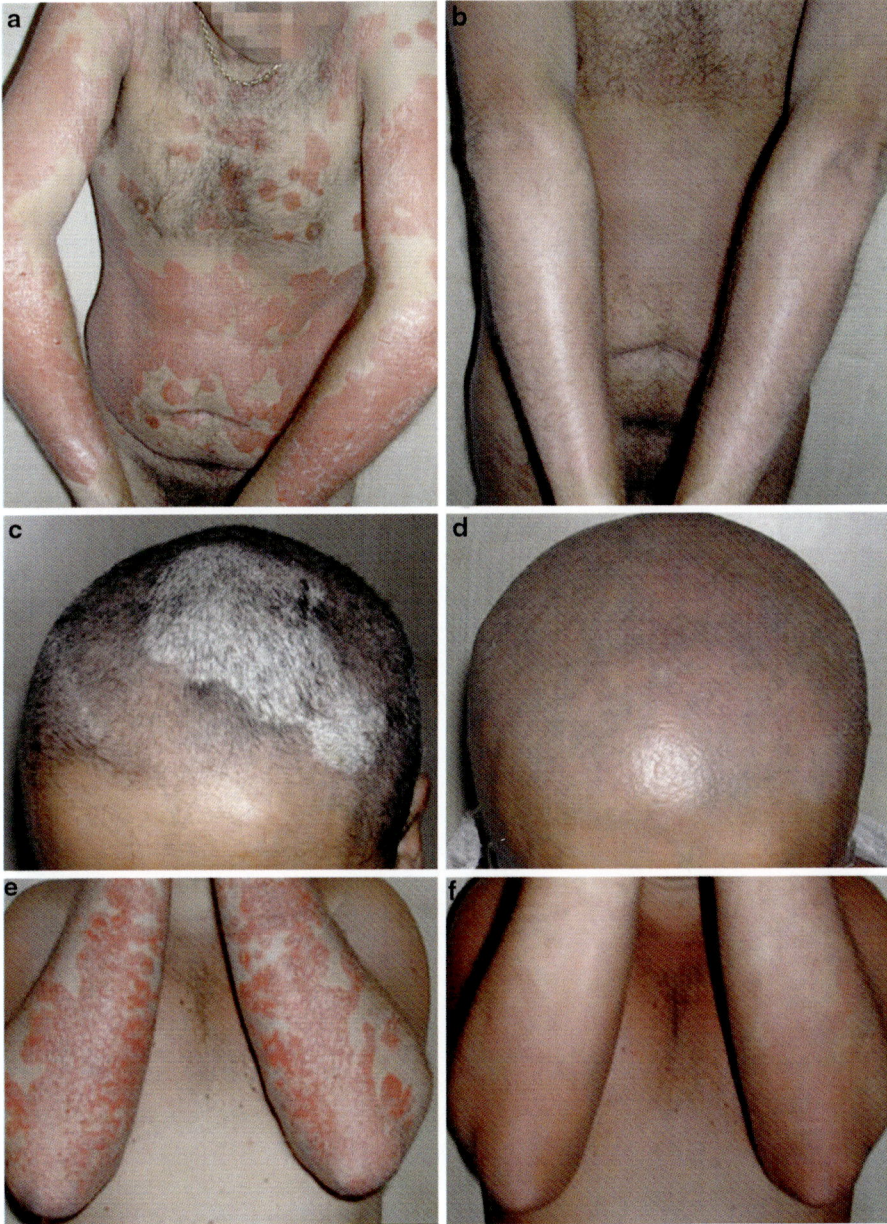

Fig. 6.6 Three patients suffering from psoriasis before (*left*) and after a 3-week course of treatment with ichthyotherapy (2 h daily fish bath) combined with UVA radiation (*right*)

unusually long remission periods. Sixty-five percent of patients stated that subsequent relapses were less severe than before treatment. No significant adverse events were reported. Mild, transient bleeding from open crusted lesions was reported by one patient with eczema, and UV radiation-related erythema by two others.

In conclusion, only two small, non-controlled clinical studies have been published to date, but both studies show significant benefits of ichthyotherapy for psoriasis.

6.5 Impact on Health Related Quality of Life

Psoriasis, like many other disfiguring skin diseases, can negatively impact patients' health related quality of life (HRQoL) (Bhosle et al. 2006). Psoriasis patients often experience problems in body image and self-esteem, with feelings of shame and embarrassment regarding their appearance (Fortune et al. 2005). Krueger et al. (2001) reported that at least 20 % of psoriasis patients had once contemplated suicide. The chronic and recurring nature of this disease often results in feelings of hopelessness (Vardy et al. 2002). The lack of control over the disease may be one of the most bothersome issues for psoriasis patients (Rapp et al. 1998).

HRQoL is a well-established measure for evaluating treatment outcome in dermatology because HRQoL-questionnaires reflect patients' evaluation of the impact of disease and its treatment on their wellbeing. HRQoL is a physician-independent evaluation of treatment outcome. Many different validated measures and indices have been used in the past to assess HRQoL in psoriasis patients (Bhosle et al. 2006). The *Skindex-29* is a self-administered questionnaire with 29 items that assess three domains: burden of symptoms, social and physical functioning, and emotional response (Chren et al. 1996). The *Psoriasis Disability Index* is designed for use in adults (i.e. over the age of 16), and is comprised of 15 questions that assess daily activities, school or work, personal relationships, leisure, and treatment (Finlay and Kelly 1987).

In a recent study, presented at the *8th International Conference on Biotherapy* in Los Angeles evaluating health related quality of life in a cohort of 82 psoriasis patients before and after ichthyotherapy using "skindex-29" and the "Psoriasis Disability Index" (PDI), Grassberger (2010) reported a significant reduction in overall and subcategory scores in both indices (Figs. 6.7 and 6.8). The results of this HRQoL study further corroborate the results of the studies from Özcelik et al. (2000) and Grassberger and Hoch (2006): ichthyotherapy is, indeed, a highly efficacious treatment option with significant improvement in health related quality of life.

6.6 Ichthyotherapy for Congenital Ichthyosis – A Case Report

The male infant presented in this previously unpublished report was diagnosed with *lamellar ichthyosis (LI)*, which belongs to the *autosomal recessive congenital ichthyoses* (ARCI), characterized by non-bullous hyperkeratosis. LI has an estimated prevalence of 1:200,000–300,000 (Oji and Traupe 2006). Recommended treatments for infants with congenital ichthyosis include topical keratolytics (with propylene glycol and alpha hydroxyl), emollients with urea, calcipotriol (a derivative of calcitriol or vitamin D), tazarotene (a topical retinoid) and exceptionally oral retinoids (Vahlquist et al. 2008).

Skindex-29 "overall" (n=82)

P<0.0001

Skindex-29 "functioning" (n=82)

P<0.0001

Skindex-29 "symptoms" (n=82)

P<0.0001

Skindex-29 "emotions" (n=82)

P<0.0001

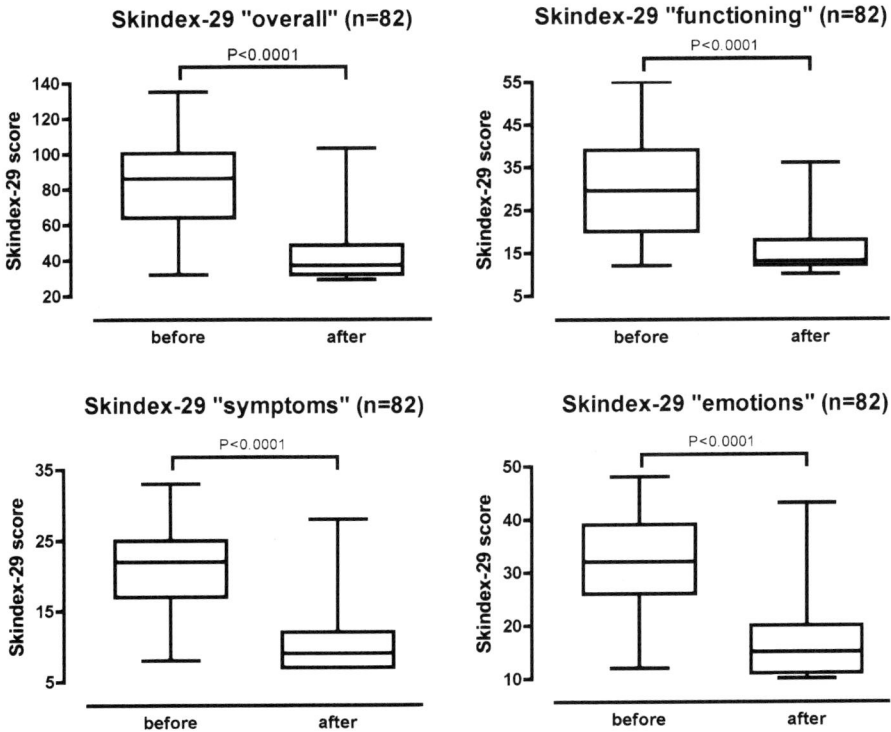

Fig. 6.7 Boxplot graphs for Skindex-29 before and after treatment for overall scores and the respective subcategories. *Horizontal lines* indicate the 75th percentile, median, and 25th percentile; *whiskers* indicate the range

Although the local side-effects of topical therapy for ichthyosis are usually minimal, patient compliance with the treatment regimen may still be poor. Topical treatment is time consuming, has to be repeated several times per day and can be quite painful. Creams or ointments may have unpleasant odors, might be difficult to apply, or might be rejected by the patient for other reasons (Vahlquist et al. 2008). As a result, these therapies are suboptimal and often fail.

6.6.1 Case

The caucasian male patient was born on term in 2001, encased in a tight shiny covering with erythroderma, referred to as collodion membrane (so-called "collodion baby"). During his first weeks of life, the membrane was gradually replaced with thick yellowish-brown plate-like hyperkeratotic scales on the whole integument, with marked palmoplantar hyperkeratosis. Soon after birth, he was diagnosed with congenital lamellar ichthyosis (Ichthyosis congenita, autosomal recessive). No known family anamnesis was detected. Subsequently the patient developed painful irregularly

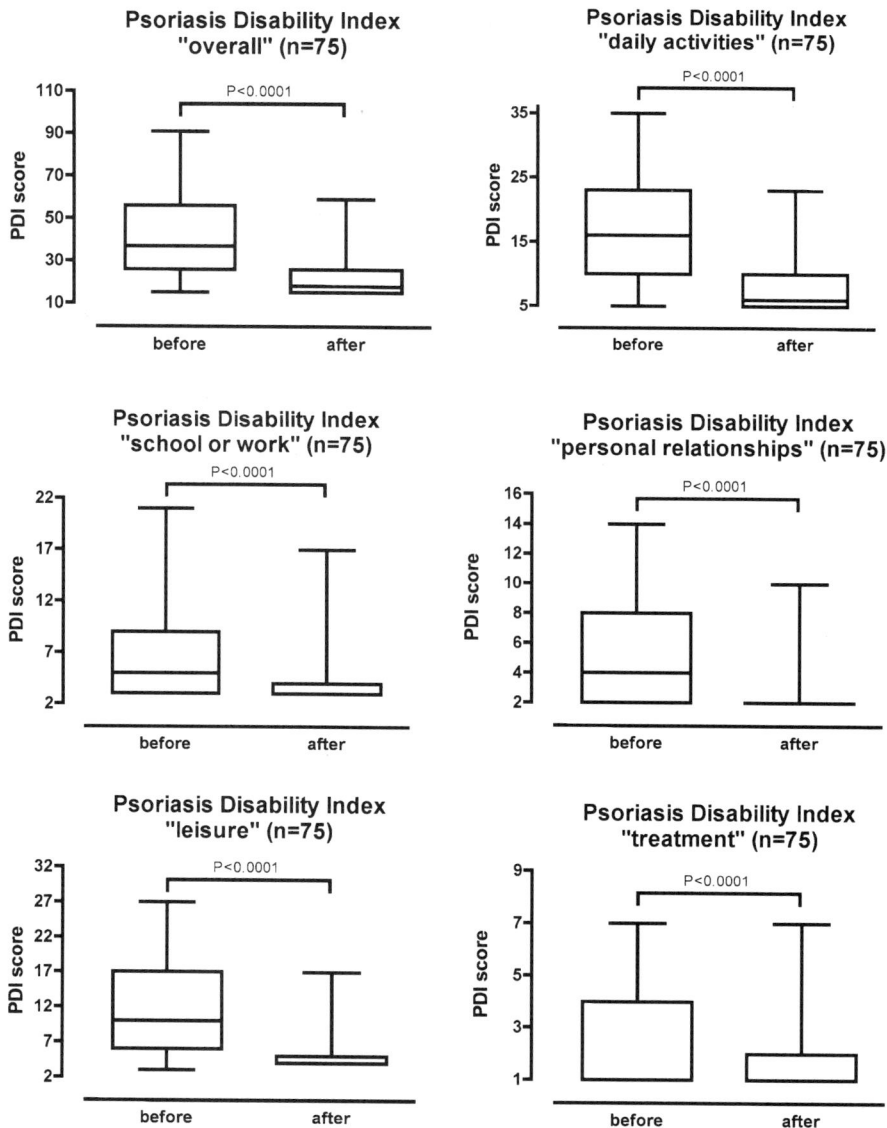

Fig. 6.8 Boxplot graphs for Psoriasis Disability Index before and after treatment for overall scores and the respective subcategories. *Horizontal lines* indicate the 75th percentile, median, and 25th percentile; *whiskers* indicate the range

branched deep cutaneous fissures with an inflammatory component causing severe pain on movement. An unpleasant and annoying smell emanated from the skin. The patient soon presented with ectropium, blepharoconjunctivitis and eclabium (Eversion of a lip). In the first year, contractures of the ischiocrural muscles and the peritrochanteric muscles ensued with already marked muscular atrophy (Fig. 6.9a).

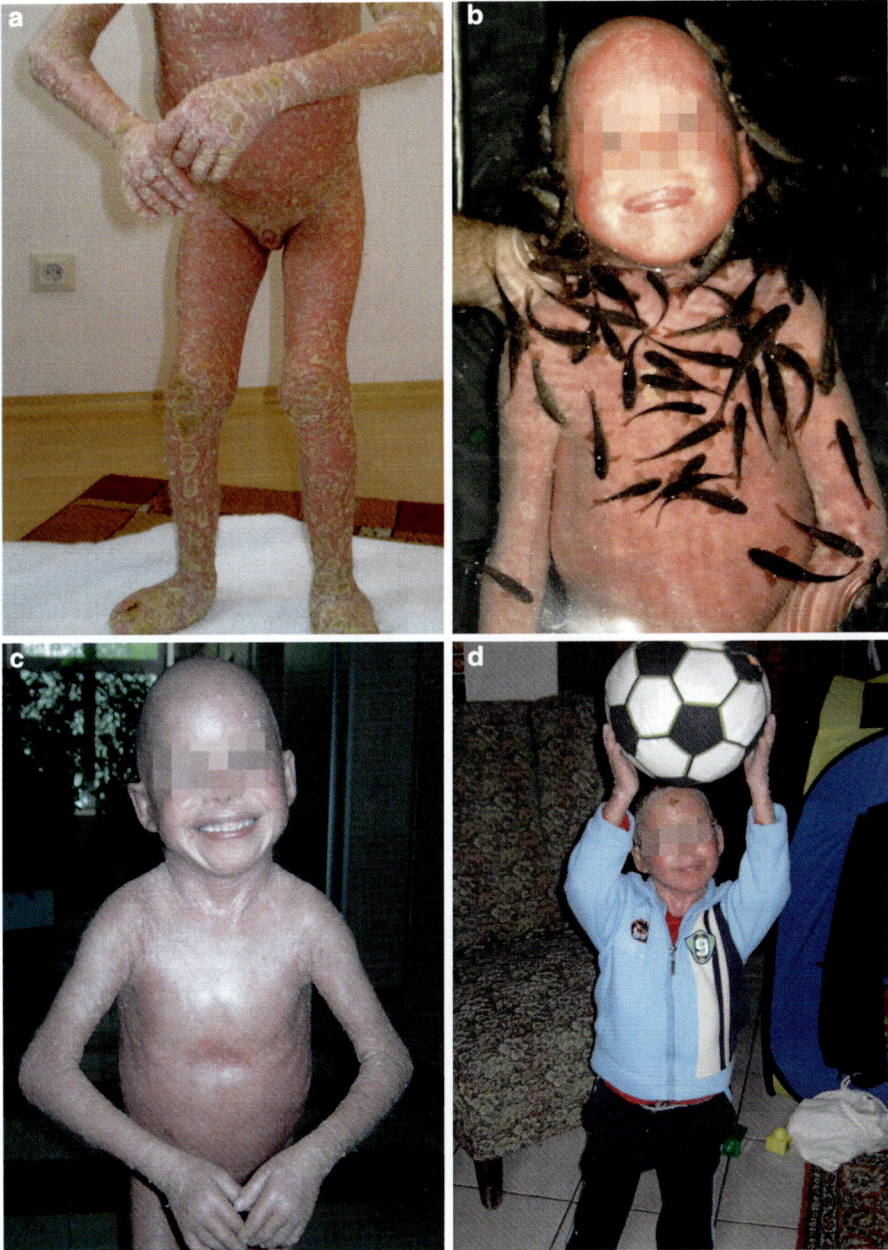

Fig. 6.9 Successful long-term treatment of congenital ichthyosis with ichthyotherapy. (**a**) Before treatment; (**b**) during treatment in fish bath; (**c**) after several courses of ichthyotherapy; (**d**) through regular fish baths the young patient was able to elevate his arms and move without pain

By the time he reached 1.5 years, the scales had a thickness of up to 1 cm. Touching and moving his fingers was now very painful. A keratolytic treatment with Polidocanol and urea (Optiderm®), almond oil and chamomile was used on a daily basis. Pediatric assessment at age 2, documented a dystrophic toddler with massive delay in neuro-motor development.

In 2003, the child was admitted to a specialized clinic, where the patient received a complex treatment regimen of medical baths twice daily and several external oint-ments. The recommended regular mechanical keratolysis with a microfiber cloth was exceptionally painful and therefore very distressing for the mother as well. Regular physiotherapeutic/ergotherapeutic exercises were performed.

After discharge, the treatments were continued, but his condition soon worsened again. The child was again admitted to the specialized clinic. Instead of immersion baths, he now received a 15-min steam bath daily. Being essentially unable to sweat, this treatment was unbearable. Just 1 day without rubbing would result in an imme-diate relapse of symptoms.

Over the ensuing years, a variety of specialists was consulted without noticeable improvement. Movement was increasingly impaired, accordingly the child could not sit, hold a pencil or cutlery, bend, or raise his arms. In 2007, at age 6, the patient was scheduled for primary school, however, writing and prolonged sitting was a severe problem and made regular school attendance impossible.

In 2008, the child had his first ichthyotherapy treatment with specimens of *G. rufa* 2-h daily in the fish bath, for 3 weeks (Fig. 6.9b). After the first week, a rapid improvement was noted in the child's skin condition. After 2 weeks, almost all thick scales were removed (Fig. 6.9c), leaving only elbows and knees with some thicker scales. By now, the patient's life changed profoundly. Being able to move painlessly, the boy started to do things he never did before, e.g. using the slide and the carousel at the playground on his own, sitting, writing with a pencil, walking on his own, play-ing with a ball and raising his arms over the head (Fig. 6.9d). This resulted in joyful parents, a happy patient and a subsequent increase in body weight. By the end of treatment, even the head became virtually free of scales. The boy stated that he was able to sweat for the first time in his life, and was no longer getting overheated.

Yet, 5 days after the end of the 3-week treatment course, the child's skin condi-tions worsened again. As a result, continuous fish treatment was recommended. With the help of a public fundraising, a treatment tub was installed at the boy's home. Routine home treatment with fish baths every other day resulted in reduced redness, reduced itching, reduction of scales, and a resolution to the unpleasant and annoying odor of old, dead skin. This enormous increase in quality of life enabled the boy to visit school without handicap.

6.6.2   Case Discussion

Although systemic retinoids (which promote shedding of the hyperkeratotic plates) have become a standard therapeutic regime in disorders of keratinization (Lacour et al. 1996) the side-effects of systemic retinoids (alterations of lipids and liver function tests) render them inadequate in infants. Successful treatment depends upon the continuous removal of the sick skin and must be done physically. Ichthyotherapy is an elegant and painless method of reducing the hyperkeratotic scales; consequently, quality of life is improved. This case highlights the benefits of ichthyotherapy with *G. rufa* on the quality of life in a young boy suffering from congenital lamellar ichthyosis.

6.7   Possible Mechanisms of Action

Several mechanisms have been suggested regarding the observed efficacy of ichthyotherapy. One obvious mechanism is the physical contact with the fish, which feed on the desquamating skin, thus leading to a rapid reduction of superficial skin scales. Additionally many patients consistently report a pleasing micro-massage like feeling while the fish nibble at their skin (Grassberger and Hoch 2006). Interestingly, the fish seem to prefer hyperkeratotic to healthy skin, possibly because it is easier to access and remove the scales (nibble).

Another suggested mechanism is the increased direct effect of ultraviolet radiation associated with ichthyotherapy, be it natural or in the form of a sunbed. Phototherapy is a well-recognized option for patients with widespread psoriasis lesions (Paul et al. 2012), and removal of scales by the fish probably facilitates greater penetration of UV rays into the dermis.

The presence of a high level of selenium (1.3 mg/L) in the Kangal hot springs water has also been suggested as a contributing factor (Özcelik et al. 2000), although Grassberger and Hoch (2006) reported similar efficacy with relatively low levels of selenium in the water. Therefore, water selenium concentration is probably not a major factor in the observed efficacy of ichthyotherapy.

Psychological factors like stress are regarded a causal or exacerbating factors in psoriasis (Arck and Paus 2006). Since most patients refer to the fish baths as relaxing and pleasing, the stress-reducing and psychological benefits of ichthyotherapy might contribute to the observed treatment benefits.

It is also possible that the oral secretions of the fish have direct anti-inflammatory or desquamation effects. The copious epidermal mucus of fish plays an important role in host defense, particularly in the prevention of colonization by pathogens. Several studies have shown that the surface mucus or "slime" of fish contains a variety of substances such as crinotoxins, calmodulin, pheromones and a variety of antimicrobial substances like fatty acids, immunoglobulins, complement components, lectins, lysozyme, proteolytic enzymes and antinociceptive substances (Subramanian et al. 2008; Jais et al. 1998). A polysaccharide isolated from the mucus of *Misgurnus anguillicaudatus*, a freshwater fish in the loach family (Cobitidae), exhibited strong anti-proliferative and apoptosis-inducing properties (Zhang and Huang 2005). This is a remarkable finding, given the fact that the problem in psoriasis and ichthyosis is a very high mitotic rate of epidermal cells. However, no research has been conducted yet with *G. rufa* to determine whether a similar effect exists in this species as well.

While various providers of ichthyotherapy claim on their websites that *G. rufa* secretes a unique enzyme called dithranol (synonym of anthralin, which prevents epithelial proliferation), there is no scientific evidence to support those claims. Actually, it seems highly unlikely that fish are capable of producing this substance, which is derived from the bark of the Araroba tree of South America and has been manufactured synthetically for decades.

6.8   Indications, Contraindications and Possible Side Effects

6.8.1   Indications

Probably all hyperkeratotic skin conditions would benefit from ichthyotherapy, since the mechanical removal of excessive skin scales by the fish accomplishes the most basic treatment endpoint for these disorders. As already demonstrated, psoriasis and one form of ichthyosis have been treated successfully (Özcelik et al. 2000; Grassberger and Hoch 2006; Grassberger 2010).

6.8.1.1   Psoriasis

Psoriasis is the illness that has most often been treated with ichthyotherapy. It is a common skin disorder with a worldwide distribution. The average prevalence in Europe and the USA has been estimated at about 2–3 %. It is hypothesized to be an immune-mediated disease with a genetic component. The patients often have extensive red and/or silver scaly plaques. These are areas of inflammation and excessive skin production, often on elbows, knees, scalp and genitals. Psoriasis is a chronic, often relapsing disease. Whilst considerable advances have been made in the management of this disease in recent years, there is still no cure, and no simple, safe and invariably effective treatment. The disease carries a substantial burden even when not extensive, and is associated with widespread treatment dissatisfaction.

6.8.1.2   Ichthyosis

Ichthyoses are a heterogeneous group of cornification disorders of the skin, characterized by generalized hyperkeratosis and excessive scaling (Oji et al. 2010). The small but diverse subgroup of *congenital ichthyoses* (CI), typically present at birth with a collodion membrane and/or ichthyosiform erythroderma. CI are very rare congenital disorders that often pose a diagnostic as well as a therapeutic challenge for the caring physician.

6.8.1.3   Diabetic Foot

Diabetic patients often suffer from xerosis and callus of the plantar skin, which often complicate the situation and predispose to the formation of plantar ulcers. In addition to optimal sugar control, effective prevention and treatment measures include pressure relief ("off-loading"), and avoidance or reduction of callus formation. Therefore, the treatment of calloused feet with ichthyotherapy in patients with diabetes mellitus might result in significant benefits. However, studies addressing this issue are still lacking, and special hygienic measures would have to be considered when these often immunocompromised patients are exposed to the potentially contaminated fish tank.

6.8.1.4   Atopic Dermatitis

Whether conditions like atopic dermatitis are also a suitable indication for this kind of alternative treatment remains to be studied.

6.8.1.5   Pedicure in "Fish-Spas"

In recent years, an increased number of "fish pedicure" salons opened to the public all over the world, where customers immerse their feet in small fish tubs to mid-calf level to let *G. rufa* (and in several cases other species) remove excess and thickened skin from their feet. In mid 2011, a survey among environmental health practitioners in the United Kingdom identified 279 registered fish spas in the UK alone (Health Protection Agency 2011). The fish nibble on the thickened skin of the feet with reported satisfaction of the customers. This application of *G. rufa* does not serve therapeutic purposes and is regarded as a wellness or recreational activity. One major problem associated with this new form of fish pedicure, according to local health authorities, is the possible spread of infections.

To educate fish pedicure salon operators on health and safety issues related to this practice, the British Health Protection Agency issued a manual for "guidance on the management of the public health risks from fish spas," based on available evidence and expert consensus (Health Protection Agency 2011). This manual was conceived and produced by representatives of the Health Protection Agency, Health Protection Scotland, the Health and Safety Laboratory and local authorities; it was agreed upon by the Department of Health, Social Services and Public Safety in Northern Ireland and Public Health Wales. The authors concluded that, on the evidence identified and the consensus view of experts, the risk of infection as a result of a fish pedicure is likely to be very low, but cannot be entirely excluded. In order to reduce the risk even further, the document provides operators of fish spas with practical recommendations for safe use.

Despite the fact that there is so far no scientific evidence for spread of infection through *G. rufa*, fish pedicures and ichthyotherapy facilities have been banned by health authorities in several countries (including many U.S. states and Canadian provinces) because of safety concerns. These bans are mainly based on the following assumptions and circumstances: the fish used as "instruments" can neither be disinfected nor sterilized; animals (in this case fish) are prohibited from doctor's offices and pedicure salons; and animal welfare concerns, in general.

6.8.2 Contraindications

Based on general medical considerations, certain patient groups are likely to be at increased risk for infections when undergoing treatments like ichthyotherapy. Therefore, they should be discouraged from undergoing such treatment, especially if they have obvious breaks in the skin. Those at increased risk include severely immunocompromised patients or patients with immune deficiency, whether induced or due to an underlying illness.

Patients that have a known infection with a blood-borne virus, such as hepatitis B and C or HIV should be excluded from the treatment with fish reused on other patients. However, a home based treatment tub for single patient use can be an alternative in such cases. Generally, broken skin, bleeding wounds or infectious skin conditions are also considered contraindications. The transmissibility of contagious skin infections (i.e. bacterial skin infections, herpes viruses and warty human papilloma virus lesions) via ichthyotherapy is unknown, and at this point, individuals should not immerse active lesions into shared tanks in order to minimize potential risks to the fish and to other human clients.

The risk of contracting an infection can probably be reduced by use of disease-free fish reared in controlled facilities under high standards of husbandry and welfare (Verner-Jeffreys et al. 2012).

6.8.3 Side Effects

In the Turkish study by Özcelik et al. (2000), the only side effect reported was sunburn in two patients. In the first Austrian study by Grassberger and Hoch (2006), no severe side effects were recorded during the treatment period. Mild, transient bleeding from open crusted lesions was reported in one patient with eczema, and UV-radiation-related mild erythema in two others. Ichthyotherapy in combination with UVA-treatment was generally very well tolerated.

Given the scarcity of scientific studies on ichthyotherapy there is too little data to draw valid conclusions on possible side-effects and their frequency. Therefore, it seems prudent to consider the theoretical risks of infection when using living animals like fish for therapeutic purposes. However, based on available data and experience thus far, the observed benefits appear to far outweigh the minimal side-effects thus far encountered.

6.9 Hygienic Approach

6.9.1 Possible Risks

The major concern with ichthyotherapy is risk of infection. In the context of ichthyotherapy, there are three potential routes of pathogen transmission:

- fish to person (i.e. Zoonotic disease)
- water to person (i.e. Infection through contaminated water) and
- person to person (either via equipment or through the fish).

If the patient has an underlying condition with reduced defense mechanisms (i.e. immunocompromised patients) or open skin lesions the theoretical risk of infection is increased.

6.9.2 Transmission from Fish to Person (Zoonotic Disease)

Any disease or infection that is naturally transmissible from vertebrate animals to humans and vice-versa is classified as a zoonotic disease (zoonosis). In general, humans contract fish-borne bacterial diseases either through ingestion of contaminated fish tissue or water, or by injection of the organism into puncture wounds or abrasions. This mode of infection also includes transmission via fish tank surface or treatment tubs to patients.

6.9.2.1 Bacterial Infections

Exposures to fish-borne bacteria can result in asymptomatic or mild episodes of gastroenteritis or in localized infections of the skin and the underlying tissue (Nemetz and Shotts 1993). A few bacteria are highly pathogenic, and the status of the human host immune system plays a vital role in the severity of disease.

Gram-negative bacteria are the major cause of fish infectious diseases, both in terms of severity and incidence (Nemetz and Shotts 1993). A few of these microbes, along with some gram-positive species, also cause disease in humans (mostly of an opportunistic nature). According to Lehane and Rawlin (2000), the main pathogens acquired cutaneously from fish (through spine puncture or open wounds) are *Aeromonas hydrophila*, *Edwardsiella tarda*, *Erysipelothrix rhusiopathiae*, *Mycobacterium marinum*, *Streptococcus iniae*, *Vibrio vulnificus* and *Vibrio damsela*. The spectrum of manifestations is wide, varying from cases of mild cellulitis, to severe life-threatening necrotizing fasciitis requiring radical surgery, to sepsis and death (Finkelstein and Oren 2011).

In the setting of ichthyotherapy, the following organisms are probably of greatest concern: gram-positive bacteria like *Streptococcus* sp. and *Staphylococcus* sp., the latter causing erysipelas and cellulitis (a diffuse inflammation of connective tissue with severe inflammation of the dermal and subcutaneous layers of the skin). *Erysipelothrix rhusiopathiae* and *Streptococcus iniae* are associated with handling fish, but zoonotic spread to humans occurs only rarely. *E. rhusiopathiae* can cause erysipeloid or "fish rose", with septicemia and endocarditis as potential sequel. *Streptococcus iniae* causes a high mortality rate and rapid death in infected fish. In humans, infection is clearly opportunistic with all cases to date associated with direct infection of puncture wounds during preparation of contaminated fish, generally in elderly or immunocompromised individuals (Agnew and Barnes 2007).

Mycobacteria like *Mycobacterium marinum*, *M. fortuitum*, *M. piscium* and *M. chelonei* can cause cutaneous lesions (so-called "fish tank granuloma" or "swimming pool granuloma") at skin contact sites, through contamination of lacerated or abraded skin, especially in immunocompromised patients. *M. marinum* has been identified in fish as well as in biofilms on inanimate surfaces (Aubry et al. 2002), and often infects home aquarium hobbyists. Infections with other non-tuberculous Mycobacteria (e.g. *M. mageritense*) have been reported in association with footbaths at nail saloons, and shaving prior to the footbaths has been identified as a risk factor (Winthrop et al. 2002; Gira et al. 2004). Since Mycobacteria are typically associated with pools and fish tanks, these pathogens and associated risk factors (e.g. broken skin and biofilms), deserve special attention in the ichthyotherapy setting.

Infections with gram-negative bacteria of the species *Plesiomonas shigelloides* cause gastroenteritis, followed by septicemia in immune deficient patients (Wadstrism and Ljungh 1991).

Aeromonas spp. such as *A. hydrophila*, *A. sobria* and *A. caviae* are capable of causing gastroenteritis and diarrheal disease when ingested, and localized wound infections when there is invasive contact with the skin, such as with water-related trauma (Janda and Abbott 2010). Interestingly, Aeromonas infections have been reported in association with the use of medicinal leeches (Snower et al. 1989; Whitaker et al. 2011), which also live in sweet waters. Skin infections may be superficial or may progress to cellulitis, deep muscle necrosis or septicemia, especially in the immunocompromised host. *Aeromonas sobria* has recently been reported in association with *G. rufa* by Majtan et al. (2012), however, reports of serious human infections are rare.

Other bacteria of relevance are *Pseudomonas* spp. and Enterobacteriaceae of the genera *Escherichia*, *Salmonella*, *Klebsiella and Edwardsiella*. Infection with *Edwardsiella tarda* is possible by ingestion or through a penetrating wound. Infected fish or contaminated water are possible sources of infection (Janda and Abbott 1993). A variety of disease may develop, including gastroenteritis, localized infections or septicemia. Salmonellae have also been reported in association with fish tanks and tropical fish.

Since most of these infections are associated with ingestion, patients should be educated to avoid swallowing the tub water and to wash their hands after the treatment or after hand contact with the fish.

6.9.2.2   Parasitic Infections

Fish parasites like *Diphyllobothrium latum* (fish tapeworm) pose no danger in well-controlled ichthyotherapy settings since the fish are only fed with commercially available fish food in addition to the patient's skin scales. Therefore, the risk of fish tapeworm infestation can be ruled out, based on the absence of a suitable interme-diate host (normally freshwater crustaceans). Fish flukes (trematodes in the phy-lum Platyhelminthes) can only be transmitted to man by eating raw or undercooked fish. Although potentially zoonotic species of *Giardia* and *Cryptosporidium* have been found in fish, there is no evidence that these could be transmitted via the mouths of *G. rufa*, nor via the water, as ingestion will not occur (Health Protection Agency 2011).

6.9.2.3   Viral and Fungal Infections

To the best of our knowledge, no human infections with fish specific viruses or fungi have been reported in the medical literature so far.

6.9.3   Transmission from Water to Person

In addition to the above-mentioned bacterial pathogens found in fish and fish-water, the following bacteria may be found in bathtub water not associated with fish: *Legionella pneumophila*, *Escherichia coli* and *Pseudomonas aeruginosa*. *Legionella pneumophila* can cause respiratory infections from mild to fatal pneumonia, but the risk of contracting a *Legionella*-associated respiratory tract infection is considered very low since substantial aerosols as in whirlpools and hot tubs are normally not generated during a fish bath. *Pseudomonas aeruginosa* is known to colonize bio-films on underwater surfaces and to cause whirlpool-associated dermatitis and fol-liculitis, which usually manifests as self-limiting pustular rash (Hudson et al. 1985; Ratnam et al. 1986; CDC 2000).

6.9.4   Transmission from Person to Person

In the ichthiotherapy setting, person-to-person spread of infection can occur either via direct patient contact, through the water, the equipment and/or the fish them-selves. Health authorities are particularly concerned about the transmission of blood borne viruses that might occur as a result of sharing a fish bath.

6.9.4.1   Viral Infections

Although the risk of transmission of hepatitis B virus has been described as particu-larly high in athletes in contact and collision sports (Kordi and Wallace 2004) and survival of hepatitis B virus in the environment has been reported for 7 days on dry surfaces (Bond et al. 1981), there is no data available for survival in water. Ciesek et al. (2010) found that at 37 °C and in a moist environment, hepatitis C virus was inactivated after just 2 days. It is generally considered that viral pathogens contaminating the fish's mouth are not likely to remain there in a sufficient manner to cause transmission. Heistinger et al. (2011), using a mammalian model virus, the Equine Herpes Virus 1 (EHV 1) as model for an enveloped DNA virus and Equine Rhino Virus (ERV) as model for a non-enveloped RNA-virus, have addressed this issue in *G. rufa*. Results showed that mammalian DNA virus did not survive on fish tissue, whereas mammalian RNA virus did survive on fish tissue for about 5 min. Although this study investigated only non-human viruses, there is no scientific record for cold-blooded animals like fish transmitting human viral disease, because viruses usually cannot survive in or on these animals. However, given the scarcity of publications, studies investigating the survival in fish and the possible transmission of blood-borne viruses are still important areas of future research.

Although the theoretical risk can never be completely excluded, and despite the lack of rigorous safety studies, it is noteworthy that the British Health Protection Agency (2011) has come to the conclusion that based on the available evidence to date, the risk of infection with a blood-borne virus as a result of a fish pedicure bath is likely to be extremely low.

6.9.4.2 Bacterial Infections

An infection with human pathogenic bacteria such as *Staphylococcus aureus* is more likely from skin contact with surfaces outside the water, with the risk being similar to that in a gym, since dilution by water makes water-borne transmission very unlikely (Health Protection Agency 2011).

6.9.4.3 Fungal Infections

Fungal infections like athlete's foot contracted via surviving organisms on surfaces such as floors and towels, are not unique to ichthyotherapy and are found frequently in areas like public pools, foot spas and sports clubs, where people walk barefoot. Following general standards of hygiene are therefore mandatory to prevent such infections.

6.9.5 Hygiene Guidelines for Ichthyotherapy

Apart from the guidance paper for fish foot spas by the British Health Protection Agency (2011), approved detailed hygiene standards for the treatment of skin diseases with ichthyotherapy have not yet been established. However, the variety and abundance of treatment centers and recent regulatory problems in many communities make it necessary to discuss possible health risks for patients and provide minimal requirements for hygiene when using this therapy. Based on the afore mentioned

theoretical modes of transmission, the following recommendations regarding maintenance of treatment tubs and fish should be followed as minimum requirements for
ichthyotherapy facilities.

6.9.5.1   Hygienic Measures

In light of the potential risks of infections, predominantly from bacterial pathogens, the
most important steps to establish a hygienic and safe ichthyotherapy or fish spa are:

- the use of an open freshwater system, which renews the bathing water constantly
  or at regular intervals, with an estimated complete water exchange 3–4 times a day,
- preventing floating debris and contaminants through the use of a cascade filters
  or similar physical barriers,
- use of a filter pump to constantly clean the recycling water, thereby removing
  fish excrements and immersed particles,
- use of an ultraviolet (UV-C) water sterilization device,
- oxygen enrichment of the water and
- use of an oxygen-separating water and surface disinfectant that is approved for
  use in aquaculture.

Ideally, all of the above-mentioned measures are taken by a fully automated digi-
tally controlled system.

Plants, rocks or sand in the treatment tubs are neither necessary nor appropriate,
since these features are prone to biofilm formation. However, corresponding to the
fish's natural habitat, the floor of the tubs should provide hiding places for the fish
to comply with animal welfare requirements.

6.9.5.2   Water Disinfection

Unlike other pool facilities, the water in the treatment tubs cannot be treated with
classical chemical disinfectants, e.g. chlorine or quaternary ammonium compounds,
since these substances are not tolerated by the fish. Therefore, an oxygen separating
substance, approved for use in aquaculture (e.g. Sanosil®) has proven useful in disinfecting the water to the required level, whilst maintaining fish viability. The active
substances of Sanosil® are stabilized hydrogen peroxide, and traces of ionic silver.
Both substances work synergistically and are highly effective against a broad spectrum of viruses, bacteria and fungi, are active against biofilm formation and – in the
correct concentrations – are non-toxic to patients and fish.

6.9.5.3   Surface Disinfection

Most common nosocomial pathogens may well survive and persist on inanimate
surfaces for weeks or even months, making these surfaces a continuous source of

transmission, if no regular preventive surface disinfection is performed (Kramer et al. 2006). Therefore, general guidelines for surface disinfection in specific patient care areas, apart from the treatment tubs, are mandatory for ichthyotherapy facilities. The treatment tubs should be thoroughly cleaned and disinfected at regular intervals with an oxygen-separating agent, preferably after each 3-week treatment course.

6.9.5.4   Water and Fish Samples

For surveillance of microbiological parameters, water samples should be drawn at regular intervals to monitor quality. Official regulations from local health authorities for water quality in public pools or whirlpools may be followed as guidance and should provide a minimal framework. To control potential zoonotic infections, scheduled fish sampling by accredited veterinarian specialists is recommended, with special focus on the presence of non-tuberculous Mycobacteria and *Aeromonas* spp. A fish health certificate should always be obtained.

6.9.5.5   Patient Requirements

We recommend that patients should be tested for blood-borne viruses (BBV) such as HIV and Hepatitis B and C prior to fish treatment, since many people are unaware of their BBV status. Additionally, patients should be under medical supervision throughout the entire course of treatment, and should be excluded from ichthyotherapy when an infectious skin disease is suspected or diagnosed (see also "contraindications"). After providing patients with all relevant information on ichthyotherapy, including contraindications and possible side-effects, they should be asked to sign an informed consent form prior to treatment, to ensure that they understood the information given and are not aware of any contraindications to the treatment.

In addition to the above recommendations, one must always comply with the local regulations when performing ichthyotherapy.

6.10   Practical Application

6.10.1   Fish Species

The few scientific studies published so far employed the species *G. rufa*. Therefore, no other fish should be used for this treatment unless safety and efficacy data are available for those species. The fish, preferably reared under veterinary supervision in an adjacent breeding facility (Fig. 6.10a), should be at least 3 cm in length

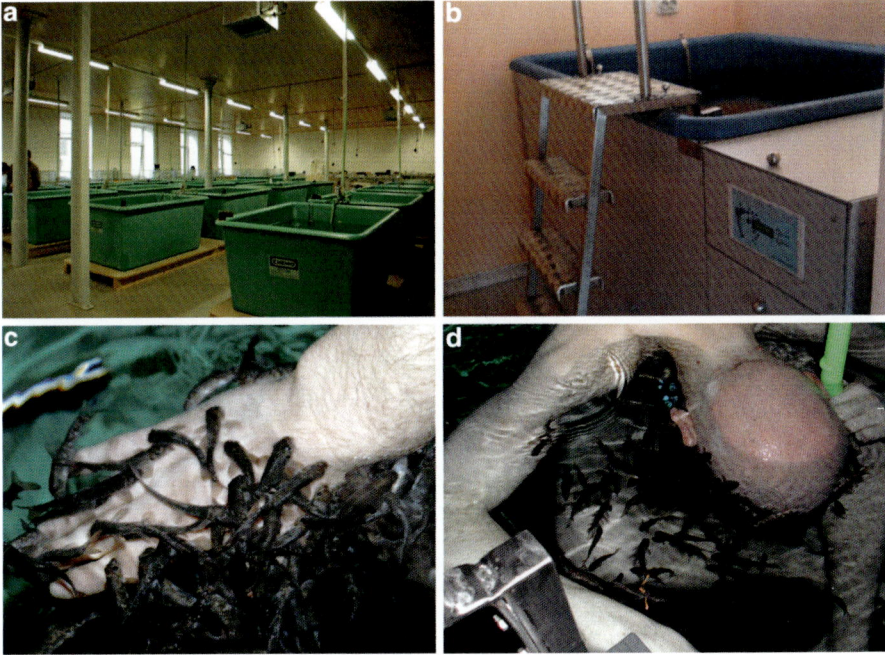

Fig. 6.10 (**a**) In conjunction with an ichthyotherapy facility, a larger scale adjacent breeding facility is necessary, especially when home users with their own tubs have to be supplied; (**b**) treatment tub with built in fully automated control unit; (**c**) healthy fish are a prerequisite for a successful treatment; (**d**) in some cases special procedures are necessary, i.e. shaving the head and using a snorkel to treat scalp lesions

(approx. 1.5 years old) when used for treatment. The fish must always be fed nutritious (i.e. commercially available) fish food daily after the treatment sessions, since feeding on patients' skin scales does not provide most of the required nutrients. There should never be any shortness in food supply and it is not necessary to starve the fish in order to be therapeutically active.

6.10.2 Tubs and Number of Fish

Every patient should be allocated his own treatment cabin equipped with a tub. The treatment tubs, made from food-safe plastic usually have a capacity of 700–1,000 l (Fig. 6.10b). Between 250 and 400 fish are used on each patient, depending on the size and severity of the skin lesions. Ideally, the bath tubs are equipped with an elaborate hygiene and fresh water system, to ensure healthy active fish (Fig. 6.10c) and patient safety. The bath tubs should be set to a comfortably warm temperature (about 36 °C).

6.10.3   Duration of Treatment

According to Grassberger and Hoch (2006) a treatment course with a duration of approximately 3 weeks with a daily 2-h fish bath is sufficient for a satisfying result in psoriasis. Ongoing treatments might be necessary thereafter to maintain the treatment result and keep the patient symptom free, i.e. in remission. Daily treatments lasting for 1–2 h makes ichthyotherapy acceptable for most patients, and considerably improves compliance (Grassberger and Hoch 2006).

6.10.4   UV-Radiation (Phototherapy)

Patients with no contraindication to UV exposure may additionally use a commercially available stand-up tan facility after each bath session according to skin type and after consultation with a dermatologist.

6.10.5   Emollients

In the Austrian ichthyotherapy study (Grassberger and Hoch 2006), patients applied a generic skin lotion (emollient) containing Shea butter and *Aloe vera* extract after fish and UV exposure. Whether this is a major contribution to the observed effect remains to be studied.

6.10.6   Special Procedures

If psoriasis lesions severely involve the scalp, the patient's head can optionally be shaved before treatment and then immersed while the patient breathes through a snorkel (Fig. 6.10d). Sensitive skin areas might be clothed or taped to avoid irritation. However, it should be emphasized that bleeding skin lesions present a contraindication in most treatment centers.

6.11   Future Research

6.11.1   Efficacy

Based on the few but well documented clinical studies, ichthyotherapy combined with a short course of UVA treatment can be regarded as a relatively safe and effective treatment option for patients with disfiguring hyperkeratotic skin conditions like psoriasis and ichthyosis. Prospective randomized controlled trials are now warranted to validate the efficacy of this unusual and apparently highly effective biotherapeutic modality, and to compare ichthyotherapy with controls (e.g., the conventional standard regimens or water with the UV therapy alone). Additionally, the aesthetic or recreational use of *G. rufa* in the form of foot spas and pedicure parties should be clearly separated from the medical or therapeutic application of these fish.

6.11.2   Hygiene

Based on the above given hygienic standards, "Best Practice Guidelines" regarding minimal requirements in hygiene should be established and subsequently published to provide health authorities with an aid to assess commercial ichthyotherapy facilities and to harmonize treatment standards throughout the world. Larger safety studies might uncover some uncommon side effects, or might help to allay fears that serious problems will soon declare themselves. Although no experiment can prove that an event will never occur, laboratory models can help demonstrate the likelihood (or unlikelihood) of various theoretical infectious complications. Given the regulatory restrictions that currently exist in many countries, and given the large numbers of people who could potentially benefit from ichthyotherapy if it were more widely available, such studies will certainly be undertaken in the near future.

6.11.3   Mechanisms of Action

The underlying mechanisms of ichthyotherapy are not yet fully understood. The most noticeable effect of the fish is removal of excess skin scales. However, the observed dramatic reduction of the inflammatory component in psoriasis patients suggests additional, possibly molecular mechanisms. Biochemical studies should be undertaken to identify and characterize the properties of *G. rufa* mucus in this context.

References

Abdoli A (2000) The inland freshwater fishes of Iran (in: Farsi). Iranian Museum of Natural and Wildlife, Tehran, 378 pp

Agnew W, Barnes AC (2007) *Streptococcus iniae*: an aquatic pathogen of global veterinary significance and a challenging candidate for reliable vaccination. Vet Microbiol 122(1–2):1–15

Arck P, Paus R (2006) From the brain-skin connection: the neuroendocrine-immune misalliance of stress and itch. Neuroimmunomodulation 13(5–6):347–356

Aubry A, Chosidow O, Caumes E, Robert J, Cambau E (2002) Sixty-three cases of *Mycobacterium marinum* infection. Arch Intern Med 162:1746–1752

Berg LS (1949) Freshwater fishes of Iran and adjacent countries. Tr Zool Inst Akad Nauk SSSR 8:783–858

Bhosle MJ, Kulkarni A, Feldman SR, Balkrishnan R (2006) Quality of life in patients with psoriasis. Health Qual Life Outcomes 6(4):35. doi:10.1186/1477-7525-4-35

Bond WW, Favero MS, Petersen NJ, Gravelle CR, Ebert JW, Maynard JE (1981) Survival of hepatitis B virus after drying and storage for one week. Lancet 1(8219):550–551

Carlin CS, Feldman SR, Krueger JG, Menter A, Krueger GG (2004) A 50 % reduction in the Psoriasis Area and Severity Score (PASI 50) is a clinically significant endpoint in the assessment of psoriasis. J Am Acad Dermatol 50:859–866

CDC (2000) Pseudomonas dermatitis/folliculitis associated with pools and hot tubs – Colorado and Maine 1999–2000. MMWR 49(48):1087–1091

Chren MM, Lasek RJ, Quinn LM, Mostow EN, Zyzanski SJ (1996) Skindex, a quality-of-life measure for patients with skin diseases: reliability, validity and responsiveness. J Invest Dermatol 107:707–713

Ciesek S, Friesland M, Steinmann J, Becker B, Wedemeyer H, Manns MP, Steinmann J, Pietschmann T, Steinmann E (2010) How stable is the hepatitis C virus (HCV)? Environmental stability of HCV and its susceptibility to chemical biocides. J Infect Dis 201(12):1859–1866

Coad BW (2012) Freshwater fishes of Iran. http://www.briancoad.com. Accessed 29 Oct 2012

Finkelstein R, Oren I (2011) Soft tissue infections caused by marine bacterial pathogens: epidemiology, diagnosis, and management. Curr Infect Dis Rep 13(5):470–477

Finlay AY, Kelly SE (1987) Psoriasis: an index of disability. Clin Exp Dermatol 12:8–11

Fishbase.org "*Garra rufa* (Heckel, 1843)" http://www.fishbase.org/summary/Garra-rufa.html. Accessed 29 Oct 2012

Fortune DG, Richards HL, Griffiths CE (2005) Psychologic factors in psoriasis: consequences, mechanisms, and interventions. Dermatol Clin 23:681–694

Gira AK, Reisenauer AH, Hammock L, Nadiminti U, Macy JT, Reeves A, Burnett C, Yakrus MA, Toney S, Jensen BJ, Blumberg HM, Caughman SW, Nolte FS (2004) Furunculosis due to *Mycobacterium mageritense* associated with footbaths at a nail salon. J Clin Microbiol 42(4):1813–1817

Grassberger M (2010) Ichthyotherapy for skin disease. Presented at the 8th international conference on biotherapy, Los Angeles, 11–14 Nov 2010

Grassberger M, Hoch W (2006) Ichthyotherapy as alternative treatment for patients with psoriasis: a pilot study. Evid Based Complement Alternat Med 3(4):483–488

Health Protection Agency (2011) Guidance on the management of the public health risks from fish pedicures. http://www.hpa.org.uk/webc/HPAwebFile/HPAweb_C/1317131045549. Accessed 25 Oct 2012

Heistinger K, Heistinger H, Lussy H, Nowotny N (2011) Analysis of potential microbiological risks in Ichthyotherapy using Kangal fish (*Garra rufa*). Egypt J Aquat Biol Fish 15(3):525–537

Hudson PJ, Vogt RL, Jillson DA, Kappel SJ, Highsmith AK (1985) Duration of whirlpool-spa use as a risk factor for Pseudomonas dermatitis. Am J Epidemiol 122(5):915–917

Jais AM, Matori MF, Kittakoop P, Sowanborirux K (1998) Fatty acid compositions in mucus and roe of Haruan, *Channa striatus*, for wound healing. Gen Pharmacol 30(4):561–563

Janda JM, Abbott SL (1993) Infections associated with the genus *Edwardsiella*: the role of *Edwardsiella tarda* in human disease. Clin Infect Dis 17(4):742–748

Janda JM, Abbott SL (2010) The genus Aeromonas: taxonomy, pathogenicity, and infection. Clin Microbiol Rev 23(1):35–73

Kordi R, Wallace WA (2004) Blood borne infections in sport: risks of transmission, methods of prevention, and recommendations for hepatitis B vaccination. Br J Sports Med 38(6):678–684

Kramer A, Schwebke I, Kampf G (2006) How long do nosocomial pathogens persist on inanimate surfaces? A systematic review. BMC Infect Dis 6:130

Krueger G, Koo J, Lebwohl M, Menter A, Stern RS, Rolstad T (2001) The impact of psoriasis on quality of life: results of a 1998 National Psoriasis Foundation patient-membership survey. Arch Dermatol 137:280–284

Kullander SO, Fang F (2004) Seven new species of *Garra* (Cyprinidae: Cyprininae) from the Rakhine Yoma, southern Myanmar. Ichthyol Explor Freshw 15(3):257–278

Lacour M, Mehta-Nikhar B, Atherton DJ, Harper JI (1996) An appraisal of acitretin therapy in children with inherited disorders of keratinization. Br J Dermatol 134(6):1023–1029

Lehane L, Rawlin GT (2000) Topically acquired bacterial zoonoses from fish: a review. Med J Aust 173(5):256–259

Majtan J, Cerny J, Ofukana A, Takac P, Kozanek M (2012) Mortality of therapeutic fish *Garra rufa* caused by Aeromonas sobria. Asian Pac J Trop Biomed 2(2):85–87

Menon AGK (1964) Monograph of the cyprinid fishes of the genus *Garra* Hamilton. Mem Indian Mus 14:173–260

Nemetz TG, Shotts EB (1993) Zoonotic diseases. In: Stoskopf MK (ed) Fish medicine. WB Saunders Company, Philadelphia, pp 214–220

Oji V, Traupe H (2006) Ichthyoses: differential diagnosis and molecular genetics. Eur J Dermatol 16(4):349–359

Oji V, Tadini G, Akiyama M, Blanchet Bardon C, Bodemer C, Bourrat E, Coudiere P, DiGiovanna JJ et al (2010) Revised nomenclature and classification of inherited ichthyoses: results of the First Ichthyosis Consensus Conference in Sorèze 2009. J Am Acad Dermatol 63(4): 607–641

Özcelik S, Polat HH, Akyol M, Yalcin AN, Ozcelik D, Marufihah M (2000) Kangal hot spring with fish and psoriasis treatment. J Dermatol 27(6):386–390

Paul C, Gallini A, Archier E, Castela E, Devaux S, Aractingi S, Aubin F, Bachelez H, Cribier B, Joly P, Jullien D, Le Maître M, Misery L, Richard MA, Ortonne JP (2012) Evidence-based recommendations on topical treatment and phototherapy of psoriasis: systematic review and expert opinion of a panel of dermatologists. J Eur Acad Dermatol Venereol 26(Suppl 3):1–10

Rapp SR, Feldman S, Fleischer AB Jr, Reboussin DM, Exum ML (1998) Health related quality of life in psoriasis: a biopsychosocial model and measures. In: Rajagopalan R, Sherertz EF, Anderson R (eds) Care management of skin diseases life quality and economic impact. Marcel Dekker, New York, pp 125–145

Ratnam S, Hogan K, March SB, Butler RW (1986) Whirlpool-associated folliculitis caused by *Pseudomonas aeruginosa*: report of an outbreak and review. J Clin Microbiol 23(3):655–659

Snower DP, Ruef C, Kuritza AP, Edberg SC (1989) *Aeromonas hydrophila* infection associated with the use of medicinal leeches. J Clin Microbiol 27(6):1421–1422

Subramanian S, Ross NW, MacKinnon SL (2008) Comparison of antimicrobial activity in the epidermal mucus extracts of fish. Comp Biochem Physiol B Biochem Mol Biol 150(1):85–92

Teimori A (2006) Preliminary study of freshwater fish diversity in Fars province. M.Sc. thesis, Shiraz University, Shiraz

Teimori A, Esmaeili HR, Ansari TH (2011) Micro-structure consideration of the adhesive organ in Doctor Fish, *Garra rufa* (Teleostei; Cyprinidae) from the Persian Gulf Basin. Turk J Fish Aquat Sci 11:407–411

Timur M, Çolak A, Marufi M (1983) A study on the systematic identification of the Balikli thermal spring (Sivas) fish and the curative effects of the fish on dermal diseases. Vet Fakült Dergisi (Ankara) 30:276–282

Vahlquist A, Gånemo A, Virtanen M (2008) Congenital ichthyosis: an overview of current and emerging therapies. Acta Derm Venereol 88:4–14

Vardy D, Besser A, Amir M, Gesthalter B, Biton A, Buskila D (2002) Experiences of stigmatization play a role in mediating the impact of disease severity on quality of life in psoriasis patients. Br J Dermatol 147:736–742

Verner-Jeffreys DW, Baker-Austin C, Pond MJ, Rimmer GSE, Kerr R, Stone D et al (2012) Zoonotic disease pathogens in fish used for pedicure [letter]. Emerg Infect Dis 18(6):1006–1008, http://dx.doi.org/10.3201/eid1806.111782

Wadstrism T, Ljungh A (1991) *Aeromonas* and *Plesiomonas* as food and waterborne pathogens. Int J Food Microbiol 12:303–312

Warwick D, Warwick J (1989) The doctor fish – a cure for psoriasis? Lancet 335:1093–1094

Whitaker IS, Josty IC, Hawkins S, Azzopardi E, Naderi N, Graf J, Damaris L, Lineaweaver WC, Kon M (2011) Medicinal leeches and the microsurgeon: a four-year study, clinical series and risk benefit review. Microsurgery 31(4):281–287

Winthrop KL, Abrams M, Yakrus M, Schwartz I, Ely J, Gillies D, Vugia DJ (2002) An outbreak of mycobacterial furunculosis associated with footbaths at a nail salon. N Engl J Med 346(18):1366–1371

Zhang CX, Huang KX (2005) Apoptosis induction on HL-60 cells of a novel polysaccharide from the mucus of the loach, *Misgurnus anguillicaudatus*. J Ethnopharmacol 99(3):385–390

Chapter 7
Helminth Therapy

David E. Elliott, David I. Pritchard, and Joel V. Weinstock

7.1 Introduction

The best solutions address root causes of a problem. The therapeutic use of helminths (parasitic worms) proposes to treat a root cause of autoimmune disease, loss of exposure to these organisms due to modern hygienic lifestyle. There are many types of helminths. Two are being explored for potential medical application. The first is *Trichuris suis* or porcine whipworm. The second is *Necator americanus*, a human hookworm. There are more than 80 different autoimmune diseases, which afflict people in highly-developed industrialized countries. Most of these diseases are rare in tropical lesser-developed countries, where helminth exposure is common. Diseases currently being studied for treatment by helminths include Crohn's disease, ulcerative colitis, multiple sclerosis, celiac disease, rheumatoid arthritis, and autism.

D.E. Elliott, M.D., Ph.D. (✉)
Department of Internal Medicine, Division of Gastroenterology/Hepatology,
University of Iowa Carver College of Medicine, 200 Hawkins Dr.,
Iowa City, IA 52242, USA
e-mail: david-elliott@uiowa.edu

D.I. Pritchard, Ph.D.
Faculty of Science, University of Nottingham, Room D07
Boots Science Building, University Park, NG7 2RD Nottingham, UK
e-mail: david.pritchard@nottingham.ac.uk

J.V. Weinstock, M.D.
Division of Gastroenterology, Tufts Medical Center, Boston, MA 02493, USA
e-mail: jweinstock2@tuftsmedicalcenter.org

M. Grassberger et al. (eds.), *Biotherapy - History, Principles and Practice:*
A Practical Guide to the Diagnosis and Treatment of Disease using Living Organisms,
DOI 10.1007/978-94-007-6585-6_7, © Springer Science+Business Media Dordrecht 2013

7.2 General Historical Aspects

Our immune system protects us from invading organisms like viruses and bacteria. To do this, the immune system must identify an invader and then mount a defense. If the immune system inappropriately identifies normal cells or commensal bacteria as an invader, the misguided defense will injure healthy tissue. Diseases caused by misguided immune reactions are named autoimmune and immune-mediated inflammatory diseases. These types of diseases were very rare before the 1930s but now afflict more than 10 % of the population in the United States and Western Europe. The dramatic increase in autoimmune and immune-mediated disease that occurred over the last 75 years, suggests that environmental changes, associated with advanced industrialization, result in an increased risk for immune-mediated disease. Indeed, there is now a great deal of data showing that as lesser-developed countries become more developed, autoimmune and inflammatory diseases emerge.

Many lifestyle changes occur as countries become more developed. One change with profound immunologic impact is the loss of natural exposure to helminths. Helminths influence their host's immune system to promote anti-inflammatory immune regulatory responses. Loss of exposure to helminths means loss of this immune regulatory influence. In theory, loss of the helminthic immune regulatory influence permits development of an autoimmune or immune mediated inflammatory disease. If this theory is true, then exposure to helminths should serve to protect individuals from these types of diseases.

We and others have used animal models to test the theory that helminth exposure protects individuals from developing immune-mediated disease. An animal model is a well-defined experimental setting where an animal, usually mice or rats, develops an illness that mimics a human disease. Mice or rats given helminths are protected from developing illnesses that mimic inflammatory bowel disease (Elliott et al. 2000, 2003, 2004; Khan et al. 2002; Reardon et al. 2001), multiple sclerosis (La Flamme et al. 2003; Sewell et al. 2003), type 1 diabetes (Cooke et al. 1999; Zaccone et al. 2003; Saunders et al. 2007; Liu et al. 2009; Espinoza-Jimenez et al. 2010), rheumatoid arthritis (Salinas-Carmona et al. 2009; Osada et al. 2009; He et al. 2010; McInnes et al. 2003), and asthma (Wilson et al. 2002; Kitagaki et al. 2006; Mangan et al. 2006).

The theory that helminth infection protects from autoimmune disease can be tested in people by comparing the prevalence of a disease in helminth highly-exposed and less- or non-exposed groups. These epidemiologic studies show that wheezing or allergic skin reactions (Scrivener et al. 2001; Araujo et al. 2004; van den Biggelaar et al. 2000, 2004; Endara et al. 2010; Flohr et al. 2010), multiple sclerosis (Correale and Farez 2007), and inflammatory bowel disease (Kabeerdoss et al. 2011) are less common or much milder in people often exposed to helminths. In patients with multiple sclerosis, removal of naturally acquired helminths results in worsening neurologic symptoms (Correale and Farez 2011). In a case report of ulcerative colitis, treatment of pinworm resulted in a flare of the inflammatory bowel disease (Büning et al. 2008). These observations suggest that exposure to helminths does protect individuals from developing immune-mediated illness.

7.3   General Mechanisms of Action

Biological organisms need to respond to challenges quickly in order to remain alive. To accomplish this task, most systems within an organism are maintained in a state of balance so they can move in either a positive or a negative direction depending on need. After meeting a challenge, the system moves back to a baseline or neutral position. This state of dynamic balance is called homeostasis. The immune system demonstrates active homeostatic balance. Many different cell types make up the immune system. Each cell type has members that pull toward a maximal response and members that push to reduce (regulate) that response. The interactions of these varied components result in a homeostatic network that responds to invading bacteria and viruses, while ignoring normal cells and helpful commensal organisms. Autoimmune and immune-mediated inflammatory diseases occur when this immune regulatory network becomes dysfunctional. Helminths need to live in their host, while causing as little damage as possible. Therefore, they have evolved mechanisms that augment immune regulatory responses. In our laboratories, we are studying how helminths suppress inflammatory responses and strengthen immune regulatory networks.

A major immune regulatory network is composed of "helper" T lymphocyte cells. These cells recognize and respond to proteins displayed on antigen-presenting cells (APC; dendritic cells, macrophages, and B lymphocytes). To sense proteins displayed by APC, T helper cells use a surface component named CD4 and are often called CD4+ T cells. There are several different subtypes of T helper cells known as Th1, Th2, Th17, Th3, Tr1, and Treg cells (Fig. 7.1).

These subtypes are identified by the hormone-like molecules (called cytokines or interleukins (IL)) that they produce. Th1 cells make IFNγ, which instructs macrophages to make toxic substances used to kill intracellular bacteria and viruses. Th2 cells make IL-4, which helps the growth of B cells that make antibodies, while Th17 cells make IL-17, which promotes growth of neutrophils to kill extracellular bacteria. Th3 cells make TGF-β that helps suppress activity by Th1 and Th2 cells. Tr1 cells make IL-10 that suppresses activity by Th1, Th2, and Th17 cells. Tregs suppress growth of Th1, Th2, and Th17 cells. How the immune system responds to a challenge or to normal cells depends on the mix of Th1/Th2/Th17/Th3/Tr1/Treg lymphocytes. Animals exposed to helminths have reduced Th1 and Th17 cell activity but augmented Th2, Th3, Tr1, and Treg activity as compared to helminth naïve animals (Elliott and Weinstock 2012). Similar patterns are seen in people from regions where helminth infection is commonplace (highly endemic) (Elliott et al. 2000; Borkow et al. 2000; Sabin et al. 1996; Bentwich et al. 1996). Exposure to helminths also increases regulatory subtypes of CD8+ T cells, dendritic cells, macrophages, and B cells (Elliott and Weinstock 2012).

There is strong evidence that helminth exposure results in changes to the immune system, which decreases risk for developing immune disorders. Therefore, exposure to helminths may prevent onset of the immune-mediated diseases that have become common in developed countries. In addition, previously established colitis improves

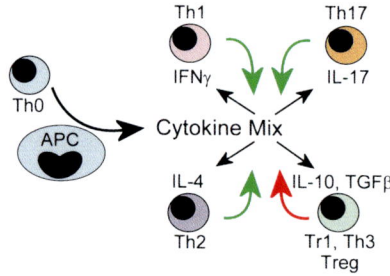

Fig. 7.1 T helper cell subtypes. Naïve T cells (Th0) respond to proteins displayed by antigen-presenting cells (*APC*) can develop into different subtypes (Th1, Th2, Th17, Tr1, Th3, Treg) depending on the mix of different cytokines/interleukins they are exposed to. Th1 cells make cytokines that promote differentiation of more Th1 cells. Th2 cells make cytokines that promote differentiation of more Th2 cells. Th17 cells make cytokines that promote differentiation of more Th17 cells. Tr1, Th3, and Treg cells suppress differentiation and function of Th1, Th2, and Th17 cells to help the immune system stay in homeostatic balance

in mice treated with helminths (Elliott et al. 2004). This suggests that helminths may be used to treat patients that have already developed autoimmune and immune-mediated inflammatory illnesses.

Investigators are working to identify how helminths are able to alter their host immune systems. For example, helminths make products that can alter how the cells of the immune system react (McInnes et al. 2003; Schnoeller et al. 2008). There are many different types of helminths and each type likely has a unique set of products that shape immunity. Helminths are divided into three major groups: round worms (nematodes), flukes (trematodes), and tapeworms (cestodes). The ability to live inside and at the expense of another organism (parasitism), developed independently in each of these groups. Furthermore, many of these parasites live in different parts of the body (niche) or use different ways of gaining access to their host. Therefore, the exact methods used by parasitic worms to alter host immune systems likely varies between these groups and probably even varies by parasite species within a group. In addition, helminths are very complex multicellular organisms that have influenced human genetic variation (Goncalves et al. 2003; Fumagalli et al. 2009). We expect that a given helminth utilizes several mechanisms to evade and suppress host immunity.

7.4   Helminths Currently Being Studied to Treat Human Diseases

Left untreated, autoimmune and immune-mediated inflammatory diseases cause significant damage to patient's tissues. However, many of the medications we use to combat these diseases have significant risks. Commonly used medications include: glucocorticoids (e.g. prednisone) that can cause infections, diabetes, osteoporosis, cataracts, muscle weakness, avascular necrosis of the large joints, and growth arrest; methotrexate that can cause low blood counts, lung scaring, liver inflammation, mouth sores, diarrhea, birth defects, and abortion; purine anti-metabolites (e.g. azathioprine, mercaptopurine) that can cause low blood counts, hair loss, pancreatic inflammation, liver inflammation and lymphoma; mitoxantrone which can cause hair loss, nausea, vomiting and severe heart damage; TNFα blockers (e.g. infliximab, adalimumab, certolizumab, etanercept), which can cause severe viral, bacterial or fungal infection, liver failure, heart failure, demyelinating disease (similar to multiple sclerosis), and lupus-like disease; and natalizumab, which is rarely associated with an irreversible brain infection called progressive multifocal leukoencephalopathy (PML). When medications fail, we are often left with attempts at surgical repair of damage caused by immune mediated disease. With current medical therapy, about 50 % of patients with Crohn's disease will require at least one bowel resection. A similar percentage of patients with ulcerative colitis require colon removal either due to ongoing inflammation or colon cancer associated with longstanding disease. Patients with rheumatoid arthritis often require joint replacement when feasible. Each of these surgeries carries its own set of risks, therefore, alternative approaches to treat these diseases are needed.

Experiments using animal models and studies of people from endemic areas suggest that helminth exposure can prevent and treat immune-mediated disease. There are many species of helminths that can infect people. One challenge to using helminths therapeutically is to identify species with little risk for causing disease on their own. Although most infections cause no symptoms, some helminth species can cause significant problems to their hosts over time. Examples of these helminths include *Wuchereria bancrofti* and *Brugia malayi*, which cause elephantiasis from severe swelling of the lower extremities. While the biologic behavior of most helminths does not change in patients with altered immune systems, some are able to multiply by autoinfection in immune compromised patients. The usually asymptomatic helminth *Strongyloides stercoralis* can exponentially multiply in patients placed on glucocorticoid medications causing a potentially fatal disease named "fulminant strongyloidiasis". Some helminth larvae have the capacity to migrate to parts of the body where they can cause damage. For example, instead of laying eggs to reproduce, adult *Onchocerca volvulus* release microfilaria that migrate to the skin where they are ingested by biting flies that then transmit infection to other people. *O. volvulus* microfilariae carry a symbiotic bacteria called *Wolbachia pipientis*. When microfilariae die in the skin before they are ingested by a fly, they release bacterial products that causes a mild itch. Microfilariae also can inadvertently migrate into the eye. If they die while in the eye, the release of bacterial products prompts an inflammatory response that can cause irreversible blindness. Helminths that have the ability to cause significant disease are poor candidates for therapeutic use. Yet, there are many helminths that produce few symptoms and little or no disease.

Compared with current medications and surgeries, the clinical use of selected helminths has very low risk. We have performed early clinical studies using *Trichuris suis* or *Necator americanus* to treat patients with active immunologic disease. These helminths are discussed individually below.

7.4.1   Trichuris suis: *Porcine Whipworm*

7.4.1.1   Choice and Biology of *Trichuris suis*

Trichuris suis is a helminth that is closely related to the human whipworm, *Trichuris trichiura* (Cutillas et al. 2009). People can briefly become colonized with *T. suis* (Beer 1976) though there are no reports of illness due to occupational exposure in farmers. *Trichuris* species have a simple lifecycle. The helminth is acquired by swallowing microscopic eggs (ova) that contain an infective worm larva. The ova hatch in the intestine to release larvae that attach to the lining of the intestine and mature into adult worms. Adult worms mate and lay immature ova that are deposited in the stool and are passed with a bowel movement. Once out of the body, the ova slowly mature (embryonate) in the soil over several weeks. Mature ova can survive for years in the soil waiting to be swallowed by a new host.

  T. suis provides some unique advantages for use as a therapeutic agent. Helminths need to live in hosts to reproduce. Although closely related to *T. trichiura*, *T. suis* can be grown in specially bred pigs. This permits production of the significant numbers of mature ova needed to treat many patients with immune-mediated disease. The ova are easy to isolate in pure form from infected pigs. The ova are very stable permitting storage and distribution through normal pharmaceutical channels. *T. suis* worms do not multiply in their host permitting accurate dosing of exposure. Because the ova take weeks to mature in the environment, people cannot easily transmit the *T. suis* to close contacts. In humans, *T. suis* colonization is transient. Ova can be taken orally every 2 or 3 weeks to maintain exposure. These characteristics make *T. suis* an unusually robust choice for clinical use (Fig. 7.2).

7.4.1.2   Potential Indications, Contraindications and Side Effects

Many if not most autoimmune and immune-mediated inflammatory diseases could improve with helminthic therapy. These diseases result from inadequate or dysfunctional regulation of the immune response. Helminths induce immune regulatory responses that restrain immune-medicated disease. Illnesses currently targeted for study of "helminthotherapy" include: Crohn's disease, ulcerative colitis, multiple sclerosis, Type 1 diabetes, rheumatoid arthritis, psoriasis, food allergy, and autism.

  There are no human diseases attributed to *T. suis*. In pigs, *T. suis* increases infections caused by known bacterial pathogens like *Campylobacter jejuni* (Mansfield et al. 2003). Therefore, *T. suis* should not be given to people with known pathogenic intestinal bacterial infections (eg. *Clostridium difficile*, *E. coli* O157, *Klebsiella oxytoca*, *C. jejuni*, *Shigella* sp., *Salmonella* sp.). It is currently standard of care to test for these infections before starting medications in patients with intestinal symptoms. Helminths are resistant to most antibiotics and are probably not affected by generally used medications. There are current plans to test pigs colonized with *T. suis* with common medications to determine if they influence helminth viability.

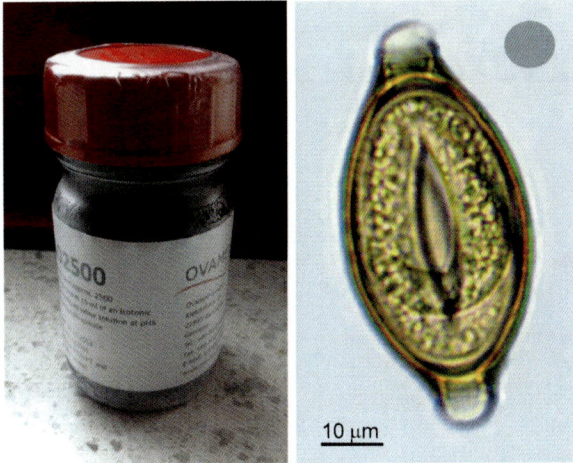

Fig. 7.2 *Trichuris suis* ova (TSO). (*Left*) Bottle containing 2,500 microscopic embryonated ova in 30 ml of fluid. (*Right*) Close up micrograph of an embryonated egg. Each *Trichuris suis* ova is about 25–30 μm wide and 65–75 μm long. For comparison, the *gray circle* is the size one human red blood cell

Helminth infections usually cause no symptoms. No side effects were noted in initial studies of *T. suis* in Crohn's disease and ulcerative colitis. Studies in allergic rhinitis and multiple sclerosis suggest that some patients may experience soft stools or diarrhea that resolves without stopping continued dosing with *T. suis* ova (Bager et al. 2010; Fleming et al. 2011). However, these mild, transient side effects are not apparent in patients with inflammatory bowel disease.

7.4.1.3 Diseases Studied

The earliest clinical studies of *T. suis* were "open label" meaning that the patients and health care team knew who was receiving helminth eggs. We performed a small trial of four patients with Crohn's disease and three patients with ulcerative colitis. Each patient received one dose of 2,500 embryonated *T. suis* ova. Each patient had improvement in his/her disease symptoms (Summers et al. 2003). We then did a second open label study that tested repeated dosing with embryonated *T. suis* ova. Approximately 2,500 *T. suis* ova were given to 29 patients with Crohn's disease every 3 weeks for 24 weeks (Summers et al. 2005a). After this period, 79 % responded with a drop in their disease activity scores, and 72 % had resolution of their Crohn's symptoms (Summers et al. 2005a). There were no side effects or complications attributable to *T. suis* colonization. These safety and early efficacy trials suggested that patients with inflammatory bowel disease may improve with exposure to *T. suis* and that this helminth is safe to use in people with intestinal inflammation.

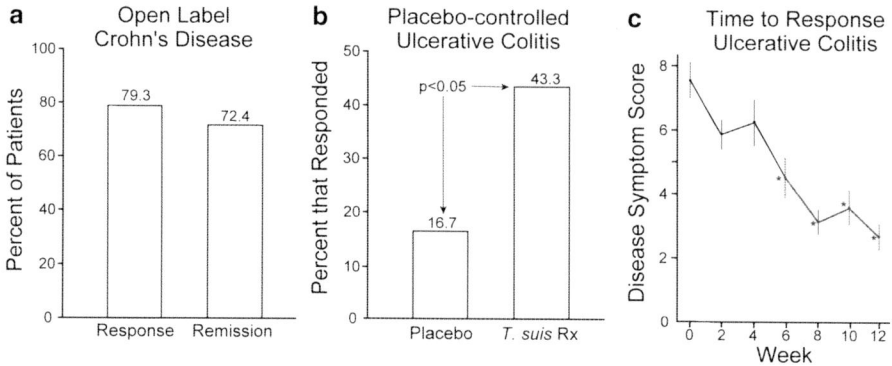

Fig. 7.3 Results of clinical studies in IBD. (**a**) Open label study of *T. suis* every 3 weeks for 24 weeks in Crohn's disease. Response is a drop in Crohn's disease activity index (CDAI) of 100 or more points. Remission is a CDAI below 150 points at end of study period. (**b**) Double blind placebo controlled study of *T. suis* every 2 weeks for 12 weeks in ulcerative colitis. Response is a drop in the ulcerative colitis disease activity index of 4 or more points at end of study period. (**c**) Time to response. Disease activity (Simple index) was measured serially in ulcerative colitis patients receiving *T. suis*. Significant (p<0.05) sustained drop in disease activity occurred about 6–8 weeks after starting treatment. Please see text for citations

Open-label studies show if a treatment is safe. However, to show that a treatment is effective, randomized double-blind placebo-controlled studies are used. In these types of studies, patients are randomly assigned to groups that are given either the treatment being studied or an inactive look-a-like (i.e. placebo control). Neither the patient nor members of the health care team know, which group a patient is in until the study is completed (i.e. double blind). We performed randomized double-blind placebo-controlled trial of *T. suis* in 54 patients with ulcerative colitis (Summers et al. 2005b). Patients received either 2,500 *T. suis* ova or placebo every 2 weeks for 12 weeks. Many of the patients repeatedly exposed to *T. suis* had significant improvement in disease activity compared to those given placebo (43 % vs. 17 %, p<0.05, Fig. 7.3b). The study also included a crossover phase, according to which, after the initial 12-week study, patients with continued symptoms that were originally on placebo were switched to *T. suis* and those on *T. suis* were crossed-over to placebo. The patients and team members remained "blinded" to the treatment groups. In the crossover phase, 56 % of the patients given *T. suis* responded compared to 13 % of patients given placebo (p=0.02) (Elliott et al. 2005). It took about 6–8 weeks for ulcerative colitis patients to respond to *T. suis* treatment (Fig. 7.3c). This study suggested that exposure to *T. suis* is safe and effective for reducing symptoms and disease activity in patients with ulcerative colitis.

A randomized double-blind placebo-controlled study testing the effect of *T. suis* exposure on grass pollen-induced allergic rhinitis was conducted in Denmark (Bager et al. 2010). During allergy season, 96 rhinitis patients received placebo or 2,500 *T. suis* ova (TSO) every 3 weeks for a total of up to eight doses. However, test subjects received between two and five (median 3) doses of TSO or placebo by the peak of allergy season. By the end of the trial, 24 % of the patients had not received the full eight doses. Patients that received *T. suis* eggs had increased eosinophil counts and titers of anti-*T. suis* antibody as compared to those in the placebo group. Although patients in the *T. suis* group used fewer medications, symptom scores and response to skin prick tests did not differ between the two groups; suggesting that exposure to *T. suis*, as administered in this study, may not help seasonal allergy symptoms. This study has been criticized for timing of exposure to *T. suis* in relation to the onset of allergy season and for assuming that the same dose that reduces symptoms in inflammatory bowel disease would be effective in rhinitis (Hepworth et al. 2010; Summers et al. 2010). Circulating allergen-specific IgE antibodies bind to mast cells causing these cells to release histamine and other mediators upon allergen exposure and triggering the symptoms of allergic rhinitis. Thus, it is reasonable to assume that *T. suis* therapy will require long-term administration well before the allergy season to allow the naturally occurring allergen-specific IgE levels to wane. But, this study did show that *T. suis* exposure influences some immunologic responses in people. Also the agent appeared safe though some patients receiving helminths reported more loose stools, upper abdominal discomfort and flatulence than did those who received placebo. The reported symptoms dissipated quickly. Further studies are required to determine if *T. suis* is effective in allergic rhinitis.

Helminthic therapy of multiple sclerosis is also being investigated. A series of publications from South America suggest that natural infection with helminths arrests the further development of brain lesions in patients with multiple sclerosis. This has led to independent clinical trials both in the United States and Germany testing *T. suis* in patients with this disease. An open label trial of repeated *T. suis* exposure in five patients with relapsing-remitting multiple sclerosis showed that helminth colonization resulted in less neurological symptoms and development of fewer CNS lesions as measured by magnetic resonance imaging (Fleming et al. 2011). Disease activity returned after stopping *T. suis* administration. These researchers found that exposure to helminths increased the peripheral blood eosinophil count, increased production of anti-*T. suis* antibodies, and increased the level of circulating IL-4 and IL-10 cytokines. It was found that repeated doses of *T. suis* were well tolerated but some patients had a brief period of low grade intestinal symptoms and loose stools. A group working in Germany performed an open-label study of *T. suis* on four patients with progressive multiple sclerosis (Benzel et al. 2011). Clinical symptoms stabilized in three of the four patients. Helminth exposure resulted in slight increase in eosinophil counts and suppression of pro-inflammatory cytokine (IFNγ) production by peripheral blood T cells. Similar to scientists in the USA, they also found that helminth exposure was well tolerated by patients with multiple sclerosis.

7.4.1.4 Case Report

A patient with unresponsive ulcerative colitis, whose symptoms did not improve with medications, went into remission after traveling to Thailand and purposefully acquiring *T. trichiura* (human whipworm) infection (Broadhurst et al. 2010).

However, the patient's symptoms returned when the helminths died off. The person then reacquired *T. trichiura* and went back into remission. It should be noted that human whipworm is closely related to porcine whipworm (*T. suis*).

7.4.1.5   Future Clinical Trials with *T. suis*

The pharmaceutical industry presently is conducting large multi-center randomized double blind trial testing this agent in patients with inflammatory bowel disease. There are trials underway in Crohn's disease in Europe and in the United States (ClinicalTrials.gov Identifier NCT01279577). A large multi-center trial for ulcerative colitis is slated for the United States. Also underway are, a single center investigator-initiated study of *T. suis* in peanut food allergy (NCT01070498), a single center investigator-initiated study of *T. suis* in autism (NCT01040221), and three independent single center investigator-initiated studies of *T. suis* in multiple sclerosis (NCT00645749, NCT01006941, NCT01413243). Many of these studies also plan to start evaluating how *T. suis* may alter human immune responses. For one study of patients with ulcerative colitis, that is the major focus of the investigation (NCT01433471).

Over the next few years, *T. suis* treatment will be tested for efficacy in many of the major autoimmune and immune-mediated inflammatory diseases like psoriasis and Type 1 diabetes. Alongside those clinical trials, translational studies in people and experiments in animal models of human disease will define the mechanisms by which helminths manipulate their host to suppress inflammation. Hopefully, we eventually will identify helminth-derived factors that reproduce at least some of these effects.

7.4.2   Necator americanus*: Human Hookworm*

7.4.2.1   Choice and Biology of Organism

Necator americanus, a hookworm, was chosen because of its propensity to re-infect immunologically primed individuals indicating the ability to moderate the immune system. This was further demonstrated in epidemiological studies, where hookworm infection was associated with a reduction in allergy, and supported by the identification of a number of immune-modulatory molecules in parasite secretions. This work has recently been comprehensively reviewed (Pritchard et al. 2012).

Our current understanding is that this hookworm species induces an allergic phenotype in the infected host soon after infection, a phenotype, which affords the host some protection. However, the phenotype would appear to be transient, implying counter immune regulation by the parasite.

Nevertheless, this parasite exhibits a dark side to its biology; *Necator* enters the body trans-cutaneously, causing a pruritus in the process. This is called "ground itch" in the tropics, and is a key feature of experimental infection during trial. Once in the skin, the worms migrate to the lungs, enter the air space, to be coughed up and swallowed. The potential for lung damage is real, hence the requirement to conduct short–term, low infection intensity safety trials. Once in the gut, the worms feed on mucosal tissue and blood, again a cause for concern for patients and physicians alike.

Consequently, the early human trials with this parasite were safety orientated, with small numbers of worms resident for a short period of time (12 weeks). The trials were designed to choose an asymptomatic dose that did not adversely affect lung or gut function.

Physicians are currently confident that doses of up to 25 larvae can be safely administered. This dose will induce a natural immunological phenotype, with what one would hope to be a parallel natural immune-regulatory response, conducive to disease suppression.

7.4.2.2   Diseases Studied

Following on from safety trials, *N. americanus* is currently being used in an attempt to moderate relapsing remitting multiple sclerosis. The immunology of this disease suggests that it would be amenable to moderation by a parasite, which promotes a T helper 2 phenotype. Patients will be infected with 25 larvae during remission, as indicated in the Worms for Immune Regulation of Multiple Sclerosis (WIRMS) clinical trial (NCT01470521) documentation, and the worms will remain in residence for the duration of the trial. Disease progression will be monitored by MRI, and patient immunology screened in parallel.

7.4.2.3   Future Research

Views on the direction of future research are outlined in a recent opinion piece (Pritchard 2012). Should TSO or *Necator* prove to be of benefit in any of the medical conditions described above, there will be a natural tendency towards experiments, which will help clinicians to understand the molecular and cellular mechanisms of immune-moderation. Our view is that parasite immunologists should also mine proven therapeutic parasites for non-proteinaceous and non-immunogenic immune suppressants. This view is based on the hypothesis that parasites need such molecules to evade host-immunity, i.e., to survive. Immunogenic parasite secretions that elicit targeted immune responses would not benefit the parasite for prolonged colonization (Pritchard et al. 2012). Any therapeutic effect of parasites in immunological disease could be the result of bystander effects of parasite-derived molecules on an over-active dysregulated immune system. On the other hand, helminths, which are complex animals, can certainly use more than one molecule to evade host immunity. It seems likely that the whole living organism, rather than a single molecular product, will remain most effective at controlling immune mediated disease.

References

Araujo MI, Hoppe BS, Medeiros M Jr, Carvalho EM (2004) *Schistosoma mansoni* infection modulates the immune response against allergic and auto-immune diseases. Mem Inst Oswaldo Cruz 99: 27–32

Bager P, Arnved J, Ronborg S et al (2010) *Trichuris suis* ova therapy for allergic rhinitis: a randomized, double-blind, placebo-controlled clinical trial. J Allergy Clin Immunol 125:123–130

Beer RJ (1976) The relationship between *Trichuris trichiura* (Linnaeus 1758) of man and *Trichuris suis* (Schrank 1788) of the pig. Res Vet Sci 20:47–54

Bentwich Z, Weisman Z, Moroz C, Bar-Yehuda S, Kalinkovich A (1996) Immune dysregulation in Ethiopian immigrants in Israel: relevance to helminth infections? Clin Exp Immunol 103:239–243

Benzel F, Erdur H, Kohler S, Frentsch M, Thiel A, Harms L, Wandinger KP, Rosche B (2011) Immune monitoring of *Trichuris suis* egg therapy in multiple sclerosis patients. J Helminthol 86(3):339–347

Borkow G, Leng Q, Weisman Z, Stein M, Galai N, Kalinkovich A, Bentwich Z (2000) Chronic immune activation associated with intestinal helminth infections results in impaired signal transduction and anergy. J Clin Invest 106(8):1053–1060

Broadhurst MJ, Leung JM, Kashyap V, McCune JM, Mahadevan U, McKerrow JH, Loke P (2010) IL-22+ CD4+ T cells are associated with therapeutic *Trichuris trichiura* infection in an ulcerative colitis patient. Sci Transl Med 2:60ra88

Büning J, Homann N, von Smolinski D, Borcherding F, Noack F, Stolte M, Kohl M, Lehnert H, Ludwig D (2008) Helminths as governors of inflammatory bowel disease. Gut 57(8):1182–1183

Cooke A, Tonks P, Jones FM, O'Shea H, Hutchings P, Fulford AJ, Dunne DW (1999) Infection with *Schistosoma mansoni* prevents insulin dependent diabetes mellitus in non-obese diabetic mice. Parasite Immunol 21(4):169–176

Correale J, Farez M (2007) Association between parasite infection and immune responses in multiple sclerosis. Ann Neurol 61:97–108

Correale J, Farez MF (2011) The impact of parasite infections on the course of multiple sclerosis. J Neuroimmunol 233(1–2):6–11

Cutillas C, Callejón R, de Rojas M, Tewes B, Ubeda JM, Ariza C, Guevara DC (2009) *Trichuris suis* and *Trichuris trichiura* are different nematode species. Acta Trop 111(3):299–307

Elliott DE, Weinstock JV (2012) Helminth-host immunological interactions: prevention and control of immune-mediated diseases. Ann N Y Acad Sci 1247:83–96

Elliott DE, Urban JF Jr, Argo CK, Weinstock JV (2000) Does the failure to acquire helminthic parasites predispose to Crohn's disease? FASEB J 14:1848–1855

Elliott DE, Li J, Blum A, Metwali A, Qadir K, Urban JF Jr, Weinstock JV (2003) Exposure to schistosome eggs protects mice from TNBS-induced colitis. Am J Physiol Gastrointest Liver Physiol 284(3):G385–G391

Elliott DE, Setiawan T, Metwali A, Blum A, Urban JF Jr, Weinstock JV (2004) *Heligmosomoides polygyrus* inhibits established colitis in IL-10-deficient mice. Eur J Immunol 34:2690–2698

Elliott DE, Summers RW, Weinstock JV (2005) Helminths and the modulation of mucosal inflammation. Curr Opin Gastroenterol 21(1):51–58

Endara P, Vaca M, Chico ME, Erazo S, Oviedo G, Quinzo I, Rodriguez A, Lovato R, Moncayo AL, Barreto ML, Rodrigues LC, Cooper PJ (2010) Long-term periodic anthelmintic treatments are associated with increased allergen skin reactivity. Clin Exp Allergy 40(11):1669–1677

Espinoza-Jimenez A, Rivera-Montoya I, Cardenas-Arreola R, Moran L, Terrazas LI (2010) *Taenia crassiceps* infection attenuates multiple low-dose streptozotocin-induced diabetes. J Biomed Biotechnol 2010:850541

Fleming JO, Isaak A, Lee JE, Luzzio CC, Carrithers MD, Cook TD, Field AS, Boland J, Fabry Z (2011) Probiotic helminth administration in relapsing-remitting multiple sclerosis: a phase 1 study. Mult Scler 17(6):743–754

Flohr C, Tuyen LN, Quinnell RJ, Lewis S, Minh TT, Campbell J, Simmons C, Telford G, Brown A, Hien TT, Farrar J, Williams H, Pritchard DI, Britton J (2010) Reduced helminth burden increases allergen skin sensitization but not clinical allergy: a randomized, double-blind, placebo-controlled trial in Vietnam. Clin Exp Allergy 40(1):131–142

Fumagalli M, Pozzoli U, Cagliani R, Comi GP, Riva S, Clerici M, Bresolin N, Sironi M (2009) Parasites represent a major selective force for interleukin genes and shape the genetic predisposition to autoimmune conditions. J Exp Med 206(6):1395–1408

Goncalves ML, Araujo A, Ferreira LF (2003) Human intestinal parasites in the past: new findings and a review. Mem Inst Oswaldo Cruz 98(Suppl 1):103–118

He Y, Li J, Zhuang W, Yin L, Chen C, Li J, Chi F, Bai Y, Chen XP (2010) The inhibitory effect against collagen-induced arthritis by *Schistosoma japonicum* infection is infection stage-dependent. BMC Immunol 11:28

Hepworth MR, Hamelmann E, Lucius R, Hartmann S (2010) Looking into the future of *Trichuris suis* therapy. J Allergy Clin Immunol 125(3):767–768

Kabeerdoss J, Pugazhendhi S, Subramanian V, Binder HJ, Ramakrishna BS (2011) Exposure to hookworms in patients with Crohn's disease: a case-control study. Aliment Pharmacol Ther 34:923–930

Khan WI, Blennerhasset PA, Varghese AK, Chowdhury SK, Omsted P, Deng Y, Collins SM (2002) Intestinal nematode infection ameliorates experimental colitis in mice. Infect Immun 70(11): 5931–5937

Kitagaki K, Businga TR, Racila D, Elliott DE, Weinstock JV, Kline JN (2006) Intestinal helminths protect in a murine model of asthma. J Immunol 177(3):1628–1635

La Flamme AC, Ruddenklau K, Backstrom BT (2003) Schistosomiasis decreases central nervous system inflammation and alters the progression of experimental autoimmune encephalomyelitis. Infect Immun 71:4996–5004

Liu Q, Sundar K, Mishra PK, Mousavi G, Liu Z, Gaydo A, Alem F, Lagunoff D, Bleich D, Gause WC (2009) Helminth infection can reduce insulitis and type 1 diabetes through CD25- and IL-10-independent mechanisms. Infect Immun 77(12):5347–5358

Mangan NE, van Rooijen N, McKenzie AN, Fallon PG (2006) Helminth-modified pulmonary immune response protects mice from allergen-induced airway hyperresponsiveness. J Immunol 176(1):138–147

Mansfield LS, Gauthier DT, Abner SR, Jones KM, Wilder SR, Urban JF (2003) Enhancement of disease and pathology by synergy of *Trichuris suis* and *Campylobacter jejuni* in the colon of immunologically naive swine. Am J Trop Med Hyg 68:70–80

McInnes IB, Leung BP, Harnett M, Gracie JA, Liew FY, Harnett W (2003) A novel therapeutic approach targeting articular inflammation using the filarial nematode-derived phosphorylcholine-containing glycoprotein ES-62. J Immunol 171:2127–2133

Osada Y, Shimizu S, Kumagai T, Yamada S, Kanazawa T (2009) *Schistosoma mansoni* infection reduces severity of collagen-induced arthritis via down-regulation of pro-inflammatory mediators. Int J Parasitol 39:457–464

Pritchard DI (2012) Worm therapy: how would you like your medicine? Int J Parasitol Drugs Drug Resist 2:106–108

Pritchard DI, Blount DG, Schmid-Grendelmeier P, Till SJ (2012) Parasitic worm therapy for allergy: is this incongruous or avant-garde medicine? Clin Exp Allergy 42:505–512

Reardon C, Sanchez A, Hogaboam CM, McKay DM (2001) Tapeworm infection reduces epithelial ion transport abnormalities in murine dextran sulfate sodium-induced colitis. Infect Immun 69:4417–4423

Sabin EA, Araujo MI, Carvalho EM, Pearce EJ (1996) Impairment of tetanus toxoid-specific Th1-like immune responses in humans infected with *Schistosoma mansoni*. J Infect Dis 173:269–272

Salinas-Carmona MC, de la Cruz-Galicia G, Pérez-Rivera I, Solís-Soto JM, Segoviano-Ramirez JC, Vázquez AV, Garza MA (2009) Spontaneous arthritis in MRL/lpr mice is aggravated by *Staphylococcus aureus* and ameliorated by *Nippostrongylus brasiliensis* infections. Autoimmunity 42(1):25–32

Saunders KA, Raine T, Cooke A, Lawrence CE (2007) Inhibition of autoimmune type 1 diabetes by gastrointestinal helminth infection. Infect Immun 75:397–407

Schnoeller C, Rausch S, Pillai S, Avagyan A, Wittig BM, Loddenkemper C, Hamann A, Hamelmann E, Lucius R, Hartmann S (2008) A helminth immunomodulator reduces allergic

and inflammatory responses by induction of IL-10-producing macrophages. J Immunol 180:
4265–4272

Scrivener S, Yemaneberhan H, Zebenigus M, Tilahun D, Girma S, Ali S, McElroy P, Custovic A,
Woodcock A, Pritchard D, Venn A, Britton J (2001) Independent effects of intestinal parasite
infection and domestic allergen exposure on risk of wheeze in Ethiopia: a nested case-control
study. Lancet 358:1493–1499

Sewell D, Qing Z, Reinke E, Elliot D, Weinstock J, Sandor M, Fabry Z (2003) Immunomodulation
of experimental autoimmune encephalomyelitis by helminth ova immunization. Int Immunol
15:59–69

Summers RW, Elliott DE, Qadir K, Urban JFJ, Thompson R, Weinstock JV (2003) *Trichuris suis*
seems to be safe and possibly effective in the treatment of inflammatory bowel disease. Am J
Gastroenterol 98:2034–2041

Summers RW, Elliott DE, Urban JF Jr, Thompson R, Weinstock JV (2005a) *Trichuris suis* therapy
in Crohn's disease. Gut 54:87–90

Summers RW, Elliott DE, Urban JF Jr, Thompson RA, Weinstock JV (2005b) *Trichuris suis* therapy for active ulcerative colitis: a randomized controlled trial. Gastroenterology 128:825–832

Summers RW, Elliott DE, Weinstock JV (2010) *Trichuris suis* might be effective in treating allergic rhinitis. J Allergy Clin Immunol 125:766–767

van den Biggelaar AH, van Ree R, Rodrigues LC, Lell B, Deelder AM, Kremsner PG, Yazdanbakhsh
M (2000) Decreased atopy in children infected with *Schistosoma haematobium*: a role for
parasite-induced interleukin-10. Lancet 356:1723–1727

van den Biggelaar AH, Rodrigues LC, van Ree R, van der Zee JS, Hoeksma-Kruize YC, Souverijn
JH, Missinou MA, Borrmann S, Kremsner PG, Yazdanbakhsh M (2004) Long-term treatment
of intestinal helminths increases mite skin-test reactivity in Gabonese schoolchildren. J Infect
Dis 189(5):892–900

Wilson MS, Taylor MD, Balic A, Finney CA, Lamb JR, Maizels RM (2002) Suppression of allergic
airway inflammation by helminth-induced regulatory T cells. J Exp Med 202(9):1199–1212

Zaccone P, Fehérvári Z, Jones FM, Sidobre S, Kronenberg M, Dunne DW, Cooke A (2003)
Schistosoma mansoni antigens modulate the activity of the innate immune response and prevent
onset of type 1 diabetes. Eur J Immunol 33:1439–1449

Chapter 8
Phage Therapy

Elizabeth M. Kutter, Guram Gvasalia, Zemphira Alavidze, and Erin Brewster

8.1 Introduction

Phage therapy is the use of bacteriophages – viruses that can only infect bacteria – to treat bacterial infections. In some parts of the world, phages have been used therapeutically since the 1930s.

Phage therapy was first developed at the Pasteur Institute in Paris early in the twentieth century and soon spread through Europe, the US, the Soviet Union and other parts of the world, but with mixed success. Since the advent of chemical antibiotics in the 1940s, it has been largely ignored in the West, while still being used to varying degrees in some countries, with major claims of success. Today, however, the resurgence of bacteria that are resistant to most or all available antibiotics is precipitating a major health crisis, and interest is growing in the potential use of phage to complement antibiotics as a way to fight infection. This chapter has been written to put phage therapy into historical and ecological perspective and to briefly explore the very interesting early research in France, the US and Eastern Europe as

E.M. Kutter, Ph.D. (✉)
PhageBiotics and the Evergreen State College, Olympia, WA, USA
e-mail: kutterb@evergreen.edu

G. Gvasalia, M.D.
Department of General Surgery, Tbilisi State Medical University,
33 Vaja-Pshavela Avenue, Tbilisi 0160, Georgia
e-mail: guramgvasalia@yahoo.com

Z. Alavidze
LTD Bacteriophage Production Facility, Gotua Str 3, Tbilisi 0160, Georgia
e-mail: zemal@caucasus.net

E. Brewster
Research Associate, Bacteriophage Lab, Evergreen State College,
Olympia, WA 98505, USA
e-mail: brewster.erin@gmail.com

M. Grassberger et al. (eds.), *Biotherapy - History, Principles and Practice:*
A Practical Guide to the Diagnosis and Treatment of Disease using Living Organisms,
DOI 10.1007/978-94-007-6585-6_8, © Springer Science+Business Media Dordrecht 2013

well as growing recent studies worldwide. Here, we discuss the nature of phage and the mechanisms of phage infection of bacteria, examine an existing body of research indicating the potential for a widespread application of phage use as treatment and prevention of various pathologies and present details of specific clinical phage applications. Phages specific for virtually every well-studied bacterial species have now been isolated and characterized. It has become clear that phages play the major role in maintaining the bacterial balance throughout nature; for example, at any given time a substantial fraction of the bacteria in the oceans are infected with replicating phage, which are thus key in the cycling of nutrients as well as in preventing overgrowth by any particular bacterial species. The reported results of early phage therapy work worldwide and of more recent French and Eastern European therapeutic phage applications are very encouraging in terms of such factors as lack of side effects and interactions with other medications as well as of efficacy, particularly as our understanding of phage biology and ecology has grown enormously. However, since most clinical application reports involved individualized applications to infections recalcitrant to all other available treatments rather than the double-blind clinical trials so prized by Western medicine, they have often been disregarded.

There have been encouraging recent developments, many of them discussed here. In 2006, the US FDA and the EU both approved phage preparations targeting *Listeria monocytogenes* on ready-to-eat foods (Lang 2006). Currently, phage therapy is part of standard medical practice in the Republic of Georgia and is fairly readily available in Russia, Poland and other eastern European countries. It is occasionally applied elsewhere, such as in Australia and France, on a compassionate-use basis. In the American states of Washington and Oregon, licensed naturopathic physicians have the specifically expressed right to use natural products that have been approved and long used in other countries; under this provision, a few are beginning to use phages from Tbilisi, particularly the commercial Eliava Institute product *Intestiphage* for irritable bowel syndrome and other intractable gastrointestinal problems, on a special trial basis. Most excitingly, the Nestlé Corporation is currently carrying out an extensive project in Dhaka, Bangladesh, to study the safety and efficacy of phage therapy in treating *Escherichia coli*-induced diarrhea in children (clinical trials.gov; Brüssow 2012). In parallel tracks, two phage preparations – a novel cocktail of their own well-characterized T4-like phages isolated from infant Bangladeshi patients and a commercially available Russian anti-*E. coli* phage cocktail (Microgen) – are being added to the current standard oral rehydration solution and are being controlled with a double-blind placebo. This work, discussed in detail in Sect. 8.6.4, builds on very extensive genomic and mouse-gut studies and is the first phage therapy study to demonstrate all of the key elements of large-scale modern clinical trials. Bruessow (2012) and Pirnay et al. (2012) make clear the vast challenges for companies trying to run such phage therapy trials within the scope of the Western pharmaceutical paradigm.

A wide variety of resources on the subject are now available, both in fairly general terms (Häusler 2003, 2006; Kutter 2008, 2009; Sulakvelidze et al. 2001; Summers 1999, 2001) and with substantially more technical detail (Kutter et al. 2010, 2013; Abedon et al. 2011; Alavidze et al. 2007; Brüssow 2007; Górski et al. 2006, 2009;

Kutter and Sulakvelidze 2005; Chanishvili 2012; Górski et al. 2012; Merabishvili et al. 2009; Miedzybrodzki et al. 2012). Phage therapy discussions now play a substantial role in the biennial Evergreen International Phage Meetings. The topic is also a key component of the new biennial Viruses of Microbes meetings, the first of which were held at the Pasteur Institute, Paris in 2010 and in the Military Academy in Brussels in 2012, with the latter drawing substantial governmental, medical and military as well as scientific and commercial interest. In light of the increasingly desperate problems with antibiotic resistance, it is to be strongly hoped that broad public-private-academic collaborations continue to develop to make phage therapy available for external applications much more broadly, while regulators and researchers deal with the challenges to getting funding for full-scale clinical trials of more invasive approaches.

8.1.1 Basic Bacteriophage Biology and Ecology

Viruses are like spaceships that are able to transfer their genomes from their cell of origin to a new cell where they can reproduce. In the case of bacteriophages, the targets are bacterial cells. Each phage is specific for certain strains of bacteria; some hit only a relatively few strains of a single species, while others can infect much broader ranges of strains, in some cases even strains from more than one related species. No bacteriophages can multiply in the cells of eukaryotes, and none are known which can infect both Gram-negative and Gram-positive bacteria.

Every virus consists of a protein shell into which is packed a long strand of genetic material (DNA or RNA) containing the genetic information that determines all of the properties of that virus (Fig. 8.1). Unlike any of the viruses that infect human, animal, plant or yeast cells, most cultured bacteriophages have tails, the tips of which have the ability to bind to one or more of the specific molecules on the surface of their target bacteria. The long phage DNA molecule (Fig. 8.1c) then travels rapidly through the tail and on into the host cell, where it directs the production of progeny phage, providing both the template for the new phage DNA and the information for transforming the host cell into a factory for making that particular phage. Using the electron microscope (EM), each phage family can be seen to have its own specific shape and range of sizes. With recent advances in DNA sequencing techniques, our understanding of various phages is exploding in powerful and important ways; the genomes of several thousand, infecting a wide variety of organisms, have now been sequenced, revealing a remarkable variety of types and properties that can be very important when considering potential therapeutic applications. In general appearance, 95 % of the studied phages (and all of those used therapeutically) belong to one or another of three general tailed morphotypes: the stubby-tailed podoviridae, the siphoviridae with long, often flexible tails; and the myoviridae, with tails composed of an inner tube and an outer contractile sheath attached to a complex baseplate (Fig. 8.1). All three of these morphotypes are clearly very ancient, and groups of each type can be found infecting

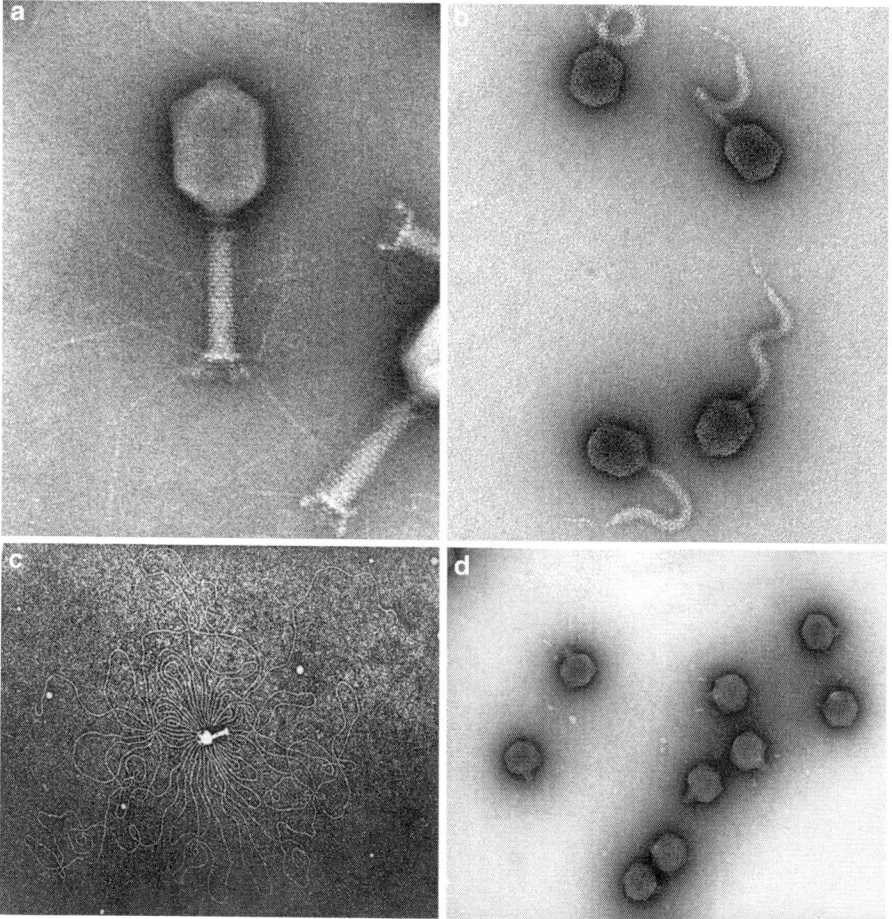

Fig. 8.1 (**a**) Electron micrographs of myovirus bacteriophage T4; (**b**) of siphovirus coliphage gamma; (**c**) of osmotically shocked T4 phage showing the entire linear molecule of its DNA as released from the central phage particle and (**d**) of podovirus coliphage T3. These micrographs were a parting gift from Michael Wurtz to E. K. and the phage community on his retirement from the Biozentrum, Universität Basel, Switzerland

Gram-negative bacteria, Gram-positive bacteria and even some Archae, but members of each phage genus generally only infect members of relatively closely related species of bacteria.

Understanding of the interactions between bacteria and phages during this lytic infection process was greatly aided by one-step growth curve experiments. These demonstrated an eclipse period, during which the DNA began replicating and no free phage could be found in the cell; a latent period of accumulation of intracellular phages; and a precisely-timed lysis process that released the phage to go in search of new hosts. This phage infection cycle is illustrated in Fig. 8.2 for coliphage T4, which does a particularly effective job of shutting off all host functions. With T4, about 300 progeny phage are produced in just half an hour within each bacterial cell

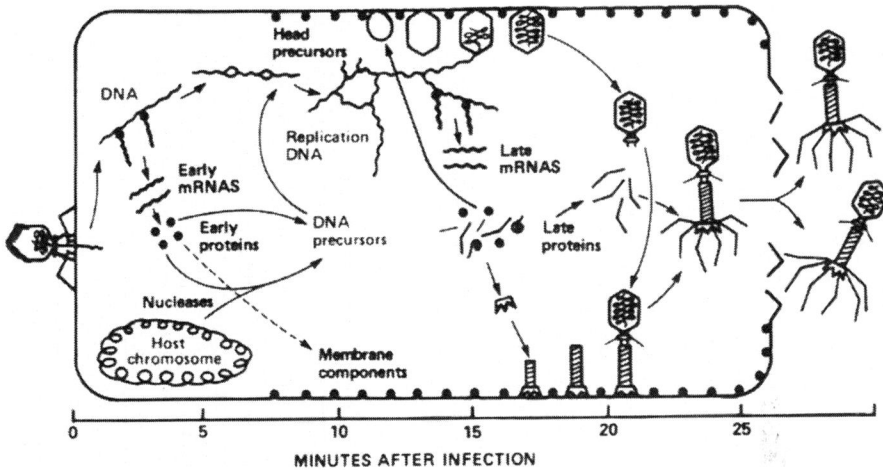

Fig. 8.2 Bacteriophage intracellular growth cycle. Noteworthy features: nucleolytic action on the host chromosome together with phage nucleotide-synthesizing enzymes furnish the nucleotides for phage DNA synthesis; replicating phage DNA is a concatemer of multiple DNA molecules; a number of phage-coded proteins become associated with the host membrane; maturation of the phage head and tail occur at membrane sites; specifically-timed phage lysozyme action makes holes in the peptidoglycan layer, releasing phage to carry on the infection cycle

under optimum conditions. Relatives of T4 are very widespread in nature, having developed a variety of sophisticated tricks for making their hosts into particularly efficient phage factories. For example, a majority of the phages infecting cyanobacteria in the oceans are members of this subfamily; some of those actually carry genes for certain somewhat unstable elements of the cyanobacterial photosynthetic system, enabling them to both turn off synthesis of all host proteins and be sure that the host energy production will be functional throughout the 24-h infection cycle of these phages. T4-related phages infecting enteric bacteria are major components of commercial therapeutic cocktails targeting gut pathogens, such as "Intestiphage".

8.1.1.1 The Challenges of Temperate Phages

Through the first half of the twentieth century, arguments raged over whether the lytic principle termed *bacteriophage* required regular reinfection by an external agent – a virus – as claimed by discoverer Felix d'Herelle, whose early work is discussed below, or simply reflected an inherent property of the specific bacteria. During the 1930s and 1940s, it gradually became clear that in some senses both were true – that there were in fact two quite fundamentally different groups of phages – but both are viral in nature. "Professionally lytic" or "obligatorily lytic" phages have only one option: infect from outside, reprogram the host cell, and release a burst of phage through lysing the cell after a fixed short period of time.

The so-called "temperate phages" like bacteriophage λ (lambda), on the other hand, also have another option. First described by Esther Lederberg in 1950 when she isolated λ from *E. coli* K12, most temperate phages are able to integrate their DNA into the host DNA, much as HIV integrates the DNA copy of its RNA, and to remain stably integrated for many generations as "prophages". Here, they are replicating in the protected environment of the bacterium itself while a phage-encoded repressor protein keeps most of their own genes in a dormant state. Occasionally one of these so-called "lysogenic" bacteria will suddenly stop making the repressor, switch into the normal lytic cycle for that phage, and lyse after a fixed short time to release them; this can happen either spontaneously or in response to some external trigger such as mitomycin, UV light or a particular environmental change.

The genetic regulatory circuitry of bacteriophage λ was the first complex regulatory circuit of any life form to be analyzed in molecular detail and has been very extensively studied (Friedman and Court 2001; Campbell 2006; Little 2006). Some bacteria, such as spirochetes and bacteria living in extreme environments, are infected almost exclusively by temperate rather than professionally lytic phages, while others, like *E. coli* and *Pseudomonas*, are targeted by a wide range of both types of phages. For several reasons, such temperate phages are not appropriate for therapeutic applications. They generally make the host resistant to related phages, blocking treatment efficacy, and also may carry genes that actually increase the pathogenicity of the host; very specific prophages are a key factor in such diseases as cholera and diptheria (Boyd 2005; Waldor et al. 2005).

There are both obligatorily lytic and temperate families of phages that have each of the three basic morphotypes, so other kinds of data are also needed to tell whether a newly isolated phage is temperate or professionally lytic. The sequence of the phage genome is especially useful here.

8.1.1.2   Phage Isolation

Every kind of bacterium has its own phages, which can generally be isolated wherever that bacterium grows – from sewage, feces, soil, even ocean depths and hot springs – but finding phages suitable for therapeutic applications is much easier for some species than for others; this relates significantly to our ability to grow the host in the laboratory. The surface of each bacterial cell is made up of complex arrays of protein, carbohydrate, and lipopolysaccharide molecules. These molecules are involved in forming pores, in selective uptake of needed nutrients, in motility, in protecting the cell surface, and in binding of the bacteria to particular surfaces, and are characteristic for each strain of bacteria. Each such molecule can act as a receptor for particular phages. Development of resistance to a particular phage most often reflects mutational alteration or loss of that phage's specific receptor. This loss sometimes has negative effects on the bacterium, such as making such mutants less virulent, and does not confer protection against the many other phages that use different receptors.

Phage isolation is quite straightforward for phages targeting many of the best-studied pathogens. The process starts with putting together a substantial relevant bacterial strain collection to use in isolating the phages and in determining the host

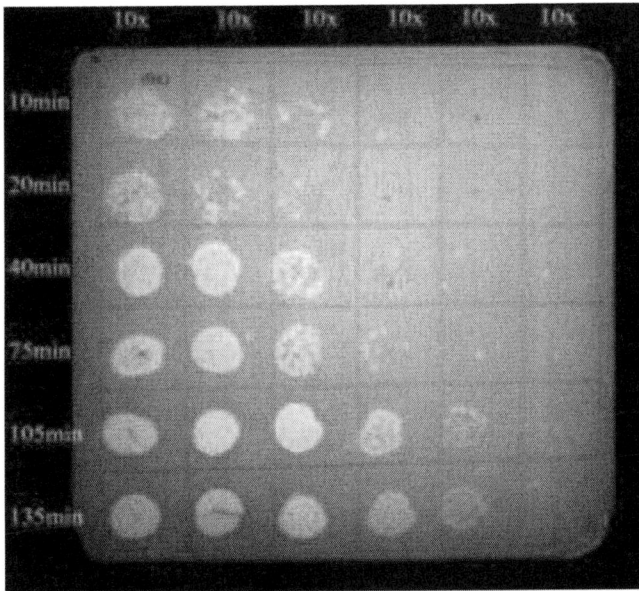

Fig. 8.3 Phage titering. Each row presents 10 μl samples from a series of tenfold dilutions of a given sample – the number of individual spots in the first square where these "plaques" are countable, multiplied times the number of dilutions to that point times 100, gives the concentration in phage per ml. In this case, a series of time points during the infection of *E. coli* by bacteriophage T4 is shown, where the samples have been treated with chloroform to release any fully-formed intracellular phage. One can see the decrease in free phage initially as most of them inject their DNA into host cells, but with many still left unadsorbed, and then the increase in phage numbers as the new phage particles are completed and released

range for each phage that is isolated. Then a relevant environmental sample, such as feces or clarified sewage, is added to an appropriate nutrient-fortified solution, inoculated with several of the targeted bacterial strains, and incubated over night. A series of 10-fold dilutions is prepared in phage buffer and 10 μl of each dilution is spotted on a nutrient-agar plate spread with a single strain of the targeted bacterium (Fig. 8.3). The next day, a dense bacterial lawn is seen, in successful cases with cleared spots and, at some dilution, dotted with individual small clear *plaques*.

Each such plaque contains about a billion phages, all of them progeny of a single initial phage. Individual plaques are plucked, diluted, plated and replucked multiple times to be sure that a single kind of phage is present. Each such isolate is then used to infect a sensitive, preferably non-pathogenic host strain to give a homogeneous stock of that phage, whose host range on the bacterial strains of interest can then be studied. For each bacterial target, a large number of phages are isolated in this fashion, and a set of phages that, between them, give the best coverage of the targeted strains are selected for further study of their other properties, such as stability, lytic properties and burst size. All of the process to this point was well laid out already by Felix d'Herelle, co-discoverer of phage, in the very useful appendix to his 1938 book; this appendix was recently translated from the French (Kuhl and Mazure 2011). Today, the characterization also includes virion morphology (as seen using a transmission electron microscope) as well as DNA size and pattern of sensitivity to a few restriction enzymes (measured by pulse field gel electrophoresis, or PFGE). It is currently expected that those phages chosen for the final therapeutic cocktails will also have their DNA sequence determined, to be sure that they carry no genes for toxins or other genes related to host pathogenesis or lysogeny. However, the process of isolating phages is much more challenging for the many pathogens that are hard to cultivate *in vitro*. While phage therapy is clearly useful for many of the most common bacterial pathogens, phages cannot be viewed as a potential universal substitute for antibiotics.

8.2   Historic Context

8.2.1   Discovery and Early Research

In 1915 and 1917 respectively, Edward Twort and Felix d'Herelle independently reported the isolation of filterable entities that were able to lyse bacterial cultures and to make small cleared areas, called by d'Herelle as "plaques" on plated bacterial cultures when they were sufficiently diluted, implying that discrete particles were involved. It was d'Herelle (Fig. 8.4), a Canadian working at the Paris Pasteur Institute, who gave the name *bacteriophage* to these entities that he isolated from the stools of soldiers recovering from dysentery. Soon thereafter, he isolated phages targeting avian typhosis and tested them during an outbreak in French chickens, showing that ¾ of the untreated chickens died but all of the phage-treated infected chickens survived. Dysentery was a severe problem in Paris at the time, and he accepted a request to treat a child battling dysentery at the Paris Hospital des Enfants-Malades; the day before, he and several assistants swallowed far more phage than they would be administering, thus carrying out the first known phase 1 clinical trial. Treatment of this first child was fully successful and he treated several additional children, but then turned to intense study of phages before publishing the results or carrying out further therapeutic applications. The first known publication discussing successful phage therapy came from two Belgians, Bruynoghe and Maisin (1921), who used phage to treat staphylococcal skin infections.

D'Herelle did an amazingly thorough job of studying phage biology. In a 300-page book, *The Bacteriophage: Its Role in Immunity*, d'Herelle (1922) carefully described such fundamental phage concepts as host specificity of phage adsorption and multiplication, the formation and make-up of plaques on lawns of susceptible bacteria, infective centers (which include phage-pregnant bacteria), the lysis process whereby the progeny phage are released from host cells, the dependence of phage production on the precise state of the host, isolation of phages from various sources, and factors which control the stability of a free phage. He quickly became

Fig. 8.4 Canadian biologist
Felix d'Herelle, who
discovered the bacteriophage
phenomenon in Paris in 1917
while working to better
understand dysentery and
soon applied phages in
disease treatment

fascinated with the role of phages in the natural control of microbial infections, noting that phages isolated from recuperating patients were frequently very specific for the disease organisms the patients were combatting and that phage populations often showed rather rapid variations over time. Throughout his life, he worked to develop the therapeutic potential of properly selected phages against the most devastating health problems of the day, traveling to many parts of the world, teaching at Yale from 1928 to 1933, and establishing his own Laboratoire du Bactériophage in Paris, run by his son in law, Theodore Mazure, which produced the first commercial phage cocktails: Bacté-Coli-Phage, Bacté-Intesti-Phage, Bacté-Dysentérie-Phage, Bacté-Pyo-Phage and Bacté-Rhino-Phage.

Although France is clearly a western country, most reviews of phage therapy have left out the successful continuation of phage therapy in France until the early 1990s, where physicians largely used well-made phage preparations produced by d'Herelle's lab or by the Bacteriophage Service of the Lyon and Paris branches of the Institute Pasteur. Sarah Kuhl, M.D., has translated much of the relevant French literature, and summarized in a detailed review of many aspects of phage therapy (Abedon et al. 2011).

One key to d'Herelle's many successes was that he focused intensely on understanding phage biology and on applying that knowledge in his production and application of phages, including careful ongoing quality control, close work with physicians and development of appropriate treatment modalities. He wrote several more detailed books on phage and phage therapy, two of which were also translated into English. The depth of his insights into the practice of phage therapy became more accessible via the recent publication of a translation of the appendix of his 1938 book [d'Herelle 1938, in Bacteriophage 1 (2011)]. Through much of the time, he worked closely with George Eliava, director of the Georgian Institute

Fig. 8.5 The Eliava Institute, built in 1933 specifically to be the world center of phage therapy, with wings both for production of therapeutic cocktails and for basic phage research and ongoing refinement of those cocktails. It still continues that mission, though it is now surrounded by the tall buildings of a bustling modern city

of Microbiology, who had seen bactericidal action of the water of the Koura River in Tbilisi (Tiflis) that he could not explain until he became familiar with d'Herelle's work while working at the Pasteur Institute from 1920 to 1921.

Over the years, the two developed plans to found an Institute of Bacteriophage Research in Tbilisi as a world center of phage therapy, including scientific and industrial facilities and supplied with its own experimental clinics. A large wooded campus on the river Mtkvari was allotted for the project in 1926, and d'Herelle sent large amounts of supplies, equipment, and library materials. In 1934–1935, he visited Tbilisi for 6 months, set up his laboratory in the main building of what is now the Eliava Institute (Fig. 8.5), and worked closely with the scientists studying phage and developing new therapeutic cocktails. While there, he also wrote a book entitled *The Bacteriophage and the Phenomenon of Recovery*, which was translated into Russian by Eliava and had a very strong impact on Stalin and thus on the implementation of phage therapy in the Soviet Union.

D'Herelle intended to move to Georgia; in fact, the cottage built for him still stands on the institute's extensive grounds. However, in 1937 Eliava was arrested as a "people's enemy" by Beria, then head of the KGB in Georgia and soon to direct the Soviet KGB as Stalin's much-feared henchman. Eliava was soon executed, sharing the tragic fate of many Georgian and Russian progressive intellectuals of the time, and d'Herelle, disillusioned, never returned to Georgia. However, their Bacteriophage Institute survived under the leadership of a group of women well trained by Eliava and d'Herelle. The Institute was put under the People's Commissary of Health of Georgia in 1938 and was transferred to the All-Union Ministry of Health in 1951, taking on the leadership role in providing bacteriophages for therapy and bacterial typing throughout the former Soviet Union.

8.2.2   Early Attempts at Commercialization

Phage therapy was explored extensively in the pre-antibiotic era, with many successes being reported for a variety of diseases, including dysentery, typhoid and paratyphoid fevers, pyogenic and urinary tract infections, and cholera. Phages have been given orally, through colon infusion, and as aerosols as well as infused directly into lesions. They have also been given as injections: intradermal, intramuscular, intraduodenal, intraperitoneal, into the pericardium and arteries leading to infected areas, and even intravenously. The early strong interest in phage therapy is reflected in some 800 papers published on the topic between 1917 and 1956. The reported results were quite variable. Many of the physicians, entrepreneurs and pharmaceutical firms who initially became very excited by the potential clinical implications jumped into application efforts with very little understanding of phages, microbiology in general, or the basic scientific process. Thus, many of the studies were anecdotal and/or poorly controlled; many of the failures were predictable, and some of the reported successes did not make sense in light of current knowlege. Too often, uncharacterized phages at unknown concentrations were given to patients without prior specific bacteriologic diagnosis, often apparently without follow-up or untreated controls. Much of the understanding gained by d'Herelle was ignored in this early work, and inappropriate methods of preparation, "preservatives," and storage procedures were often used. On one occasion, d'Herelle reported testing 20 preparations from various companies and finding that not one of them contained isolatable phages! In general, there was little of the essential quality control except in a few research centers such as those of d'Herelle, the French Pasteur Institutes, and the Institute d'Herelle had helped establish in Tbilisi. Large clinical studies were rare, and the results of those few that were carried out were largely inaccessible outside of what was then the Soviet Union.

8.2.3   Specific Problems of Early Phage Therapy Work

There are some who erroneously still believe that phage therapy was proven not to work; however, this was mainly based on phages which were produced by unqualified, poorly controlled and financialy biased companies, while the serious work done by qualified researchers was not widely known. Thus, it is important to carefully consider the reasons for the early problems and questions:

- Lack of understanding of the heterogeneity and ecology of phages and bacteria involved
- Lack of availability or reliability of laboratories for carefully identifying the pathogens involved (which is particularly important considering the relative specificity of phage therapy) – the tools for such identification were still very primative
- Use of too few different, complementing phages in infections that involved mixtures of different bacterial species and strains
- Failure, all too often, to properly test phage virulence against the target bacteria in vitro before using them in patients
- Emergence of phage-resistant bacterial strains through selection of resistant mutants (especially in cases where only one phage strain was used against a particular bacterium) or through lysogenization (if temperate phages were used)
- Failure to appropriately characterize or titer phage preparations, many of which, even from major companies, were shown to be totally inactive (Straub and Appelbaum 1933)
- Failure to neutralize gastric pH prior to oral phage administration
- Inactivation of phages by both specific and nonspecific factors in bodily fluids

All of these factors need to be dealt with, taking full advantage of modern tools, as we work to formally document and enhance phage efficacy and to integrate phage into medical practice worldwide to help deal with the growing crisis in antibiotic resistance.

8.3   Key Therapy-Relevant Properties of Phages

8.3.1   Professionally Lytic Phages and the Development of Molecular Biology

In 1943, an event occurred that had a major impact on the orientation of phage research in the United States and much of western Europe, strongly shifting the emphasis from practical applications to basic science and building strongly on special properties of bacteriophage T4 and its closest relatives. Physicist Max Delbruck joined with Alfred Hershey and Salvador Luria to form the "Phage Group" and establish an annual phage course and meeting at Cold Spring Harbor, Long Island. A major factor contributing to the success of phage as model systems for working out fundamental biologic principles at the molecular level was that Delbruck persuaded most U. S. phage biologists to focus on one bacterial host (*E. coli* B) and seven of its highly lytic phages, arbitrarily named types T1 through T7, making it possible to effectively work together. Fortuitously, T2, T4, and T6 were quite similar to one another, defining the family of myoviridae now called the *T-even phages*, which were key in demonstrating that DNA is the genetic material, that viruses can encode enzymes, that gene expression is mediated through special "messenger RNA," that the genetic code is triplet in nature, and other fundamental concepts. Much of this work was made possible by the fact that the T-even phages encode enzymes that make the cell incorporate an unusual base, hydroxymethyl-cytosine (HMdC) rather than cytosine into their progeny DNA, and then take advantage of that substitution to completely block reading of the genetic information from any cytosine-containing DNA, including that of the host. Until then, the existence of such an entity as transient messenger RNA (mRNA) had been thoroughly masked by the very large quantity and stable nature of ribosomal RNA; with that inhibited, the presence of a fraction of RNA that complemented the phage DNA and kept changing over the course of the phage infection, could be clearly documented.

The negative side of this focus on a few phages growing on one host under rich laboratory conditions was that there was very little study or awareness of the ranges, roles, and properties of bacteriophages in the natural environment, or of potential applications. On the positive side, most of the strongly lytic phages selected for therapeutic applications targeting enteric bacteria have turned out to belong to one or another of these three very well-studied families of professionally lytic phages – myovirus T4, siphovirus T5 and podovirus T7 – as confirmed by sequencing data as well as morphological details. This is very useful as the target is to assure their safety and to understand the physiological properties involved. Here again, the HMdC in the DNA of T-even phage plays a role in protecting their DNA from most of the restriction enzymes that their hosts use to protect themselves against foreign DNA, including those of other phages.

The temperate phages generally encode repressors to turn off most of their own genes when they are in the lysogenic prophage state, integrases to let them insert themselves into the DNA of their hosts, and excision enzymes to eventually cut their DNA back out of some progeny to enter a lytic cycle. Professionally lytic phages, in contrast, generally have much more extensive mechanisms for shutting off the host, often involving nucleases and transcription factors as well as a variety of small proteins made under the control of strong promoters very early in infection to re-direct cellular metabolism; T4, for example, encodes at least 50 such "monkey-wrench" proteins.

8.3.2   *Phage Interactions in the Body*

A number of early experiments involving phage injection into animals intravenously led to widespread belief that phage therapy could not succeed because the phages were too rapidly cleared by the innate immune system. Early experiments in rabbits, rats and mice showed rapid disappearance of phage from the blood and organs, but long-term survival in the spleen (Appelmans 1921; Evans 1933). The same phenomenon was seen in the first modern clinical trial, involving injection of 5×10^6

phage targeting vancomycin-resistant enterococcus (VRE) into 12 volunteers on days 1, 3 and 9 of the trial (Richard Carlton, personal communication 2013). After the first injection, the phage level remained, as expected, at about 10^3/ml blood. However, when a second dose was given on day 3, the phage titer quickly dropped by about 10-fold, and after a third similar dose on day 9, only a few phage were detectable shortly after the injection.

It should be pointed out that all of these experiments were done in the absence of host bacteria in which the phage could multiply and were carried out by the unnatural mode of intravenous injection, exposing the phage almost immediately to the reticuloendothelial system. Many further studies make it clear that phage *are* seen in the mammalian circulatory system for prolonged periods when they are entering it from some sort of reservoir in other tissues and the host is dealing with infection by a bacterium in which they can replicate – precisely the way phage therapy is currently practiced in various countries. This pattern is particularly clear in research published in 1943 by noted Harvard bacteriologist René Dubos et al. (1943). They injected mice intracerebrally with a pathogenic *Shigella dysenteriae* strain at a dose sufficient to kill more than 95 % of the control mice in 2–5 days thereafter. When mice were infected with the pathogenic bacterium and concomittantly injected intraperitoneally with a mixture of phages which had been isolated from the New York City sewage, grown in the same bacteria, and purified only by sterile filtration, 46 of 64 (72 %) of the mice given 10^7–10^9 viable phages survived.

Pharmacokinetic studies on the mice showed that when phages were given to uninfected mice, they appeared in the blood stream almost immediately, but the levels started to drop within hours and very few were seen in the brain (Fig. 8.6). However, in infected animals, brain levels of peritoneally or muscularly injected viable phage targeting those bacteria soon far exceeded blood levels; around 10^7–10^9 phage/g tissue were often seen between 8 and 114 h after administration, with the level dropping after 75–138 h. After the first 18 h, blood levels were far lower than brain levels, but phages were still present in blood at 10^4–10^5 phage/ml in those animals in which the brain levels were still high.

Equally well-controlled 1943–1945 experiments by Henry Morton and Enrique Perez-Otero at the University of Pennsylvania supported those of Dubos et al. (1943) and further showed the lack of any protection when phage with inappropriate host specificities were used. These results clearly established that:

1. phages themselves, and not any other component in the lysate that stimulated normal immune mechanisms, were responsible
2. phages could rapidly find and multiply in foci of infection anywhere in the body, including crossing the blood–brain barrier, and
3. phages could be maintained in the circulation as long as there was a reservoir of infection where phages were continually being produced.

A final review authorized by the US Council on Pharmacy and Chemistry discussed the major advantages of phages, such as the ability to replicate into problem areas and treat localized infections that are relatively inaccessible via the circulatory system, and also emphasized the fact that the high specificity of phages greatly

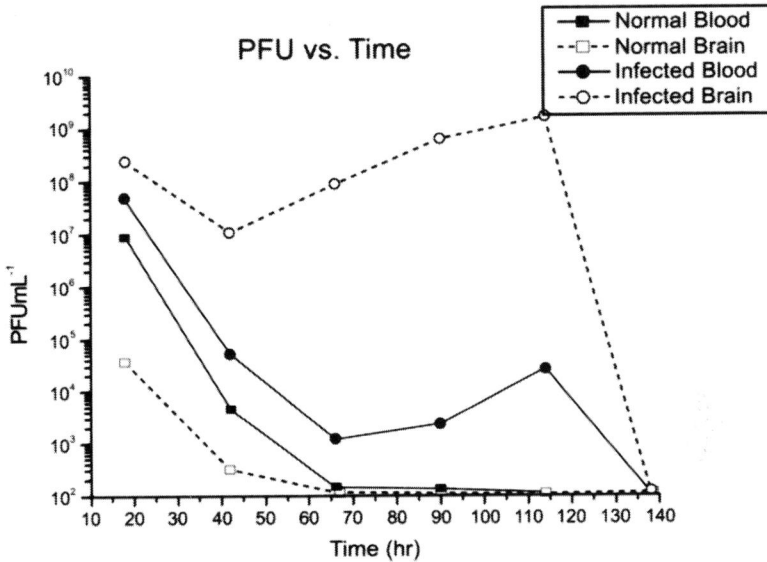

Fig. 8.6 This figure, based on the data in the 1943 mouse studies of Rene Dubos et al., provides significant insight into why phage therapy works well even in treating infections that antibiotics can't reach. When mice were injected intraperitoneally with 10^9 phage, they quickly appeared in the blood stream and some even crossed the blood–brain barrier, but they were rapidly cleared. However, if mice were also injected intracerebrally with *Shigella dysenteriae*, the host for these phages, 46/64 of the mice survived (as compared with 3/84 in the absence of appropriate viable phage) and the brain level of phage climbed to over 10^9 per gram. Once the bacteria were cleared, the level of phage dropped below detection limits

aided in reducing later problems of developing resistance, as was already being seen with antibiotics (Morton and Engely 1945). The review also concluded that, under the limitations of the times, most of the earlier phage research had been so poorly conceived and/or carried out that it offered no proof either for or against the promise of phages as antibiotics, so the negative conclusions of the earlier AMA reviews were neither unexpected nor very relevant to the potential for eventual success. It is very important to keep this in mind as there is a strong move in re-introducing phage as one important component in our battles with pathogenic bacteria.

The context of these studies sheds enlightening insights into the historical course of phage therapy in the US (Häusler 2003, 2006). In 1942, both *The Lancet* and the *British Medical Journal* published editorials about the apparently successful use of anti-dysentery phages by the Soviet military in the Middle and Far East. By November 1942, the U.S. National Research Council Committee on Medical Research (NRC/CMR) began supporting the research offered by anti-dysentery phages for dealing with this perpetual scourge of armies (including the above mouse studies) in top U.S. bacteriology labs new to phage work, initially requiring them to keep the results secret. This promising work ended in 1944, at the end of World War II, when penicillin was generally available. The military secrecy, the end of the war emergency funding, the rapid rise in antibiotic availability and their broad-spectra "wonder-drug" status, and Max Delbruck's success in persuading the phage community to shift its focus to basic molecular research involving a few model systems all contributed to the fact that there was little U.S. follow-up to these interesting and successful results. Although the results were published in major medical journals, few people knew about them, or about highly successful ongoing human applications, until they were recently dug out by Thomas Häusler (2003, 2006).

Penicillin, the first widely distributed antibiotic, works only against Gram-positive bacteria, so could not treat typhoid fever – a major health problem in the first half of the last century. Early phage efforts against typhoid also had only mixed success. Then it was found that the major pathogenic strains of *Salmonella typhi* all carried one particular surface antigen, a protein named Vi (for "virulence"). In 1936, Canadian scientists identified a number of phages specific against the Vi antigen, which could be used as "typing phages" in rapidly identifying and following typhoid outbreaks. In the late 1930s and early 1940s, physicians at the Los Angeles County Hospital were using phage treatments to help deal with repeated serious outbreaks of typhoid that were killing one in five of those afflicted. Walter Ward (1943) tested the Vi-specific phages against mouse typhus and found that the death rate fell to 6 %, versus 93 % in the controls. When they were then used to treat patients with typhoid, only 3 of 56 treated patients died, compared with the 20 % mortality for the other treatments available at the time (Knouf et al. 1946). Most impressively, they reported that the patients receiving phage therapy rapidly changed from being largely comatose to full of vigor, with renewed appetite, within 24–48 h. In 1948–1949, Desranleau treated nearly 100 patients with dysentery in the vicinity of Montreal by giving them an intravenous cocktail of six Vi-specific phages, and the death rate dropped from 20 to 2 %. However, by then chloramphenicol had been shown to work well against typhoid, and it was much easier for pharmaceutical companies to deal with, which most probably lead to the end of phage clinical trials in the Western hemisphere for many decades. The high specificity of phages still plays a strong role in the Phage Typing sets used for detecting and following problem strains of bacteria such as *Shigella*, *Salmonella* (including the typhoid pathogen), and *E. coli*, and, at last, phage therapy itself is beginning to stage a comeback.

8.4 Phage Therapy in the Age of Antibiotics

Although phage therapy research waned with the growing adoption of antibiotics in America and elsewhere in the Western world, phage therapy continued in some places. For example, in Vevey, Switzerland, the small pharmaceutical firm Saphal made "Coliphagine," "Intestiphagine," "Pyophagine," and "Staphagine" in drinkable and injectable forms, salves, and sprays into the 1960s (Häusler 2006). The owner, Harrmann Glauser, had been encouraged and trained by d'Herelle's old colleague Paul Hauduroy, who had become a professor of microbiology at the

University of Lausanne during WW II. These preparations were officially approved and were paid for by insurance.

The growing understanding of phage biology had the potential to facilitate more rational thinking about the therapeutic process, the selection of the best groups of therapeutic phages for particular applications, and ways to maximize their effectiveness. However, for a long time there was little interaction between those who were so effectively using phages as tools to understand molecular biology and those working on phage ecology and therapeutic applications. Many in the latter group were strongly spurred on by a concern about the rising incidence of nosocomial infections and of bacteria resistant against most or all known antibiotics, as well as by the fact that phage are far more effective than antibiotics in areas of the body where circulation is poor and they do not disrupt normal flora. This strong sense of the potential importance of phage therapy was particularly seen in France, Poland and the Soviet Union; in the latter two, use of therapeutic phages was never abandoned and there has been much ongoing research and clinical experience. Phage therapy continued to be used extensively in many parts of eastern Europe as a regular part of clinical practice. Today, companies in Russia make phage preparations for this purpose, selling them both in some pharmacies and on the internet. However, most of the research and much of the phage preparation evolved under the direction of key centers in Tbilisi, Georgia, and Wroclaw, Poland. In both cases, as discussed below, the close interactions between research scientists and physicians play an important role in the high degree of success obtained, just as appears to have been the case for d'Herelle's early work.

8.4.1 Pasteur Institute in France

In France, phage therapy continued with appreciable vigor into the early 1990s side by side with antibiotics. Commercial preparations were still available until 1978 from the company started by d'Herelle that first made Bacté-Pyo-Phage, Bacté-Dysentérie-phage, etc. and into the 1990s from the Pasteur Institute in Lyon and Paris, accompanied by many clinical studies. However, since virtually all publications were in French this work has been little known or credited outside of France. Many of these applications are discussed in substantial detail in Abedon et al. (2011), thanks to translations of many early and more recent French articles by co-author Dr. Sarah Kuhl.

Dr. Jean-François Vieu led the therapeutic phage efforts in the Service des Enterobacteries of the Paris Pasteur Institute until his retirement some 15 years ago. He built strongly on the widespread French work first documented in a 1936 monograph issue of La Médicine devoted to phage treatment of typhoid fever, acute colitis, peritonitis, prostate and urinary tract infections, furunculosis, sepsis, and otolaryngology. In a review, Vieu (1961) summarized the prevalent fundamental principles of French phage therapy, which was found to be particularly useful in the treatment of Staphylococcus, Pseudomonas, Proteus, and coliforms. For example, only virulent phages that completely lysed the bacteria in culture were used. The liquid lysates, containing 10^8–10^9 virions per mL, were filtered through Chamberland filters. The media that were best tolerated included broth, peptone water and appropriate micronutrients and used the least amounts of intact protein. They used both well-studied broad-spectrum cocktails, similar to those originally developed by d'Herelle and still marketed commercially in Georgia and Russia, and individual phages especially adapted for specific bacterial strains. Where necessary, efficacy could often be greatly increased by repeated passaging in the problematic bacterial strain. Local, sub-cutaneous and oral modalities of application were preferred; for intrapleural or pericardial infections, they prepared lysates in semisynthetic media as for intravenous applications. Vieu also discussed the potential for the development of phage-neutralizing antibodies after repeated injections of therapeutic phage, and concluded that extending phage treatment for long enough until antibodies become a problem shows an error in judgment, as typically phage therapy should at least begin to be effective within a few days or one should add another phage that does not cross-react serologically. Many of the details of phage administration and its effects as described in the 1936 *La Médicine* special monograph and as discussed by Vieu (1979) still appear valid in light of today's knowledge and information from other sources.

Vieu (1975; Vieu et al. 1979) includes tabulation of the 476 phages isolated and prepared by the Bacteriophage Service at the Pasteur Institute from 1969 through June 1974 for therapeutic applications, most of which targeted *Staphylococcus*, *Pseudomonas*, *E. coli*, or *Serratia*. Staphylococcal infection was reportedly the major target for phage therapy, with indications including septicemia with endocarditis, chronic osteomyelitis, suppurative thrombophlebitis, pulmonary and sinus infections, pyelonephritis, skin infections and furunculosis. Most often, phage were used for situations in which extensive antibiotic treatment had not led to clinical and bacteriological cure. The Institute Pasteur of Lyon also produced commercial therapeutic phage cocktails until the early 1990s. Its long-time director Henri de Montclos (2002) also describes how several European labs maintained an individualized production of phages by classical methods until the 1980s and discusses important safeguards when that approach was used. Successful treatment was typically achieved in a few weeks but this individualized approach did not lend itself to double-blind study establishing proof. After the AIDS crisis in the blood supply, regulations within the public health system created real challenges for continued individualized production of pharmaceuticals, including phages. A few French physicians have continued to use phages therapeutically even after the Pasteur Institute stopped making therapeutic cocktails in the mid 1990s, now generally obtaining their phages from Russia or Georgia. Dublanchet (2009) recently wrote a monograph describing his experience with phages and reporting a case study. This ties in with a major resurgence of interest in phage therapy in France, particularly for the military. In 2011, for example, a Tbilisi surgeon and an Eliava scientist were invited to a conference at a French military hospital to discuss phage therapy with a variety of military doctors and interested personnel. In January 2012, Pherecydes Pharma, which specializes in lytic phages for therapy and diagnosis, received

900,000 Euro from France's General Directorate for Armaments (DGA) to help evaluate the therapeutic efficacy, safety and pharmacodynamics of phages targeting multiresistant *Pseudomonas aeruginosa* or *E. coli* in open burn wounds. This project also involves the Institute of Genetics and Microbiology of the University of Paris XI and the Armed Services Institute of Biomedical Research. It is part of the DGA RAPID program, supporting dual innovation through projects that have great technological and commercial potential, are innovative in terms of industrial research, and have both military and civilian components. At the Pasteur Institute, scientists are doing excellent research with phage control of Pseudomonas lung infections such as those in cystic fibrosis, using mouse models. And a physician-led group called PhagEspoirs is working to make phage therapy available to patients.

There is similar enthusiasm in Belgium, where a trial on phage therapy is under way at the military hospital burn unit in Brussels (Merabishvili et al. 2009) and where a number of labs are conducting research relevant to phage therapy (Vandersteegen et al. 2012). The second biennial Viruses of Microbes meeting was held in July 2012 at the Military Academy in Brussels, drawing over 400 people. As at the similar-sized first Virus of Microbes meeting at the Pasteur Institute in Paris in 2010, one fourth of the meeting was devoted to topics related to phage therapy. The Brussels meeting was noted for the involvement of public health people as well as scientists, for candid discussions of some of the current regulatory challenges, and for a report on the progress of the major clinical trial being conducted by the Nestle company with infant diarrheal patients in Bangladesh (Brüssow 2012), described in Sect. 8.6.4.

8.4.2 Hirszfeld Institute of Immunology and Experimental Therapy (HIIET) in Poland

The most detailed publications documenting phage therapy have come from the group that was led by Stefan Slopek, long-time director of the Wroclaw Institute of Immunology and Experimental Medicine, and from his successors. An initial series of papers by Slopek et al. (1983a, b, 1984, 1985a, b, c, 1987) and Weber-Dabrowska et al. (1987) described work in 1981–1986 with 550 patients in 10 Polish medical centers. The patients ranged in age from 1 week to 86 years and venues included the Wroclaw Medical Academy Institute of Surgery, Cardiosurgery Clinic, Children's Surgery Clinic and Orthopedic Clinic; the Institute of Internal Diseases Nephrology Clinic; and the Clinic of Pulmonary Diseases. In 518 of the cases, phages were used after unsuccessful treatment with all available methods, including antibiotics, thus providing internal controls. The major categories of infections included long-persisting suppurative fistulas and abscesses, septicemia, respiratory tract suppurative infections, pneumonia and purulent peritonitis. The results were carefully analyzed and summarized with regard to factors such as nature and severity of the infection and monoinfection versus infection with multiple bacteria (Slopek et al. 1987). Rates of success ranged from 75 to 100 % for the various applications (overall 92 %), as measured by marked improvement in relevant physical condition, wound healing and disappearance of titratable bacteria; 84 % of subjects demonstrated full elimination of the suppurative process and healing of local wounds. Infants and children did particularly well, while and not surprisingly, the poorest results were obtained in elderly patients and those in the final stages of extended serious illnesses.

Appropriate individual highly-virulent phages were selected from their extensive collection. In the first study alone, 259 different phages were tested (116 for *Staphylococcus*, 42 for *Klebsiella*, 11 for *Proteus*, 39 for *Escherichia*, 30 for *Shigella*, 20 for *Pseudomonas*, and one for *Salmonella*); 40 % of them were selected to be used directly for therapy. All treatments were conducted in a research mode, with the phage prepared at the institute by standard research lab methods and tested for sterility. Treatment generally involved 10 ml of sterile phage lysate given orally half an hour before each meal, with gastric juices neutralized by consumption of basic Vichy water, baking soda, or gelatum shortly before the treatment. In addition, phage-soaked compresses were generally applied three times a day in cases of localized infections. Treatment ran for 1.5–14 weeks, averaging 5.3 weeks. For intestinal problems, short treatment sufficed, whereas long-term use was necessary for such problems as pneumonia with pleural fistula and pyogenic arthritis. Bacterial levels and phage sensitivity were continually monitored, and the phage were changed if the bacteria lost their sensitivity.

Therapy was generally continued for 2 weeks beyond the last positive test result for the bacteria. Few side effects were observed; those that were seen seemed to be directly associated with the therapeutic process (Slopek et al. 1983a, b, 1984). Various methods of administration, including oral, aerosols, and infusion either rectally or in surgical wounds, were successfully used. Intravenous administration was not recommended for fear of possible toxic shock from bacterial debris in the lysates. However, it was clear that the phages readily entered the body from the digestive tract and multiplied internally wherever appropriate bacteria were present, as measured by their presence in blood and urine as well as by therapeutic effects (Weber-Dabrowska et al. 2000). This interesting and rather unexpected finding has been replicated in many additional studies and systems (Smith and Huggins 1982, 1983, 1987; Smith et al. 1987).

The Polish work of the 1980s was all carried out in regional hospitals by the attending physicians, who kept detailed notes throughout on each patient. The final evaluating therapist also filled out a special inquiry form that was sent to the Polish Academy of Science research team along with the notes. The Computer Center at Wroclaw Technical University carried out extensive analyses of the data, using the categories established in the World Health Organization (WHO) (1977) International Classification of Diseases in assessing results. They also looked at the effects of factors such as age, severity of initial condition, type(s) of bacteria involved, and length of treatment.

After Slopek's retirement, Dr. Beatta Weber-Dabrowska carried on with the treatment work, though with far less support, and published a further, less detailed summary in English of the results for the next set of patients (Weber-Dabrowska et al. 2000). In 1998, immunologist A. Górski took over as Institute director and revived its strong involvement in phage work, with special emphasis on the immunologic consequences of phage treatment along with the Institute's other key immunological research. The Institute is also now working with the strong phage biology group of Dr. M. Lobocka in Warsaw to sequence and further characterize key phages – an important step in eventually making them available to the outside world (Lobocka et al. 2012).

In 2005, after Poland joined the European Union, the Institute opened its own outpatient Phage Therapy Unit at the IIET (http://surfer.iitd.pan.wroc.pl/pl/OTF) to support long-term treatment of chronic infections, complementing its ongoing work supplying phage and support services for physicians in regional hospitals (Międzybrodzki et al. 2012; Kutter et al. 2013). Patients are being treated in ambulatory out-patient fashion under European rules for therapeutic experiments, within a protocol named "Experimental phage therapy of drug-resistant bacterial infections, including MRSA infections", approved by the bioethics commission and registered also at ClinicalTrials.gov (ID: NCT00945087). This study deals with long-term chronic bacterial infections of the skin, subcutaneous tissues, bones and joints, fistulas, wounds, bedsores, urinary and reproductive tracts, digestive tract, middle ear, sinuses, tonsils, and respiratory tract where no surgery or hospitalization is currently required. The presence of bacteria sensitive to at least one phage in their collection must be confirmed by microbiological culture at the Institute. This collection currently consists of 524 bacteriophages, specific to *E. coli* (121 phages, 77 of which are routinely tested), *Klebsiella* (95 phages), *Enterococcus* (73), *Enterobacter* (48), *Shigella* (39) *Citrobacter* (38), *Pseudomonas* (37), *Salmonella* (32), *Stenotrophomonas* (18), *Serratia* (17), *Proteus* (17), *Morganella* (14), *Staphylococcus* (7), *Acinetobacter* (5), and *Burkholderia* (2). The phage titers in the sterile filtrates usually range between 10^6 and 10^9 pfu/ml, and are administered in a wide variety of ways. This approach, using a single phage at a time individually selected for each patient, does not lend itself to double blind experiments or to exportation to other clinics without this infrastructure. However, between January 2008 and December 2010, 157 patients with a wide range of very long-standing antibiotic-resistant bacterial infections were accepted for experimental phage therapy in the Phage Therapy Unit and the entire pool of data from these patients has now been analyzed in great depth microbiologically, immunologically and in terms of extent of healing (Międzybrodzki et al. 2012), providing a great deal of useful information for the planning and assessment of phage therapy. The clinic has also begun to provide some individualized treatment of foreign patients.

8.4.3 Eliava Institute of Bacteriophage, Microbiology and Virology (EIBMV) in Georgia

The Eliava Institute of Bacteriophage, Microbiology and Virology provides another major current model of supporting therapeutic phage treatment, in this case carried out on a very large scale, but with few details available on individual patients in recent years. During Soviet times, the Eliava Institute was the center of phage production and research for the Soviet Ministry of Health, as discussed extensively by Häusler (2003, 2006), Sulakvelidze and Kutter (2005), Chanishvili and Sharp (2009), Kutter et al. (2010) and Abedon et al. (2011). From the Bacteriophage Institute's inception, the industrial part was run on a self-supporting basis, while its scientific branch was government supported. The Institute carried out the extensive studies needed for approval by the Ministry of Health in Moscow of each new strain, therapeutic cocktail, and means of delivery. This careful study of the host range, lytic spectrum, cross-resistance, and other fundamental properties of the phages being used was a major factor in the reported successes of their phage therapy work, as were their methods for selecting highly virulent phages from among the many available against any given host. Where necessary, new cocktails were prepared with broader host ranges.

The depth and extent of the work involved was impressive, e.g. in 1983 through 1985, the Institute's Laboratory of Morphology and Biology of Bacteriophages carried out studies of growth, biochemical features, and phage sensitivity on 2,038 strains of *Staphylococcus*, 1,128 of *Streptococcus*, 328 of *Proteus*, 373 of *Pseudomonas aeruginosa*, and 622 of *Clostridium* received from clinics and hospitals across the former Soviet Union. New broader-acting phage strains were isolated and were included in a reformulation of their extensively used Pyophage preparations. A good deal of work went into developing the documentation for Ministry of Health approval of specialized new delivery systems, such as a spray for use in battlefield wounds, respiratory tract infections, in treating the incision area before surgery, and in sanitation of hospital problem areas; unfortunately, only a little of that documentation is available today.

An enteric-coated pill was also developed, building on technology first developed in Russia and using phage strains that could survive the drying process. Tablets of phage targeting dysentery accounted for the bulk of their shipments to other parts of the former Soviet Union. Much work focused on combating nosocomial infections, in which multidrug-resistant organisms have become a particularly lethal problem. Phage preparation was carried out on an increasingly enormous scale before the breakup of the Soviet Union, employing 700 people in the factory and several hundred more in the research arm of the Institute, making tons of a variety of products weekly to ship throughout the former Soviet Union. The products were available both over the counter and through physicians; 80 % went to the Soviet military for wound and burn infections and for preventing debilitating gastrointestinal epidemics. In hospitals, they were used to treat both primary and nosocomial infections, alone or in conjunction with other antimicrobials.

Throughout the 1970s and 1980s, hundreds of thousands of pathogenic bacterial samples were sent to the Eliava research arm from throughout the USSR to isolate and produce more effective phage strains and to better characterize their usefulness. A new 5-story building and huge modern fermenters provided facilities that let the 800 employees of the production unit make two tons of phage products twice a week, with 80 % of the product going to the Soviet military – tablets against dysentery, pyophage cocktails targeting wound and other purulent infections, intestiphage with phages targeting a very wide range of cultivable enteric pathogens. Much of the

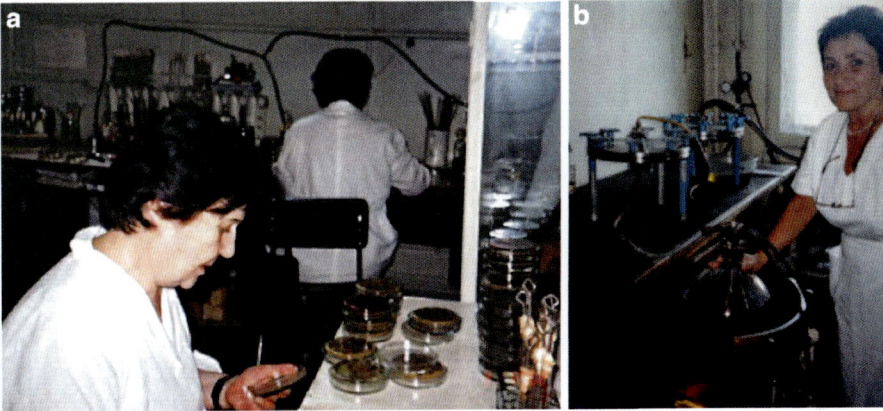

Fig 8.7 (**a**) Eliava diagnostic center, with Dr. Liana Gachechiladze checking patient samples for phage and antibiotic sensitivity and (**b**) filter sterilization of a fresh batch of Pyophage during small-scale production for local hospitals and clinics in the 1990s

research concentrated on isolating new phages, upgrading the therapeutic cocktails, isolating phages against anaerobes and defenses against potential bio warfare, while collaborating closely with military and civilian physicians to test the phage preparations and the ways of administering them. These were the golden years of the Institute, when its employees were among the best paid in Tbilisi and there were no shortages of supplies, new research targets or potential customers. In 1988, an official Scientific Industrial Union "Bacteriophage" was formed, centered in Tbilisi with branches in Gorki, Ufa, and Habarovsk, in Russia. Phage therapy thrived and blossomed through the strong collaboration between the Soviet Ministry of Health, the production facilities, the physicians, hospitals and clinics and the academic centers, to the benefit of everyone.

Much of this crashed after the 1991 break-up of the Soviet Union. The Russian branches were gradually pulled into the developing pharmaceutical giant, Microgen whose products include a wide range of phage preparations still available in pharmacies, online, and in hospitals throughout the Soviet Union and undergoing further development there. It is hard, however, to get much information about what is going on there except through individual physicians, some of whom are also involved in maggot therapy.

In Georgia, the Bacteriophage Institute (now called the Eliava Institute) lost most of its markets and funding and struggled for survival as the factory portions were stripped away and privatized, with virtually no compensation to the Institute. Phages for Georgian use were now manufactured in the research-scale production facilities within individual labs, and a new Diagnostic Center manned by the Institute's leading scientists provided some funds to keep the Institute going as well as continually supplying current problematic bacterial strains for updating the therapeutic cocktails (Fig. 8.7).

By 1996, help started becoming available from the International Science and Technology Centers programs and the Civilian Research and Development Fund, set up by NATO and the US, respectively, to give constructive opportunities to civilian scientists who had been formerly supported by the Soviet military, and from the European Union, the PhageBiotics Foundation and other sources. With hard work and outside help for the basic science and analytical work, the institute is functioning increasingly well at its original site on the Mtkvari, as further discussed below in looking at specific applications, though unfortunately none of the available US or European governmental sources have supported proper clinical trials and the programs are now winding down.

Particularly extensive work on therapeutic phages continued to be carried out by scientists at the Bacteriophage Institute, still working closely with local physicians. Phage therapy still is an accepted component of the general standard of care in Georgia, used especially extensively in pediatric, burn, and surgical hospital settings and available over the counter in local pharmacies for management of diarrheal diseases and purulent infections.

An exciting new product was licensed in 2000 by the Georgian Ministry of Health. PhagoBioDerm is a biodegradable, nontoxic polymer composite that allows the sustained release of an incorporated special version of the Pyophage cocktail. Katsarava et al. (1999) and Markoishvili et al. (2002) published a study of PhagoBioDerm involving 107 patients with ulcers that had failed to respond to conventional therapies with systemic antibiotics, antibiotic-containing ointments, and various phlebotonic and vascular-protecting agents. The ulcers were treated with PhagoBioDerm alone or in combination with other interventions during 1999 and 2000. The wounds or ulcers healed completely in 70 % of the 96 patients for whom there was follow-up data. In the 22 cases for which complete microbiologic analyses were available, healing was associated with the concomitant elimination or very marked reduction of the pathogenic bacteria in the ulcers. A newly-formulated version of PhagoBioDerm has just come on the market in Georgian pharmacies (Katsarava, personal communication 2012), which is much less expensive to produce and has additional advantages.

8.4.4 Phages in Today's Russia

After the former Soviet Union broke up, the large phage production and research facilities in places like Ufa and Novosibirsk continued to operate and supply the Russian market, but little information was available about their activities. They were all gradually acquired by Russian pharmaceutical behemoth Microgen, whose extensive on-line catalogues include a variety of phage products. One of these is currently being used as a parallel control in the study of phage targeting infant diarrhea being conducted by Nestle in Bangladesh, discussed in detail in Sect. 8.6.4. It is very difficult to find details about any research or clinical trials the Russians may be carrying out, although a few bits of information about their products occasionally appear in general phage therapy reviews by Russian phage biologists like Krylov et al. (2012) and Letarov et al. (2010).

8.5   Specific Clinical Applications

8.5.1   Purulent Surgical, Wound- and Burn-Infections

Staphylococcus aureus, both methicillin-resistant (MRSA) and methicillin-sensitive, is a common inhabitant of the human skin which, unfortunately, encodes many traits that can make it a potent pathogen when it gets into wounds, deep burns and surgical sites. MRSA is of particular health related concern worldwide, given its reduced susceptibility to antibiotic treatment, its wide prevalence in hospital- and community-acquired infections, and its potentially lethal and otherwise serious consequences such as septicemia. Georgian, Russian and French Pyophage preparations, designed to treat purulent infections, include phages targeting these bacteria, as well as ones against many strains of *Streptococcus*, *Proteus*, *Pseudomonas* or *E. coli*. The potent anti-*Staphylococcus aureus* phages are of special interest. They belong to the broad-spectrum staph phage K genus, as described by Kvachadze et al. (2011), Vandersteegen et al. (2012) and Lobocka et al. (2012). This is the only large lytic phage subfamily targeting *Staphylococcus* and the only known family of myoviruses, in contrast to what is seen for most other kinds of bacteria; Staphylococci are also targeted by a wide range of siphoviruses, but those are all temperate phages (Deghorain and van Melderen 2012). Here as elsewhere, there is no cross-resistance between phages and antibiotics; the two often function well synergistically, especially when the antibiotics are administered orally or intravenously and the phage are applied locally. Furthermore, very little development of resistance to this subfamily of phages is observed, presumably implying that the phages' still- unidentified primary receptor on the *S. aureus* surface is of substantial importance to these bacteria.

8.5.1.1   Case Reports

Case #1

A.G., a musician from Toronto, had been fighting a debilitating staph infection for 4 years. His physicians were insisting he faced death within the year if he didn't submit to amputation of his lower leg. Informing himself about phage therapy in the Republic of Georgia, and reading about an upcoming Evergreen International Phage Meeting in a New York Times Magazine article, he decided to explore that route. He attended the phage meeting, had his exudate tested for sensitivity to Eliava pyophage and went to Tbilisi half a year later to be treated when he couldn't find a doctor willing to carry out the treatment in Toronto (Fig. 8.8a). Scans were conducted to determine the nature of the cavity running through his ankle, and the sensitivity of the bacteria to individual staph phages, Pyophage and antibiotics was determined (Fig. 8.8b). Interestingly, while the IV treatment had helped keep the staph to tolerable levels in the rest of his body, it had not penetrated the reservoir in the bones of his ankle even enough to select for antibiotic resistance over the course of treatment.

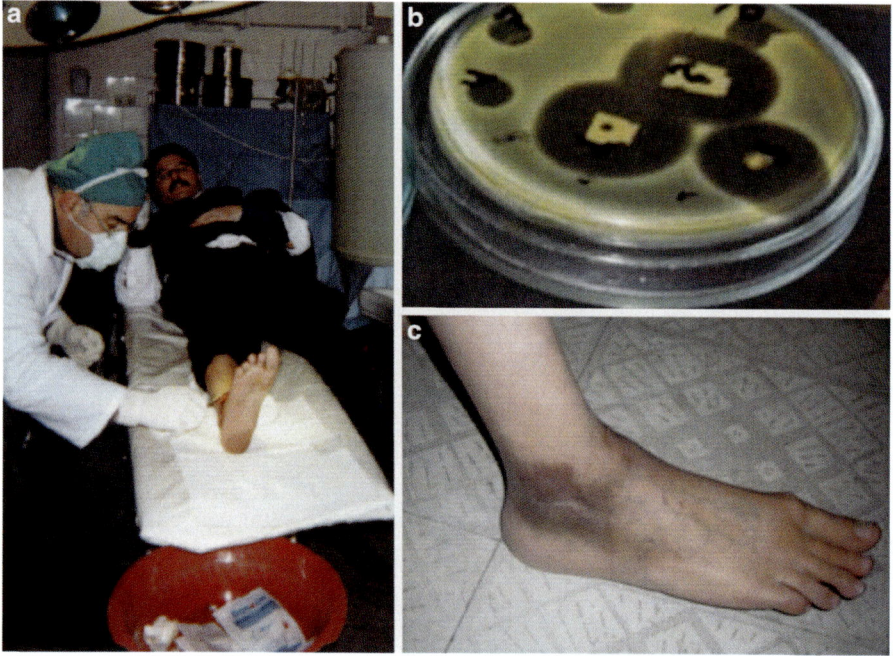

Fig. 8.8 Case 1. (**a**) *Staphylococcus aureus* was still draining from both sides of the patient's ankle after 4 years of treatment, including one full year on IV antibiotics, when he finally went to Tbilisi to be treated with phage. (**b**) The Petri dish used to test the isolated bacteria shows that they were fully sensitive to three different tested staph phages (*upper left*) and to the Pyophage in PhageBioderm (*center*). Interestingly, the bacteria draining from his ankle were also still fully sensitive to the antibiotic he had been administered IV for a full year, as seen using a standard disk for that antibiotic (*lower right of plate*). (**c**) By 2007 the ankle bones had been fused and the holes had healed; he is back to being a professional bass player and still has his foot today

In this case, pure staph was the culprit, it was not MRSA, and it still responded to the Pyophage back at home over the following years. It was clear that the staph was greatly reduced and generally kept under control by the phage but was not totally eliminated, and it remained sensitive to the phage cocktail through multiple treatments. After having undergone an ankle fusion to end the constant rubbing, the ankle holes have healed and there is no further sign of staph (Fig. 8.8c).

Case #2

An anonymous patient had suffered a very serious injury due to a mine-explosion (Fig. 8.9a). Phage therapy was initiated by infusion and spraying of the Eliava pyophage cocktail to control infection, concomitantly with a major reconstructive surgery beginning on day 3 (Fig. 8.9b). The conventional treatment continued through a period

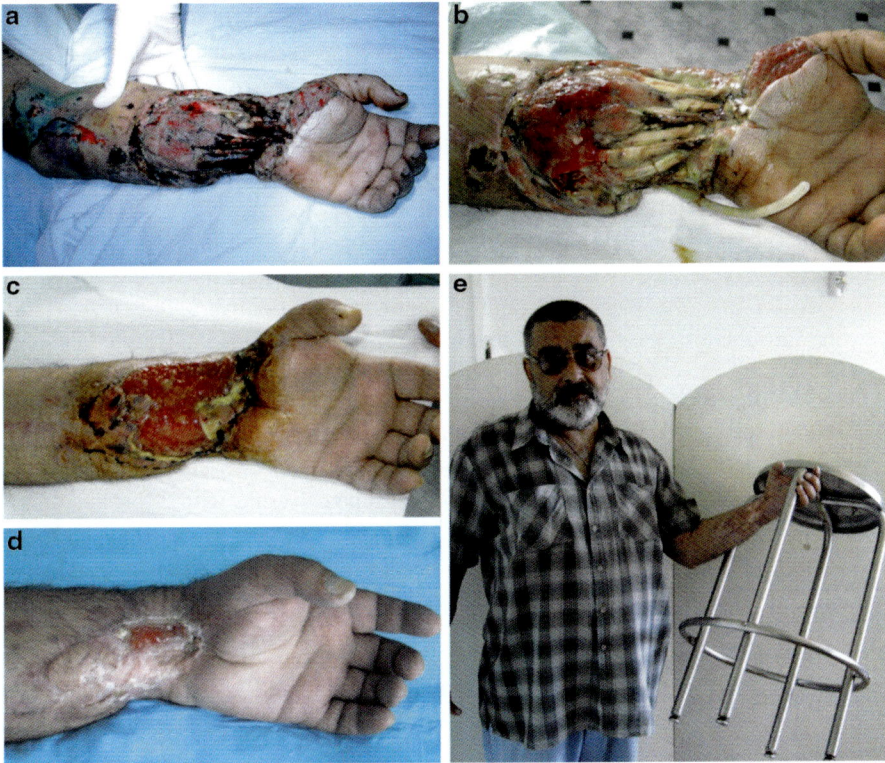

Fig. 8.9 Case 2. (**a**) Extensive injury of the left forearm due to a mine-explosion; (**b**) situation on day 3 after initiation of Phage therapy and major reconstructive surgery; (**c**) treatment was continued with daily debridement and other conventional treatments, and the wound was largely healed within one-and-half months; (**d, e**) the wound was nearly fully healed by 2 months after the injury, with full functional recovery

of extensive daily phage treatment and debridement. One-and-half months after the injury extensive healing could be seen (Fig. 8.9c). After 2 months, the wound was largely healed, and the patient was even able to lift heavy objects (Fig. 8.9d, e).

8.5.1.2 Phage Therapy and *Staphylococcus*

From the point of view of phages, MRSA is simply another strain of *Staphylococcus*. Treatment of *Staphylococcus*, including MRSA, using phages can be accomplished by external application to local infections, local injection or, if necessary, and with substantially more caution, by systemic dosing such as intra-peritoneal injection for systemic infections (MacNeal et al. 1942). Indeed, the first human phage therapy publication reported treatment of *S. aureus* skin infections (Maisin 1921), and phage targeting local *S. aureus* infections were one of the few phage applications praised as proven efficacious, in the generally critical extensive American Medical Association review by Eaton and Bayne-Jones (1934). Staph phages were used quite extensively and with high reported success in New York in the 1940s and in France up into the 1980s, as well as in Georgia, Poland and Russia (Bruynoghe and Maisin 1921; MacNeal et al. 1942; Abedon et al. 2011). A staphylococcal phage preparation for systemic application was developed at the Eliava Institute during the 1980s, including intravenous safety studies in human volunteers, with no adverse effects reported. The preparation subsequently was used to treat 653 patients, as summarized extensively by Chanishvili (2012). It was particularly effective intravenously in infants with septicemia and in immune-compromised patients, and for infusion into the urethra in cases of pelvic inflammatory disease.

Historically, questions have been raised as to whether the apparent efficacy documented in so many classic articles was due to the phage itself giving rise to bacterial lysis in situ. It has been suggested that the debris in the phage lysate, stimulating the host immune system, could be a major factor in bacterial clearance (Sulakvelidze and Barrow 2005; Kutter et al. 2010). Staph Phage Lysate (SPL), produced by Delmont laboratories, has been marketed since the 1940s as a veterinary medication, advertised as conferring resistance to staph through immune stimulation by its staphylococcal-phage induced bacterial lysis products. However, we have determined that the Delmont staph phage lysates also still contain viable phages, often at ~10^8 PFU/ml (Kutter EM, unpublished data 2004; Kuhl SJ, unpublished data 2009), a level as high as the total phage in Pyophage as determined by direct fluorescent microscopic count (Nathan Brown, personal communication 2009). These staph phage lysates initially were produced for human as well as animal use against chronic infections. However, in 1994 they were limited by the FDA to animal use pending further human efficacy trials, for which no funding has yet been found; no questions have ever been raised as to their safety.

Phages have been used in both Russia and Georgia to sanitize operating rooms and medical equipment and to prevent nosocomial infections by both Staphylococci and other bacteria (Kutter et al. 2010, 2012). A complementary approach proposed by the company Novolytics is to use "an aqueous suspension to treat nasal carriage of MRSA, thus significantly reducing the incidence of MRSA transmission" (www. novolytics.co.uk/technology.html). Leszczyński et al. (2006) described the use of oral phage therapy for targeting MRSA in a nurse who was a carrier. This individual had MRSA colonized in her gastrointestinal and urinary tracts. The result was complete elimination of cultivatable MRSA. In an earlier publication the same group argued that MRSA treatment using phages can be economically preferable to MRSA treatment using antibiotics (Miedzybodki et al. 2007). Slopek et al. (1987) reported 92.4 % positive cases for phage treatment of 550 single- and mixed-etiology infections involving *Staphylococcus aureus*. Slopek et al. (1985a) specifically address phage treatment of suppurative staphylococcal infections, with a reported 93 % "effective" rate "based on case history and data contained in a special questionnaire", while Slopek et al. (1985b) consider the treatment of various kinds of *Staphylococcus* infections of children (95.5 % positive results for the 90 children treated).

8.5.2 Urogenital Tract Infections

Phages have been applied to treat various infections of the urogenital systems either systemically, via direct injection into the bladder, or via topical application. Phage treatment of urogenital tract infections could potentially be complimented by current naturopathic protocols involving alkalinization of the urine with citrates and minerals. In their otherwise-critical 1934 AMA report, Eaton and Bayne-Jones report, were convinced that the use of phage therapy against cystitis was efficacious, as it was against staphylococcal infections. Letkiewicz et al. (2009) describe phage application rectally to target *Enterococcus faecalis* infection of the prostate, with substantial success in eliminating the target bacteria. In this case, the phages were presumed to be taken up through the rectal wall. Letarov et al. (2010) noted that rectal phage suppositories are also available on the Russian market. Slopek et al. (1985c) report 92.9 % positive cases for phage treatment of various diseases of the urogenital tract. This work has been carried out in the outpatient setting of the new Phage Therapy clinic at the Hirszfeld Institute, as discussed extensively by Międzybrodzki et al. (2012) and Kutter et al. (2013).

A good deal of work in this area has also been carried out in collaboration between the Eliava Institute and major Tbilisi hospitals. In the books of Chanishvili and Sharp (2009) and Chanishvili (2012) there are chapters on "Phage therapy in urology" and "Phage therapy in gynecology". It is reported that in cases of acute cystitis a therapeutic effect was observed within 4–5 h after the first administration and resulted in relief of pain, a decrease in the frequency of urination and a normalization of the urine parameters. Full recovery was achieved within 1–3 days in all 13 cases. However treatment of chronic forms of cystitis was less successful, with only moderate improvement observed.

Supported by a 3-year grant from the ISTC, the Eliava Institute has developed a new phage cocktail specifically targeted against a large pool of bacteria from prostatitis and urinary-tract infections (Alavidze, unpublished data). Initially, urine, sperm and/or prostate fluid from 314 patients yielded 185 *E. coli*, 112 *Staphylococcus*, 67 *P. aeruginosa*, 55 *Proteus*, 53 *Enterococcus*, 50 *Klebsiella*, and 38 *Streptococcus* isolates. The *Klebsiella*, *Enterococcus*, and *P. aeruginosa* were most antibiotic resistant. Only 147 of the 560 strains were sensitive to the current commercial Intestiphage. It is known that Pyophage contains no phages targeting *Klebsiella* or *Enterococcus*, accordingly, only 109 of the other 437 strains were sensitive to the Pyophage. These low sensitivities of bacteria from prostatitis and uterine infections presumably reflect the fact that only purulent-wound and gastrointestinal strains of the various bacteria, respectively, had been used in the formulation and production of Pyophage and Intestiphage. Therefore, many new phages were isolated against the resistant strains, studied in detail and incorporated into Urophage, attaining an activity of 86.2 % against these 560 clinical strains. Open safety and efficacy trials were carried out on 118 patients with previously recalcitrant chronic bacterial prostatitis (CBP) caused by various combinations of *Pseudomonas*, *E. coli*, *Proteus*, *Klebsiella*, *Enterococcus*, *Staphylococcus*, and *Streptococcus* species in the Urology

Department at Tbilisi's Central Clinical Hospital. The results were very encouraging both symptomatically and in lowering the bacterial concentrations, in some cases below the detection level, though in many cases some residual bacteria were still detected in prostatic fluid requiring further treatment beyond the 19-day treatment period and 10-day follow-up. Double-blind clinical trials are in the planning stage, depending on finding sources of funding.

8.5.3 Ear Infections

Chronic otitis externa, known less formally as swimmer's ear, is often caused by a *P. aeruginosa* infection that resists antibiotic treatment. The British company Biocontrol developed a remedy based on anti-*P. aeruginosa* phages targeting otitis externa, after having published studies on dogs (Marza et al. 2006, Hawkins et al. 2010). In 2009, they published the results of their double-blinded phase 1/2a (safety and small-number efficacy) trial in human patients suffering from this condition (Wright et al. 2009), rendering it the first modern double-blind clinical trial to extend into phase 2. Increases in phage numbers in situ, microbiological improvements (reductions in bacterial presence) and reduction in disease symptoms in the phage-treated cohort but not the phage-negative controls, were observed, while no side effects were seen. Complete bacterial eradication was not observed, but the extent of success was particularly notable considering that only a single phage dose was administered. Biocontrol recently merged with Targeted Genetics of Seattle and Special Phage Services of Sydney, Australia to form a new joint company, AmpliPhi Biosciences, which is moving on toward phase 3 studies and applications of their extensive *Pseudomonas* and other phage collections in other settings.

Weber-Dabrowska et al. (2001) also reported phage therapy success in treated purulent otitis media, most often caused by *Streptococcus pneumoniae*, which is the leading cause of otitis in children. Slopek et al. (1987) reported 93.8 % positive results for phage treatment of 16 cases of conjunctivitis, blepharoconjunctivitis, and otitis media.

8.5.4 Respiratory Tract Infections

Respiratory infections can be caused by different microorganisms and phage therapy is limited in efficacy to those which have a bacterial etiology. Weber-Dabrowska et al. (2001) reported success in treating pneumonia in six cancer patients. Similarly, Slopek et al. (1983a, b, 1987) report 86.7 % positive cases for phage treatment of 180 "Diseases of the respiratory system". The first case studies of phage use to treat Tbilisi children with cystic fibrosis (CF) for their chronic *S. aureus* and *P. aeruginosa* infections were recently published, using Pyophage and also a fully sequenced *S. aureus* phage of theirs delivered by standard CF nebulizers (Kutateladze and Adamia 2010; Kvachadze et al. 2011). The latter includes a detailed description of the successful treatment of a *P. aeruginosa* lung infection of a 7-year-old patient using Pyophage, along with treatment of an *S. aureus* co-infection in the same patient using phage Sb-1.

The company AmpliPhi has reported work toward expanding its anti-Pseudomonas phage therapy efforts to include treatment of children with cystic fibrosis (Kutter et al. 2010). Success in treating infections in animal models of cystic fibrosis-associated infection has also been reported by Debarbieux et al. (2010) and Carmody et al. (2010), as have explorations of the utility of nebulization as a phage delivery strategy. Phage treatments of lung infections, however, can also be successful in at least some circumstances from systemic circulation, as shown in animal models (Carmody et al. 2010).

8.5.5 *Gastrointestinal Infections*

As discussed above, d'Hérelle discovered bacteriophages in association with the examination of dysentery in humans and quickly developed an interest in its treatment using phages. Experimental anti-dysentery trials also were extensively conducted over multiple decades in Eastern Europe and Georgia. Chanishvili (2012) provided a chapter reviewing the Georgian literature on phage therapy against intestinal infections. In the treatment of dysentery, highlights included phage production to densities up to 10^{12} phage/ml for some applications, the application of many ml per dose, the use of multiple doses, buffering to prevent phage loss during passage through the gastrointestinal (GI) tract, and reductions in disease symptoms. Substantial reductions in mortality were reported. One well-controlled dysentery prevention trial was conducted in Tbilisi on 30,769 children. Neighborhoods were split up with one side of each street treated prophylactically with a phage cocktail and the other with a placebo. The result was a 3.8-fold decrease in dysentery incidence with phage treatment. Similar follow up trials were undertaken with 20,000 and 5,000 children respectively as well as another trial against *Salmonella*-associated disease. In Georgia the Intestiphage formulation is sold over the counter and routinely used to deal with a wide range of gastrointestinal diseases, including food poisoning from various causes and a variety of intransigent gut dysbioses. It is also employed prophylactically to prevent nosocomial infections, especially in pediatric hospitals, where such GI infections had been particularly prevalent. Large-scale community-wide dosing was traditionally done with phages prepared in tablets, though since the break-up of the Soviet Union the tablet form has been seldom available, and has been replaced by a liquid formulation.

The most advanced clinical trials of phage therapy using modern protocols are being carried out under the leadership of Harald Brüssow of the Nestlé Corporation, Lausanne, Switzerland. The study is designed to establish the safety and efficacy of phage therapy in treating ETEC- and EPEC-induced diarrhea in children, as discussed in some detail below.

8.6 Phages and the Human Body

8.6.1 Toxicology

From a clinical standpoint, all indications are that phages are very safe, at least for most kinds of applications. This feature is not surprising, given that humans are exposed to phages from birth. This general consensus was reflected in the decision of the FDA to approve the Intralytix phage preparations targeting listeria on ready-to-eat foods and to give GRAS (Generally Regarded as Safe) status to a similar Dutch preparation. Bergh et al. (1989) reported that nonpolluted water contains about 10^8 phage/ml. A wide variety of phages are normally found in the gastrointestinal tract, skin, urine, and mouth, where they are harbored in saliva and dental plaque (Caldwell 1928; Yeung and Kozelsky 1997; Bachrach et al. 2003). They also have been shown to be unintentional contaminants of sera and hence of commercially available vaccines, which were given dispensation to be sold despite this discovery because of the general consensus that phages are safe for humans, and there was no indication that this decision led to any problems (Merril et al. 1972; Geier et al. 1975; Milch and Fornosi 1975; Moody et al. 1975).

Very extensive preclinical animal testing was required for approving new phage formulations in the former Soviet Union, but only a few of these studies are publicly available. For example, Bogovazova et al. (1991, 1992) evaluated the safety and efficacy of *Klebsiella* phages produced by the Russian company Immunopreparat. Pharmacokinetic and toxicologic studies using intramuscular, intraperitoneal, or intravenous administration of phages were carried out in mice and guinea pigs. The researchers found no signs of acute toxicity or histologic changes, even using a dose 3,500-fold higher than the projected human dose. They also evaluated the safety and efficacy of the phages by treating 109 patients. The phage preparation was reported to be nontoxic for humans and to be effective in treating *Klebsiella* infections, as manifested by marked clinical improvements and bacterial clearance in the phage-treated patients. Chanishvili (2012) published summaries of a number of early animal studies in her extensive exploration of the material available in the archives of the Eliava Institute.

Side effects such as liver pain and fever reported in the early days of Western phage therapy may have been due to bacterial byproducts in preparations used intravenously (Larkum 1929; King et al. 1934). Concerned about this possibility, the Polish group never administers their phages intravenously. The same is true for almost all of the therapeutic work carried out in Tbilisi and probably helps explain the virtually total lack of significant problems in both places; their many years of experience, careful attention to detail and supportive infrastructure are presumably also important factors. Because phages rather readily enter the blood stream after infusion in or near wounds and other sites of localized infections and travel to sites of infection throughout the body, there generally seems to be no reason for undergoing the extra risks of intravenous administration except in such cases as septicemia, where further in-depth research and carefull ongoing monitoring are warrented.

8.6.2 Drug Interactions

No negative effects on the efficacy or safety of other drugs have been reported anywhere as a result of phage administration. No systematic studies have been carried out in this regard, but phages are so specific in their actions that it is hard to see where such interactions might occur. On the other hand, at least some antibiotics can interfere with phage treatment of localized infections by killing off the most accessible of the bacteria in which the phages need to multiply as they work their way deeper into the lesion; this would be a particular problem in cases in which the phages can still attach and infect but cannot complete their replication cycle. Georgian physicians generally believe that antibiotics should never be used topically for deep-seated infections, because the decrease in antibiotic concentration as one moves deeper below the surface provides a strong selection for antibiotic resistance, a problem which does not occur with phages.

8.6.3 Dosing Strategy

Phage cocktails can be designed and administered using either of two distinctly different dosing strategies, an active one depending on the unique ability of the phage to replicate where needed and a passive mode that uses much higher amounts of phage in a way that is more comparable to standard pharmaceuticals, while relying on the fact that phage are far less likely to have side effects at higher doses than are most antibiotics or other common drugs (Abedon and Thomas-Abedon 2010; Abedon 2011). As discussed in great detail by Abedon, whether actively or passively supplied, ultimately phage therapy efficacy is dependent on the generation of peak phage concentrations that are sufficient to result in substantial bacterial eradication in the vicinity of target bacteria.

The active treatment mode, the major approach currently used in Georgia, Russia and Poland, involves totally different principles and kinetics than any other known pharmaceutical, and is less dependent on the precise dosage administered. In general, it uses relatively low concentrations of phage and depends on the self-replicating nature of the phage in the presence of relevant host bacteria to achieve therapeutically relevant concentrations of phage in exactly the places where they are needed. There, they are usually supplied in sterile-filtered phage lysates without further purification. Total phage titers used are typically about 10^8 pfu/ml. In Georgia and Russia, the complex commercial Pyophage and Intestiphage cocktails are made by combining sterile lysates of a number of different phages, so there are only about 10^6–10^7 pfu/ml of each individual phage. As was seen in

Fig. 8.6, phage have the advantage that they can be inserted into some reservoir inside the host (in that case, the peritoneal cavity) and carried throughout the system in low concentration, to multiply only where they are needed – i.e. where there are substantial concentrations of the relevant bacteria. Furthermore, even if a very complex phage cocktail is used, only the relevant phage for the bacteria present in that particular region of the body will multiply there.

In Poland, the basic principle is the same, though so far phage therapy is available only with direct involvement of the Hirszfeld Institute, which maintains stocks of several hundred different phages (each at 10^6–10^9 per ml) and tests each patient's bacteria against the subset of their collection that targets the species involved to select the specific most appropriate phage. It is then administered either in their new outpatient clinic or by the partnering physicians in a local hospital. If resistance develops, they then chose a second or even third or fourth phage from their collection to continue the treatment.

The alternative, so-called passive application, is sometimes suggested in Western discussions. It ignores the self-replicating nature of phage in favor of a more conventional dosing strategy, where sufficient numbers of phage must be applied to treat the infection in a single dose or series of doses, with little or no local multiplicative effect. The considerable complexities of carrying out dosing calculations and getting sufficient phage in the target area are discussed in great detail by Abedon (2011), but for some accessible local applications, this may have advantages. Here also, only a single hit is required for killing, but the primary challenge for the phage is simply to find the relevant bacteria within a suitable length of time when the bacteria are present in low quantity.

8.6.4   Clinical Trial Challenges

A very pertinent question remains: if phage have such strong potential, why has it been so hard to introduce them into clinical practice in the Western world? The complex process of developing and testing a modern commercial phage therapeutic is best exemplified by an ongoing trial targeting infant diarrhea in Bangladesh. Harald Brüssow has long been working on the very serious problem of 3rd-world infant diarrhea. Major progress in targeting the viral causes was made using vaccines, but they were ineffective in targeting the 37 % that were the consequence of coliform infections and the decision was made to try phages. A number of the broadest-spectrum T4-like phages (Kutter 2009) were tested against bacteria collected from infant diarrheal patients in Dakka, Bangladesh over a decade ago; some worked, but their efficiency of plating was not high enough to seem promising. Therefore, new phages were isolated from 120 stool samples of young Bangladeshi patients (Chibani-Chennoufi et al. 2004). Two different *E.coli* strains were chosen: a widespread enteropathogenic (EPEC) strain, O127:K63, and the common lab strain K803. Remarkably, none of the 18 phages isolated using the EPEC strain hit more than a few of the *E. coli* strains isolated from the Bangladeshi infants; these were all siphoviridae belonging to 3 families. In contrast, when K803 was used for the selection, all but one were T4-like and many of these had very broad host ranges on the common Bangladeshi strains and also on a large set of EPEC and EHEC bacteria reflecting 21 different O, 10 K and 10 H antigens. Only 12 out of 31 phages tested were T-even *sensu stricto*, while 19 represented more distant members of the T4 family in 4 different genera. Building on successful in-vitro tests, preliminary animal trials and human safety studies (Bruttin and Brüssow 2005), both the search for the most effective phages for Bangladesh infants and the animal studies were expanded substantially (Denou et al. 2009). They obtained 46 epidemiologically independent pathogenic *E. coli* patient isolates from the International Center for Diarrheal Disease Research in Dhaka, Bangladesh (ICDDR,B): 15 EHEC, 15 ETEC, 10 enteroaggregative (EAggEC), and 3 enteroinvasive (EIEC) strains as well as 3 verotoxin-producing *E. coli* (VTEC). These strains were tested against T4-like phages, and a set of 10 phages was found which was giving a 52 % coverage, with 16 giving a 2/3 coverage. Subsequently, all of these phages have been sequenced. They were then tested for their gut survival capacities in mice which were previously inoculated or not with 10^{10} phage-sensitive *E. coli* K12 by intragastric tube, and the concentrations in 18 segments of the gut were measured.

A major double blind, placebo-controlled clinical trial evaluating the safety and efficacy of these phages was started in August, 2009 in Bangladesh (www.clinicaltrials.gov/ct2/show/NCT00937274). The initial steps evaluated the safety of the mixture on adults and older children. The therapy is being applied as a supplement to the standard oral rehydration solution, and also involves a second arm using a commercially available Russian anti-*E. coli* phage cocktail (from Microgen) as well as a placebo (oral rehydration solution only). The preparations are being administered at a dose of 10^6 PFU/ml in standard rehydration fluid for up to 5 days to treat diarrhea due to enterotoxigenic *E. coli* (ETEC) and enteropathogenic *E. coli* (EPEC) in children aged 6–24 months. The main outcome measures for the study include duration of diarrhea, daily and cumulative stool output, volume of oral rehydration solution intake, stool frequency, time to recovery, and weight gain. Encouraging interim results were presented at the 2012 Brussels Viruses of Microbes meeting and also published by Brüssow (2012).

8.7 Conclusion

As discussed above, phages have many potential advantages:

- They are self-replicating but also self-limiting, since they multiply only in the presence of the appropriate bacteria
- They can be targeted much more specifically than most antibiotics to the problem bacteria, causing far less of the bacterial imbalance or "dysbiosis" that are major problems with antibiotics, often leading to serious secondary infections involving relatively resistant bacteria that can increase hospitalization time,

expense, and mortality. Particularly difficult resultant problems are *Pseudomonas aeruginosa* and *Clostridium difficile*, the cause of serious diarrhea and membranous colitis (Fekety et al. 1997). Phages can often be targeted to receptors on the bacterial surface that are involved in pathogenesis, so any resistant mutants are less problematic

- No serious side effects have been reported for phage therapy
- Phage therapy would be particularly useful for people with allergies to antibiotics
- Appropriately selected phages can easily be used prophylactically to help prevent bacterial disease at times of exposure or to sanitize hospitals and help protect against hospital-acquired infections
- Phages can be prepared fairly inexpensively and locally, especially for external applications, facilitating their potential applications to underserved populations
- Phages can be used either independently or in conjunction with other antibiotics to help reduce the development of bacterial resistance

The time has come to look more carefully at the potential of phage therapy for future practice, both by strongly supporting new research and by scrutinizing the research already available, such as the very interesting human anti-typhoid phage research carried out in the 1940s (Knouf et al. 1946; Desranleau 1949) as well as the extensive earlier work in France, the US, Georgia, Poland and Russia, as explored above.

With the enormous possibilities and decreasing costs of genomic analysis, it is now possible to perform genomic sequencing of the phages included in cocktails so as to know more about the phage families involved and exclude phages from temperate families, as they may possibly carry or acquire genes related to pathogenicity or toxin production. This is now the standard procedure for therapeutic phages being developed in the West. Such modern techniques are now also being applied to some of the Georgian phage preparations with the financial support of the International Science and Technology Centers (ISTC) programs (Kvachadze et al. 2011; Karumidze et al. 2012) and a new high-tech Georgian Centers for Disease Control facility, built by the US military DARPA program. This is an important step in considering the importation of such phages for topical use in the Western world. Also, a broader consideration of other kinds of data and trials is developing (Vandenbroucke et al. 2004; Pirnay et al. 2012; Brussow 2012; Verbecken et al. 2012).

Although it seems premature to broadly introduce injectable phage preparations in the West without further extensive research, their carefully implemented use in external applications and for a variety of agricultural purposes could potentially help reduce the emergence of antibiotic-resistant strains and deal with problems we have difficulty handling today. Furthermore, compassionate use of appropriate phages seems warranted in cases in which bacteria resistant to all available antibiotics are causing life-threatening illness. Phages are especially useful in dealing with recalcitrant nosocomial infections, in which large numbers of particularly vulnerable people are being exposed to the same strains of bacteria in a closed hospital setting. In these cases especially, the environment as well as the patients can be effectively treated with phage preperations.

Acknowledgments We especially appreciate the efforts of Drs Liana Gachechiladze, Amiran Meipariani, Nino Chanishvili, Mzia Kutateladze, Rezo Adamia, Ramaz Katsarava and their colleagues in Tbilisi and of Beata Weber-Dabrowski, Andre Gorski, and others in Wroclaw to further develop phage therapy and help us understand the extensive therapeutic work carried out there. We express our thanks also to the many phage biologists and health-care personnel now working to bring phage therapy back into the Western World.

References

Abedon S (2011) Phage therapy pharmacology: calculating phage dosing. Adv Appl Microbiol 77:1–40

Abedon ST, Thomas-Abedon C (2010) Phage therapy pharmacology. Curr Pharm Biotechnol 11:28–47

Abedon S, Kuhl S, Blasdel B, Kutter E (2011) Phage treatment of human infections. Bacteriophage 1:66–85

Alavidze Z, Meiphariani A, Dzidzishvili L, Chkonia I, Goderdzishvili M, Kvatadze N, Jgenti D, Makhatadze N, Gudumidze N, Gvasalia G (2007) Treatment of the complicated forms of inflamatory wounds by bacteriophages. Med J Ga 2:123–127

Appelmans R (1921) Le bacteriophage dans l'organisme. C R Seances Soc Biol Fil 85:722–724

Bachrach G, Leizerovici-Zigmond M, Zlotkin A (2003) Bacteriophage isolation from human saliva. Lett Appl Microbiol 36:50–53

Bergh O, Børsheim KY, Bratbak G, Heldal M (1989) High abundance of viruses found in aquatic environments. Nature 340:467–468

Bogovazova GG, Voroshilova NN, Bondarenko VM et al (1992) Immunobiological properties and therapeutic effectiveness of preparations from Klebsiella bacteriophages. Zh Mikrobiol Epidemiol Immunobiol 3:30–33

Boyd EF (2005) Bacteriophages and bacterial virulence. In: Kutter E, Sulakvelidze A (eds) Bacteriophages: biology and application. CRC Press, Boca Raton, pp 223–266

Bruessow H (n.d.) Antibacterial treatment against diarrhea in oral rehydration solution. http://www.clinicaltrials.gov/ct2/show/NCT00937274. Accessed 4 Mar 2012

Brüssow H (2007) Phage therapy: the Western perspective. In: Mc Grath S, van Sinderen D (eds) Bacteriophage: genetics and microbiology. Caister Academic Press, Norfolk, pp 159–192

Brüssow H (2012) What is needed for phage therapy to become a reality in Western medicine? Virology 434(2):138–142

Bruttin A, Brüssow H (2005) Human volunteers receiving *Escherichia coli* phage T4 orally: a safety test of phage therapy. Antimicrob Agents Chemother 49:2874–2878

Bruynoghe R, Maisin J (1921) Essais de thérapeutique au moyen du bactériophage du Staphylocoque. C R Soc Biol 85:1120–1121

Caldwell JA (1928) Baceriologic and bacteriophagic study of infected urines. J Infect Dis 43:353–362

Campbell A (2006) General aspects of lysogeny. In: Calendar R (ed) The bacteriophages. Oxford University Press, Oxford, pp 66–73

Carmody LA, Gill JJ, Summer EJ, Sajjan US, Gonzalez CF, Young RF, LiPuma JJ (2010) Efficacy of bacteriophage therapy in a model of *Burkholderia cenocepacia* pulmonary infection. J Infect Dis 201:264–271

Chanishvili N (2012) A literature review of the practical application of bacteriophage research. Nova Science Publishers, Inc., New York

Chanishvili N, Sharp R (2009) A literature review of the practical application of bacteriophage research. Eliava Institute, Tbilisi

Chibani-Chennoufi S, Sidoti J, Bruttin A, Dillmann ML, Kutter E, Qadri F, Sarker SA, Brüssow H (2004) Isolation of Escherichia coli bacteriophages from the stool of pediatric diarrhea patients in Bangladesh. J Bacteriol 186(24):8287–8294

Deghorain M, van Melderen L (2012) The staphylococcal phages family: an overview. Viruses 4:3316–3335

d'Herelle F (1922) The bacteriophage: its role in immunity (trans: Smith GH). Williams & Wilkins, Baltimore

d'Herelle F (1938) Appendix from: Le Phénomene de la Gueras (trans: Kuhl S, Mazure H), Bacteriophage 1(2011):55–65

Debarbieux L, Leduc D, Maura D, Morello E, Criscuolo A, Grossi O, Balloy V, Touqui L (2010) Bacteriophages can treat and prevent Pseudomonas aeruginosa lung infections. J Infect Dis 201(7):1096–1104

Denou E, Bruttin A, Barretto C, Ngom-Bru C, Brüssow H, Zuber S (2009) T4 phages against Escherichia coli diarrhea: potential and problems. Virology 388(1):21–30

Desranleau JM (1949) Progress in the treatment of typhoid fever with Vi bacteriophages. Can J Public Health 40:473–478

Dublanchet A (2009) Des virus pour combattre les infections: la phagothérapie. Favre, Lausanne

Dubos RJ, Straus JH, Pierce C (1943) The multiplication of bacteriophage in vivo and its protective effects against an experimental infection with Shigella dysenteriae. J Exp Med 78:161–168

Eaton MD, Bayne-Jones S (1934) Bacteriophage therapy: review of the principles and results of the use of bacteriophage in the treatment of infections (I). JAMA 103:1769–1776, 1847–1853 and 1934–1939

Evans AC (1933) Inactivation of antistreptococcus bacteriophage by animal fluids. Public Health Rep 48:411–446

Fekety R, McFarland LV, Surawicz CM, Greenberg RN, Elmer GW, Mulligan ME (1997) Recurrent Clostridium difficile diarrhea: characteristics of and risk factors for patients enrolled in a prospective, randomized, double-blind trial. Clin Infect Dis 24:324–333

Friedman DI, Court DL (2001) Bacteriophage lambda: alive and well and still doing its thing. Curr Opin Microbiol 4:201–207

Geier MR, Attallah AF, Merril CR (1975) Characterization of Escherichia coli bacterial viruses in commercial sera. In Vitro 11:55–58

Górski A, Wazna E, Dabrowska BW, Dabrowska K, Switała-Jeleń K, Miedzybrodzki R (2006) Bacteriophage translocation. FEMS Immunol Med Microbiol 46(3):313–319

Górski A, Miedzybrodzki R, Borysowski J et al (2009) Bacteriophage therapy for the treatment of infections. Curr Opin Investig Drugs 10:766–774

Górski A, Międzybrodzki R, Borysowski J, Dąbrowska K, Wierzbicki P, Ohams M, Korczak-Kowalska G, Olszowska-Zaremba N, Łusiak-Szelachowska M, Kłak M, Jończyk E, Kaniuga E, Gołaś A, Purchla S, Weber-Dąbrowska B, Letkiewicz S, Fortuna W, Szufnarowski K, Pawełczyk Z, Rogoż P, Kłosowska D (2012) Phage as a modulator of immune responses: practical implications for phage therapy. Adv Virus Res 83:41–71

Häusler T (2003) Gesund durch Viren. Piper, Munich

Häusler T (2006) Viruses vs. superbugs: a solution to the antibiotic crisis. Macmillan, New York

Hawkins C, Harper D, Burch D, Anggård E, Soothill J (2010) Topical treatment of Pseudomonas aeruginosa otitis of dogs with a bacteriophage mixture: a before/after clinical trial. Vet Microbiol 146:309–313

Karumidze N, Thomas JA, Kvatadze N, Goderdzishvili M, Hakala KW, Weintraub ST, Alavidze Z, Hardies SC (2012) Characterization of lytic Pseudomonas aeruginosa bacteriophages via biological properties and genomic sequences. Appl Microbiol Biotechnol 94(6):1609–1617

Katsarava R, Beridze V, Arabuli N et al (1999) Amino acid-based bioanalogous polymers. Synthesis and study of regular poly (ester amide)s based on bis(a-amino acid), alpha, omega-alkylene diesters, and aliphatic dicarboxylic acids. J Polym Sci 37:391–407

King WE, Boyd DA, Conlin JH (1934) The cause of local reactions following the administration of Staphylococcus bacteriophage. Am J Clin Pathol 4:336–345

Knouf EG, Ward WE, Reichle PA et al (1946) Treatment of typhoid fever with type specific bacteriophage. JAMA 132:134–138

Krylov V, Shaburova O, Krylov S, Pleteneva E (2012) A genetic approach for the development of new therapeutic phages to fight Pseudomonas aeruginosa in wound infections. Viruses 4. doi:10.3390/v40x00x

Kuhl S, Mazure H (2011) F. D'Herelle preparation of therapeutic bacteriophages. From: Le Phénomène de la Guérison dans les maladies infectieuses, 1938, Masson et Cie, Paris – Appendix I (translated into English by Kuhl S, Mazure H). Bacteriophage 1:3–13

Kutateladze M, Adamia R (2010) Bacteriophages as potential new therapeutics to replace or supplement antibiotics. Trends Biotechnol 28:591–595

Kutter E (2008) Phage therapy: bacteriophages as naturally occurring antimicrobials. In: Goldman E, Green LH (eds) Practical handbook of microbiology. CRC Press, Boca Raton, pp 713–730

Kutter E (2009) Bacteriophage therapy: past and present. In: Schaechter M (ed) Encyclopedia of microbiology. Elsevier, New York, pp 258–266

Kutter E, Sulakvelidze A (2005) Bacteriophages: biology and application. CRC Press, Boca Raton

Kutter E, De Vos D, Gvasalia G, Alavidze Z, Gogokhia L, Kuhl S, Abedon ST (2010) Phage therapy in clinical practice: treatment of human infections. Curr Pharm Biotechnol 11:69–86

Kutter E, Borysowski J, Międzybrodzki R, Górski A, Kutateladze M, Alavidze Z, Goderdzishvili M, AdamiaR (2013) Clinical phage therapy. In: Borysowski J et al (eds) Phage therapy: current research and applications. Caister Academic Press (in press)

Kvachadze L, Balarjishvili N, Meskhi T, Tevdoradze E, Skhirtladze N, Pataridze T, Adamia R, Topuria T, Kutter E, Rohde C, Kutateladze M (2011) Evaluation of lytic activity of staphylococcal bacteriophage Sb-1 against freshly isolated clinical pathogens. Microb Biotechnol 4(5):643–650

Lang LH (2006) FDA approves use of bacteriophages to be added to meat and poultry products. Gastroenterology 131(5):1370

Larkum NW (1929) Bacteriophage from public health standpoint. Am J Public Health 19:31–36

Leszczyński P, Weber-Dąbrowska B, Kohutnicka M, Łuczak M, Górecki A, Górski A (2006) Successful eradication of methicillin-resistant Staphylococcus aureus (MRSA) intestinal carrier status in a healthcare worker – case report. Folia Microbiol (Praha) 51(3):236–238

Letarov AV, Golomidova AK, Tarasyan KK (2010) Ecological basis of rational phage therapy. Acta Naturae 2:60–71

Letkiewicz S, Miedzybrodzki R, Fortuna W, Weber-Dabrowska B, Górski A (2009) Eradication of Enterococcus faecalis by phage therapy in chronic bacterial prostatitis – case report. Folia Microbiol (Praha) 54:457–461

Little J (2006) Gene regulatory circuit of phage lambda. In: Calendar R (ed) The bacteriophages, 2nd edn. Oxford University Press, Oxford, pp 74–82

Lobocka M, Hejnowicz MS, Dabrowski K, Gozdek A, Kosakowski J, Witkowska M, Ulatowska MI, Weber-Dabrowska B, Kwiatek M, Parasion S, Gawor J, Kosowska H, Glowacka A (2012) Genomics of staphylococcal Twort-like phages – potential therapeutics of the post-antibiotic era. Adv Virus Res 83:143–216

MacNeal WJ, Frisbee FC, McRae MA (1942) Staphylococcemia 1931–1940. Five hundred patients. Am J Clin Pathol 12:281–294

Maisin RJ (1921) Essais de therapeutique au moyen du bacteriophage du staphylocoque. C R Soc Biol 85:1120–1121

Markoishvili K, Tsitlanadze G, Katsarava R, Morris JG Jr, Sulakvelidze A (2002) A novel sustained-release matrix based on biodegradable poly (ester amide)s and impregnated with bacteriophages and an antibiotic shows promise in management of infected venous stasis ulcers and other poorly healing wounds. Int J Dermatol 41:453–458

Marza JAS, Soothill JS, Boydell P, Collyns TA (2006) Multiplication of therapeutically administered bacteriophages in Pseudomonas aeruginosa infected patients. Burns 32:644–646

Merabishvili M, Pirnay JP, Verbeken G, Chanishvili N, Tediashvili M, Lashkhi N, Glonti T, Krylov V, Mast J, Van Parys L, Lavigne R, Volckaert G, Mattheus W, Verween G, DeCorte P, Rose T, Jennes S, Zizi M, De Vos D, Vaneechoutte M (2009) Quality-controlled small-scale production of a well-defined bacteriophage cocktail for use in human clinical trials. PLoS One 4(3):e4944

Merril CR, Friedman TB, Attallah AF (1972) Isolation of bacterophages from commercial sera. In Vitro 8:91–93

Miedzybrodzki R, Fortuna W, Weber-Dabrowska B, Gorski A (2007) Phage therapy of staphylococcal infections (including MRSA) may be less expensive than antibiotic treatment. Postepy Hig Med Dosw 61:461–465

Międzybrodzki R, Borysowski J, Weber-Dąbrowska B, Fortuna W, Letkiewicz S, Szufnarowski K, Pawełczyk Z, Rogóż P, Kłak M, Wojtasik E, Górski A (2012) Clinical aspects of phage therapy. Adv Virus Res 83:73–121

Milch H, Fornosi F (1975) Bacteriophage contamination in live poliovirus vaccine. J Biol Stand 3:2307–2310

Moody EE, Trousdale MD, Jorgensen JH, Shelokov A (1975) Bacteriophages and endotoxin in licensed live-virus vaccines. J Infect Dis 131:588–591

Montclos H (2002) Les bacteriophages therapeutique: de l'emprirism ala biologie moleculaire. Pyrexie 6:77–80

Morton HE, Engely FB (1945) Dysentery bacteriophage: review of the literature on its prophylactic and therapeutic uses in man and in experimental infections in animals. JAMA 17:584–891

Pirnay JP, Verbecken G, Rose R, Jennes S, Zizl M, Huys I, Lavigne R, Merabishvili M, Vaneechoutte M, Buckling A, De Vos D (2012) Introducing yesterday's phage therapy in today's medicine. Future Virol 7:379–390

Slopek S, Durlakowa I, Weber-Dabrowska B, Kucharewicz-Krukowska A, Dabrowski M, Bisikiewicz R (1983a) Results of bacteriophage treatment of suppurative bacterial infections. I. General evaluation of the results. Arch Immunol Ther Exp 31:267–291

Slopek S, Durlakowa I, Weber-Dabrowska B, Kucharewicz-Krukowska A, Dabrowski M, Bisikiewicz R (1983b) Results of bacteriophage treatment of suppurative bacterial infections. II. Detailed evaluation of the results. Arch Immunol Ther Exp (Warsz) 31(3):293–327

Slopek S, Durlakowa I, Weber-Dąbrowska B, Dąbrowski M, Kucharewicz-Krukowska A (1984) Results of bacteriophage treatment of suppurative bacterial infections. III. Detailed evaluation of the results obtained in further 150 cases. Arch Immunol Ther Exp 32:317–335

Slopek S, Kucharewicz-Krukowska A, Weber-Dabrowska B, Dabrowski M (1985a) Results of bacteriophage treatment of suppurative bacterial infections. VI. Analysis of treatment of suppurative staphylococcal infections. Arch Immunol Ther Exp 33:261–273

Slopek S, Kucharewicz-Krukowska A, Weber-Dąbrowska B, Dąbrowski M (1985b) Results of bacteriophage treatment of suppurative bacterial infections. V. Evaluation of the results obtained in children. Arch Immunol Ther Exp 33:241–259

Slopek S, Kucharewicz-Krukowska A, Weber-Dąbrowska B, Dąbrowski M (1985c) Results of bacteriophage treatment of suppurative bacterial infections. IV. Evaluation of the results obtained in 370 cases. Arch Immunol Ther Exp 33:219–240

Slopek S, Weber-Dabrowska B, Dabrowski M, Kucharewicz-Krukowska A (1987) Results of bacteriophage treatment of suppurative bacterial infections in the years 1981–1986. Arch Immunol Ther Exp (Warsz) 35:569–583

Smith HW, Huggins RB (1982) Successful treatment of experimental E. coli infections in mice using phage: its general superiority over antibiotics. J Gen Microbiol 128:307–318

Smith HW, Huggins RB (1983) Effectiveness of phages in treating experimental E. coli diarrhoea in calves, piglets and lambs. J Gen Microbiol 129:2659–2675

Smith HW, Huggins RB (1987) The control of experimental E. coli diarrhea in calves by means of bacteriophage. J Gen Microbiol 133:1111–1126

Smith HW, Huggins RB, Shaw KM (1987) Factors influencing the survival and multiplication of bacteriophages in calves and in their environment. J Gen Microbiol 133:1127–1135

Straub ME, Appelbaum M (1933) Studies on commercial bacteriophage products. JAMA 100:110–113

Sulakvelidze A, Barrow P (2005) Bacteriophage therapy in humans. In: Kutter E, Sulakvelidze A (eds) Bacteriophages: biology and application. CRC Press, Boca Raton, pp 335–380

Sulakvelidze A, Kutter E (2005) Bacteriophage therapy in humans. In: Kutter E, Sulakvelidze A (eds) Bacteriophages: biology and application. CRC Press, Boca Raton, pp 381–436

Sulakvelidze A, Alavidze A, Morris J (2001) Bacteriophage therapy. Antimicrob Agents Chemother 45:649–659

Summers WC (1999) Felix d'Herelle and the origins of molecular biology. Yale University Press, New Haven

Summers WC (2001) Bacteriophage therapy. Ann Rev Microbiol 55:437–451

Vandenbroucke JP (2004) When are observational studies as credible as randomized trials? Lancet 363:1728–1731

Vandersteegen K, Mattheus W, Ceyssens PJ, Bilocq F, De Vos D, Pirnay JP, Noben JP, Merabishvili M, Lipinska U, Hermans K, Lavigne R (2012) Microbiological and molecular assessment of bacteriophage ISP for the control of *Staphylococcus aureus*. PLoS One 6(9):e24418

Verbeken G, Pirnay JP, De Vos D, Jennes S, Zizi M, Lavigne R, Casteels M, Huys I (2012) Optimizing the European regulatory framework for sustainable bacteriophage therapy in human medicine. Arch Immunol Ther Exp (Warsz) 60(3):161–172

Vieu JF (1961) Intérêt des bactériophages dans le traitement de staphylococcies. Vie Med 42:823–829

Vieu JF (1975) Les bacteriphages. In: Fabre J (ed) Traité de therapeutique, Serums et vaccins. Flammarion, Paris, pp 337–430

Vieu JF, Guillermet F, Minck R et al (1979) Données actuelles sure les applications therapeutiques des bacteriophages. Bull Acad Natl Med 163:61–66

Waldor M, Friedman D, Adha S (2005) Phages: their role in bacterial pathogenesis and biotechnology. ASM Press, Washington, DC

Ward WE (1943) Protective action of VI bacteriophage in *Eberthella typhi* infections in mice. J Infect Dis 72:172–176

Weber-Dabrowska B, Debrowska M, Slopek S (1987) Studies on bacteriophage penetration in patients subjected to phage therapy. Arch Immunol Ther Exp (Warsz) 35:363–368

Weber-Dabrowska B, Mulczyk M, Gorski A (2000) Bacteriophage therapy of bacterial infections: an update of our institute's experience. Arch Immunol Ther Exp 48:547–551

Weber-Dabrowska B, Mulczyk M, Górski A (2001) Bacteriophage therapy for infections in cancer patients. Clin Appl Immunol Rev 1:131–134

Weber-Dąbrowska B, Mulczyk M, Górski A (2003) Bacteriophages as an efficient therapy for antibiotic-resistant septicemia in man. Transplant Proc 35(4):1385–1386

Wright A, Hawkins CH, Anggård EE, Harper DR (2009) A controlled clinical trial of a therapeutic bacteriophage preparation in chronic otitis due to antibiotic-resistant *Pseudomonas aeruginosa*; a preliminary report of efficacy. Clin Otolaryngol 34:349–357

Yeung MK, Kozelsky CS (1997) Transfection of *Actinomyces spp.* by genomic DNA of bacteriophages from human dental plaque. Plasmid 35:141–153

Chapter 9
Animal-Assisted Therapy: Benefits and Challenges

Mary Cole and Maureen Howard

9.1 Introduction

As counsellors working in southern Alberta, Canada, we have been fortunate to practice animal-assisted therapy to supplement more traditional counselling methods. We have seen that the presence of an animal can speed up rapport building, break down communication barriers and create a meaningful platform for learning, personal growth and desired change. At the same time, we have also directly experienced the challenges and ethical considerations when working with therapy animals (Fig. 9.1).

Our clients' positive response to animals is reflected in much of the literature compiled by the psychotherapeutic community. Practitioners working with animals have long-recognized positive effects for those who experience a wide range of maladies, including dementia, depression, autism, trauma, low motivation and self-esteem, behavioural issues, and various psychiatric ailments (Chandler 2005). The human-animal interaction is also helpful in facilitating treatment compliance, easing tensions in difficult situations and providing alternative methods for teaching various skills.

This chapter will briefly review the history of animals in therapy as well as the research evaluating benefits. Ethical issues and challenges will also be addressed and directions for future research will be discussed.

M. Cole, M. Couns., CCC (✉)
Cole and Associates Counselling Services, #270 Point McKay Terrance NW,
Calgary, Alberta, T3B 4V6, Canada
e-mail: mlcole@ucalgary.ca

M. Howard, M.Sc., R. Psych.
4057 Hacking Rd., Tappen, B.C., V0E 2X0, Canada
e-mail: maureenhoward@rocketmail.com

M. Grassberger et al. (eds.), *Biotherapy - History, Principles and Practice:*
A Practical Guide to the Diagnosis and Treatment of Disease using Living Organisms,
DOI 10.1007/978-94-007-6585-6_9, © Springer Science+Business Media Dordrecht 2013

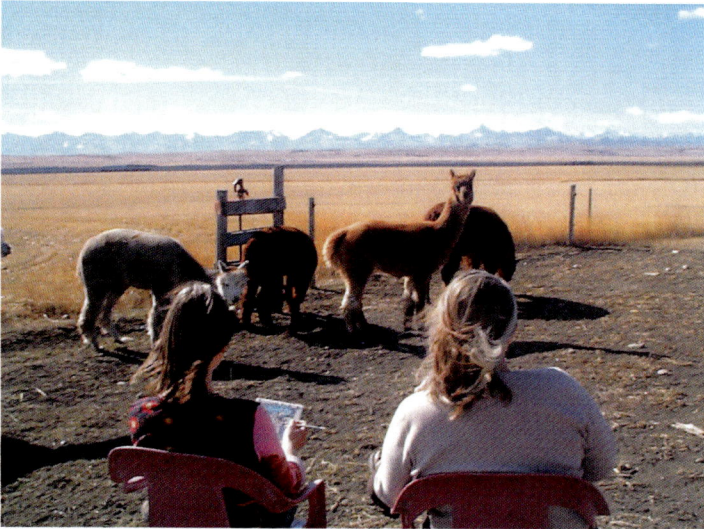

Fig. 9.1 Mary and Maureen observing therapy animals (Alpacas) in the field

9.2   History of Animals in Therapy

Historical accounts of animals contributing to the well-being of people date back to the1600s, including John Locke's discussion of the use of small animals to help cultivate empathy and responsibility in children (as cited in Fine 2000). Many other accounts beyond Locke's descriptions are also recorded in the literature. In 1792, for example, a Quaker retreat in England used farm animals to treat mental health patients in an effort to reduce the need for isolation and restraints (Baun and McCabe 2000). Over the next 100 years, mental health institutions, such as Bethel, a treatment centre in Europe for patients with epilepsy, included pet animals in their work with patients (Fine 2000) However, with the emergence of psychotropic drugs on the medical scene, references to the involvement of animals all but disappeared from the literature until the1940s, when James Bossard (as cited in Fine 2000) began to record his observations of the positive physical and emotional effects of pet ownership. Around that time, the U.S. military introduced animals into their work with veterans at a convalescent hospital in 1942.

In the1960s Boris Levinson coined the term *pet therapist* to refer to his dog, Jingles, who participated in his therapeutic work (Levinson 1964). This term marked the emergence of interest by researchers and practitioners in the psychological effects of human and animal interaction, and highlighted the critical shift to regard animals as partners in therapy rather than as tools to be exploited (Zamir 2006). Alan Beck, a well-known researcher in the field of animal-assisted therapy, and Aaron Katcher, a psychiatrist and instructor, have both contributed greatly to the development of public and professional understanding of the human-animal bond.

In the early 1980's, Beck and Katcher suggested animals could promote physical and mental health, offer companionship and even provide therapy. The second edition of their book in 1996 included research to help support their contention that human-animal relationships are not only necessary but can easily be integrated into psychological work with people (Beck and Katcher 1996). Significant growth in the field was evident in the1990s with an increasing number of written accounts outlining the benefits of the human-animal bond, and a growing number of front-line practitioners integrating animals into their practice (Taylor 2001). Currently, these anecdotal accounts (or studies with small sample sizes) outweigh rigorous experimental research available for review.

As recognition and interest in the field has unfolded, the description of how animals are included in various settings has proven to be broad and sometimes unclear, leaving both professionals and the public confused by the terminology as terms vary tremendously and are often used interchangeably in literature and on websites. To help clarify the confusion, we have chosen to refer to Pet Partners (2012), formally known as the Delta Society, to assist with the terminology. Pet Partners is an international non-profit organization established by veterinarians in the mid-1970s to increase awareness, knowledge, research, and protect animal welfare in human interactions with service and therapy animals, as well as with companion pets. They offer two broad categories of animal therapy: animal-assisted activities (AAA) and animal-assisted therapy (AAT). AAA is defined as the integration of animals into activities that facilitate education, recreation, motivation, and encourage casual interaction with no defined set of criteria or goals. AAA could include volunteers and their pets visiting a nursing home to encourage positive socialization and interaction amongst residents. While considered therapeutic in nature, there are no set goals and no planned outcomes or evaluation required.

On the other hand, Pet Partners describe AAT as intentional and therapeutic, whereby the animal's role is integral in assisting with mental health, speech, occupational therapy or physical therapy goals, and augments cognitive, physical, social and/or emotional well-being. Their criteria for AAT includes the following: the animal must meet specific criteria that fit the therapeutic goals; the animal is considered a necessary part of the treatment; therapy is directed by a qualified professional or practitioner; therapeutic intentions include physical, social, emotional, or cognitive gains; therapy can occur in group or individual sessions; and all treatment must be documented and evaluated.

Another term often used in the field is *pet therapy*. For instance, The Pet Access League Society (PALS 2012) is a well-respected non-profit organization in Alberta, Canada that utilizes animals to enhance their clients' quality of life. Through their *Pet Visitation Therapy Program*, volunteers and their animals visit facilities such as nursing homes, correctional centres and day homes (Fig. 9.2). Therefore, based on Pet Partner definitions, the pet therapy conducted by PALS would fall under the category of AAA because visits are non-directive and delivered by volunteers; animals are not involved in an intentional way to help clients reach specific therapeutic goals. In 2011, Behling et al. followed-up their 1990 survey of animal programs in long-term care facilities in Illinois, USA (Behling et al. 2011). Interestingly, this

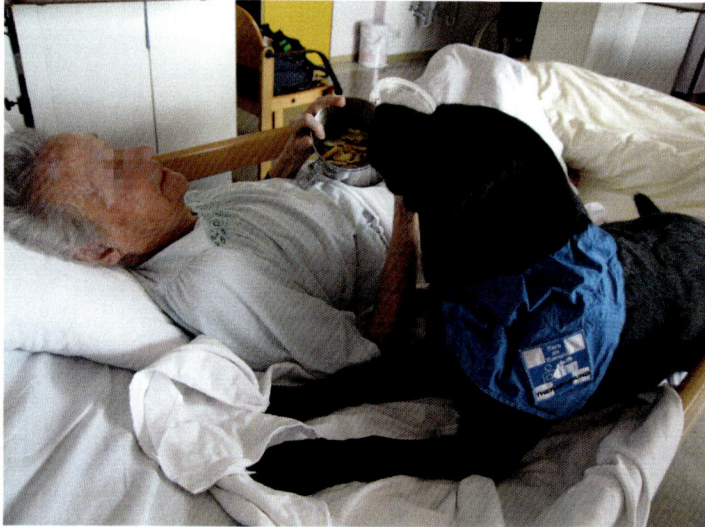

Fig. 9.2 A woman enjoying a relaxing visit from a pet therapy dog (Photo courtesy of Verein "Tiere als Therapie – TAT", Vienna, Austria)

longitudinal study revealed that the number of AAT programs had not increased since 1990 and about the same number of AAA programs have continued in these facilities. However, AAA programs are now considered a more formalized and necessary part of programming in these facilities, and respondents continue to express their support and the need for expansion of AAA.

9.3 Practical Applications

While more research is required to truly determine what makes the human-animal bond so powerful, researchers have begun to identify theoretical approaches that are best suited to the use of AAT and the characteristics of clients who seem to benefit from AAT. The following are a few common examples of accepted therapeutic approaches involving animals within the medical and psychotherapeutic communities.

9.3.1 Therapy with Horses

The term *Hippotherapy* refers to an organized and structured approach which emerged in the1960s to integrate horses into physical rehabilitation therapy (Chandler 2005). On their website, The American Hippotherapy Association (2012), known as AHA, described the therapy as a method of utilizing horse movement to complement physical, occupational and speech-language therapy sessions, with therapy not limited to one type of practitioner. For instance, a child with a physical disability may practice gross motor coordination skills by learning to balance in a saddle, or may develop expressive language skills through commands and verbal interaction with the horse and therapist.

Taylor (2001) described Equine-Facilitated Psychotherapy, referred to as EFP, as a more current approach utilizing horses in therapy. Taylor specified that EFP is derived from AAT and can only be conducted by an accredited mental health professional specifically trained in utilizing horses as part of their counseling intervention. Chandler (2005) noted that horses selected for therapy should be "well trained, calm, and friendly towards people and other horses. A therapy horse must not startle easily to noises or unfamiliar objects" (p. 31). There are a number of strengths as well as challenges to involving horses in therapy. For instance, the novelty of the horse can incite interest and involvement for many clients who may otherwise lack motivation to participate in therapy (Chandler 2005). Chandler also suggested the size and power of a horse can promote self-confidence when clients learn appropriate interaction, and the fact horses can be ridden may assist the counsellor with creating novel and interesting tasks. However, Chandler also pointed out horses require a large space and ongoing manure clean-up. As well, the potential for serious injury may outweigh the benefits and requires very experienced handlers to work with practitioners if the practitioner is not trained in the area of equine therapy. The Equine Assisted Growth and Learning Association (2012), commonly known as EAGALA, offers practical direction, education, and has developed professional standards for including horses in psychotherapy.

9.3.2   Therapy with Farm Animals

Although we refer to animal therapy as AAT in this chapter, various other terms are used across the world to describe the involvement of farm animals in practitioner's work with clients and patients, including *Green Farming*, *Green Care*, *Farming for Health*, and *Horticultural Therapy*. It was in Germany in 1987 where I (Mary Cole) first became aware of the practice of involving farm animals in therapy. I stumbled upon a small farm where several people were cleaning stalls and caring for the animals. An older gentleman milking a cow stopped his work to talk with me about how the farm was a private retreat for people recovering from a range of mental health issues. Each day was filled with chores and interactions with animals, followed by sessions with their assigned counsellor. He described the retreat as 'a break from the world', and explained how the animals, picked for their gentle qualities, were like therapists. He explained that the experience of caring for the animals encouraged him to move beyond his own reality and perhaps become more open to change.

Berget and Braastad (2011) work with the Department of Animal and Aquacultural Sciences in Norway. They describe Green Care as an umbrella term for therapeutic intervention using various stimuli, including horticulture, animals and outdoor experiences. Green Care can involve physical activities, and the emphasis is on using elements of nature to promote human health. Berget and Brastaad suggest that Green Care therapy for patients with psychiatric disorders has had promising results in helping to reduce symptoms of depression and increase self-efficacy. However, they also point out more research is required in the area. Relf (2006) offers an excellent overview of research and information about various programs involving Green Care across Europe as well as in the United States.

However, the involvement of farm animals can be both beneficial and challenging. Chandler (2005) and Mallon (1994) suggested all types of farm animals can be included in therapy as long as the handler is competent and the animal is safe to be around. We would like to add that it would also make sense that the animal is willing and interested in participating. Chandler suggested the level of training of both the animal and counsellor is dependent on the species and the type of involvement of the animal. One example of a successful farm therapy organization is Green Chimneys Children's Services in New York State, USA, a green farm residential treatment centre for youth-at-risk (Green Chimneys 2012). Green Chimneys has workshops, training for up and coming practitioners in the field, clear policy and guidelines established from many years of experience, and contributes to the overall education and research in the field of AAT. As Mallon et al. (2000) explained, the diversity of a farm experience offers much stimulation, and provides the basis for creative and varied interventions, such as providing the client with opportunities to experience and participate in nurturing activities, organizational skills, perspective-taking and problem solving.

The variety of farm animals allows the counsellor and client to choose the 'co-counsellor', or even offer a change of animal, if necessary. Concrete and meaningful daily activities may help to develop the client's sense of confidence and competence, as well as their skills. Chandler (2005) suggested disadvantages could include accessing appropriate locations, facilitating the transportation of clients, chance of serious injury, and a need for more manpower, such as an animal handler, to be involved and on-site.

9.3.3   Therapy with Dogs

Human-dog partnerships have traditionally provided a service for clients with disabilities, such as service dogs assisting people with visual impairments or with autism. Treatment was mainly focused on increasing safety and mobility for patients and clients so that they may experience more independence. Now, dogs are one of the primary animal choices for AAT (Beck and Katcher 2003) because of the relative ease of integrating dogs into various environments, the multitude of breed choices, their responsiveness and interest in humans, and their intelligence. Consequently a great deal of the human-bond and AAT research is focused on canine therapy. For instance, at the Maryland Research Institute on Aging in the United States, Marx et al. (2010) conducted research to determine whether the type of human-dog interaction would affect symptoms of patients diagnosed with Dementia. Fifty-six patients in two nursing homes viewed videos of puppies, interacted with stuffed dog toys and colouring book pictures of dogs, interacted with a real dog and even with a robotic dog. Marx and his colleagues determined that the real dog, along with the puppy video, scored the highest in improving the cognitive level of engagement of patients in their daily living activities. More information related to medical research in canine therapy is offered in the research and benefits section of this chapter.

Much like horses, dogs are expected to obey commands and to stay calm even in stressful situations. As well, it is expected that they will offer clients what is often referred to as "unconditional acceptance" (Chandler 2005, p. 28) and therefore consideration of the dog's temperament and sociability, trainability, and predictability is paramount (Urichuk and Anderson 2003). Chandler also suggested matching a dog's temperament and activity level to the client is important, along with provision for exercise, grooming, feeding and a place to defecate. To address animal welfare issues for dogs, whether in hospitals, mental health settings, schools or community spaces, a number of organizations, including Pet Partners, have developed resources, education and training as well as clear guidelines around canine selection and training.

9.3.4 Therapy with Small Animals

Small animals, such as gerbils, hamsters, guinea pigs, rabbits and even fish are utilized in psychotherapy. Referred to as *pocket pets* by Flom (2005), she noted these smaller animals often provide options in facilities with animal restrictions, such as hospital, classroom, or office settings. Most of the literature addressing pocket pets is under the category of AAA or pet therapy as these animals are often confined, can live in the facility, and can become part of the environment with relative ease (Hart 2000). Flom pointed out schools have recently imposed restrictions on various animals, such as reptiles, because of the fear of Salmonella, or on larger animals, such as dogs, because of injury risks. Flom also noted an animal in a school setting must fit the client's therapeutic needs, as well as somehow fit into the curricula. Pocket pets are also appropriate in facilities where clients would like to hold the animals, or where a larger animal may feel threatening. The shorter life span of these animals, ranging from 2 to 5 years, can be problematic, as can their tendency to be more fragile and their susceptibility to injuries and stress-related problems (Chandler 2005). Even now in 2012, little information was found in the literature regarding guidelines for integrating pocket pets into psychotherapy.

Fig. 9.3 Dogs and other animals can serve as therapeutic agents (Photo courtesy of Verein "Tiere als Therapie – TAT", Vienna, Austria)

9.3.5   AAT in Counselling

Chandler (2005) coined her own phrase for AAT in counseling as *AAT-C* to illustrate the difference in working with animals to specifically improve the psychological well-being of a client, versus all other types of therapies and activities. Chandler defined AAT-C as "the incorporation of pets as therapeutic agents into the counseling process" (p. 2) and described two methods of delivery. First, the counsellor may incorporate their own trained therapy pet, or second, they could supervise an animal handler who carries out specific interventions with the certified therapy animal (see Fig. 9.3). She explained AAT-C can occur in a variety of environments, including schools, private office settings, on acreages, and so forth, and responses to intervention are monitored and recorded to direct next steps in therapy.

In our AAT-C work with rural women diagnosed with depression for instance, we involved alpacas in various ways to supplement and reinforce each client's therapeutic goals. Easy to handle, gentle, intensely curious, trainable, and demonstrating clear non-verbals, their interest in people make alpacas ideal for working with certain clients. One client reported that observing and sometimes working with different alpacas helped her become more attuned to the connections between verbal and non-verbal information, such as the sound an alpaca makes when he is relaxed versus distressed, and how this is conveyed through the positioning of his ears, tail and face. By paying attention to these signals, she was able to more clearly identify areas for personal growth and provided us, as counsellors, with additional insight into issues that may not have come to light as quickly in a traditional therapy setting. In all cases, the women were encouraged to develop and practice positive visualization, voice and body regulation, and relaxed breathing to improve their level of calmness, control and assertiveness when working with their therapy animal. The women reported that the quietness of the country environment, the fresh air and novelty of the situation, the physical activity and the reward of their animal participating with them in the tasks, helped to reinforce the benefits of practicing these skills.

A third scenario, as described by Anderson (2008), may be to include the client's pet in the sessions if the counsellor feels competent in handling the animal and if the animal is motivated and comfortable. If it is not possible to physically include the animal, then inclusion of the pet could be achieved through methods such as storytelling, puppetry, stuffed animals, scrapbooking, metaphorical language, or photography (Chandler 2005; see also Reichart 1998).

Strategically involving animals to enhance the therapeutic experience can be daunting, particularly given the many theoretical models and approaches used by counsellors. Chandler et al. (2010) have published an informative and practical overview to guide the practitioner when matching AAT-C techniques with various counselling theories.

9.4 Research and Benefits in AAT

Over the last few decades, the medical community has laid the foundation for much of the research on the physiological benefits of the human-animal bond. For instance, a National Institute of Health (NIH) public health report (Friedmann et al. 1980) revealed an increased 1-year survival rate after a coronary attack for individuals who owned a pet. Findings in the early 1990s by Serpell (1991) reported that elderly patients who owned pets generally experienced healthier cardiovascular functioning than those who did not. In 1992, Anderson et al. reported pet ownership improved physical problems such as high blood pressure and cholesterol (Anderson et al. 1992). In 1998, Batson et al. were pioneers in studying the health benefits of integrating animals into therapy with patients afflicted with Alzheimer's disease, and they concluded that blood pressure levels improved after a participant's positive interaction with a dog (see Fig. 9.4, Batson et al. 1998). Edwards and Beck (2002) conducted an interesting study to observe whether the presence of fish tanks would improve the eating habits of patients with Alzheimer's disease. They found that the 62 patients in the study significantly improved their nutritional intake when in the calming presence of the fish, resulting in healthy weight gain and a reduced need for, and financial cost of, nutritional supplements.

In 2004, Gagnon et al. reported on Phase One of their study in Canada investigating the observable effects of a canine therapy program on children with cancer, their families and the participating nurses (Gagnon et al. 2004). The 1-year pilot project was established to address the unique issues faced by the children, including often long and painful treatments, safety and vulnerability issues, and increased stress

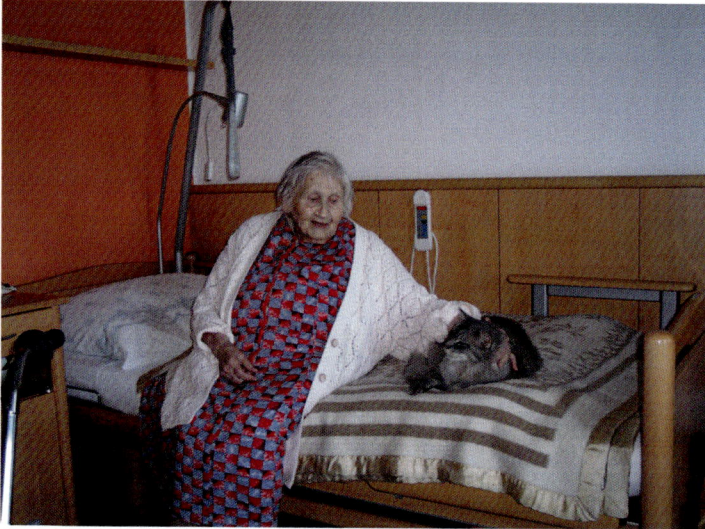

Fig. 9.4 Patients can experience positive physiological benefits when interacting with animals, as long as the patient is amenable (Photo courtesy of Verein "Tiere als Therapie – TAT", Vienna, Austria)

because of long periods of time away from home. It was designed to test the hypothesis that the presence of a dog may help to improve psychosocial conditions for all participants. As this was a tertiary care unit, there were many factors to consider and so the design was particularly rigorous. The researchers determined that the presence of a dog helped ease the stress for the children and their caregivers, as well as assist with the adjustment to the treatment program. Based on these positive results, it is now a permanent part of treatment.

Headey and Grabka (2007) concluded from their longitudinal population studies in both Germany and Australia that people who continuously own pets are the healthiest group of all as they have fewer doctor visits than non-owners or than those who have ceased to own a pet. In another study conducted by Braun et al. (2009), children hospitalized in an acute care setting were exposed to 15–20 min visits with a dog handler and a therapy dog. Pain symptoms, blood pressure and respiratory rates were compared to a control group who experienced 15–20 min of a calm setting without a dog. While both groups had lowered pain symptoms at completion, pain reduction was four times greater in the group with the dog.

As previously mentioned, there is much research investigating the efficacy of canine therapy on physiological as well as psychological well-being. In 2000 for example, Odendaal reported positive changes in neurochemicals, such as cortisol and dopamine, related directly to positive interactions with a dog (Odendaal 2000). In 2009, Miller et al. conducted a study to determine whether Oxytocin levels changed in study participants after interacting with their dog. Oxytocin is a neuro-peptide that has a role in bonding, socialization and reducing stress. The researchers found that the women in their study demonstrated statistically increased Oxytocin after interacting with their pet for 25 minutes. However, the men in the study actually experienced a decrease in the neuropeptide following interaction, and the researchers queried whether hormones, in addition to other variables such as personality traits and so forth play a role in the response of humans to canines. In another more recent example, Hamama et al. (2011) conducted a 3-month longitudinal study using a combination of cross-sectional and longitudinal designs, to determine whether canine therapy would reduce psychological stress and improve self-confidence and subjective well-being for teenage girls diagnosed with post-traumatic stress disorder (PTSD). Although the sample size was quite small, outcomes from their study suggested that by the end of the intervention all participants experienced a decline in their PTSD and depressive symptoms.

In 2012, Marcus et al. monitored outpatients at a pain clinic to see whether spending time with a therapy dog versus time in a traditional waiting room would impact their symptoms (Marcus et al. 2012). Pain, fatigue and emotional distress were found to be significantly reduced in these chronic pain patients. As well, therapy dogs had a similar effect on family and care-givers accompanying these patients. As encouraging as these results appear to be, Palley et al. (2010), note there continues to be much criticism regarding weak research design that has leading to skepticism about the effectiveness of AAT.

Occupational therapy and physiotherapy practitioners have also identified psychological benefits when using animals with their patients. For example, an occupational therapist with the U.S. Army utilized a dog named Albert to work with soldiers suffering from combat stress (Gregg 2012). He stated that not only did the presence of the dog ease tensions and discomfort when initially interviewing clients, he was able to use the dog to better teach assertiveness, communications and healthy coping.

Although the positive psychological effects of the human-animal bond have been observed and studied since the 1700s, it is only more recently that the psycho-therapeutic community has attempted to capture this information through more formalized research. They have begun to investigate how animals might aid in rapport building, increase opportunities for social chit-chat and improve clients' sense of safety and comfort in difficult situations (Beck and Katcher 2003; Chandler 2005).

Hunt et al. (as cited in Beck and Katcher 2003) described these interactions back in the early 1990s as the capacity for animals to act as *social lubricants*. This effect was highlighted in Marr et al.'s (2000) research study in which 69 men and women with a mean age of 41.5 years met together in therapy groups for one hour each day of the week for 4 weeks. Each participant had been diagnosed with a mental illness in addition to drug or alcohol abuse. Half of the patients were in a control group, the remaining participants met in the presence of a variety of animals including dogs, rabbits, guinea pigs, and ferrets. Based on the Social Behavior Scale given to both groups, Marr and his colleagues found the participants interacting with the animals were more inclined to smile and demonstrate pleasure, and were more sociable and relaxed with other participants. Prothmann et al. (2006) also found that depressed youth "may feel transported into an atmosphere that is characterized by warmth, acceptance and empathy" (p. 275), factors that promote a strong therapeutic alliance.

In her work as an AAT counsellor, Chandler (2005) finds that animals offer those clients who may not respond as well to more traditional talk therapy an opportunity for more meaningful, concrete and multisensory experiences. Katcher and Wilkins (2000) describe AAT as a positive way to maintain the attention and interest of a client who is active or struggles with focus or concentration.

Another notable theme in the research from the psychotherapeutic community is the positive effect of involving animals to enhance the therapeutic use of metaphors and symbolism. McIntosh (2002), a certified equine therapist in Alberta, Canada, states how she utilized the metaphor of 'reined in' to help resolve a conflictual parent/child relationship. Through horse work orchestrated by the counsellor, the family discovered that the horse was more compliant and responsive with a looser rein or leash. When held tight, it may well fight to gain control, or become passive and stubborn, much like a child on a tight rein. Through this experience, the clients became aware of how their behavior affected the animal's responses and consequently responses of family members.

Chandler (2005) pointed out another positive benefit of AAT in therapy is that clients with less-developed verbal skills can experience a sense of success when interacting with an animal without having to engage in using language. In 1994, Mallon summarized this concept as follows:

> Traditional forms of therapy, which rely on talking and trusting, sometimes fail children who are mistrustful of adults. The cow on the farm may in fact be the best therapist a child can have while in treatment. The cow, and other farm animals can become a companion for the child, one in whom he or she can confide all of his or her misgivings, heartaches, and pains. The cow and other farm animals can serve as the catalytic agent that brings the child and the therapist together (p. 470).

Leimer (1997) explained how unwritten rules in our communication system can make it difficult for some individuals to function well when they are not fluent in those rules, and noted that animals are often clearer because they communicate in a direct manner, both verbally and non-verbally. As Beck and Katcher (1996) pointed out, "animals do not use words, and patients can safely approach them when they cannot approach people" (p. 127) (see Fig. 9.5).

And so, while there are obvious benefits from involving animals within the medical and psychotherapeutic realms, it is still the combination of strategies and intentionality of involving an animal that dictates the success of the experience for the client (Aanderson 2008). Recently, Matuszek (2010) reinforced this notion after conducting an extensive literature review on how nurses involve animals in health care settings. She found that although there is evidence that animals are "therapeutic conduits" (p. 199), success very much depends on the situation, the patient and the reason for involving the animal. Until research clearly demonstrates that the various forms of AAT can stand alone, it continues to be recommended that AAT should be regarded as a valuable adjunct to other therapeutic modalities.

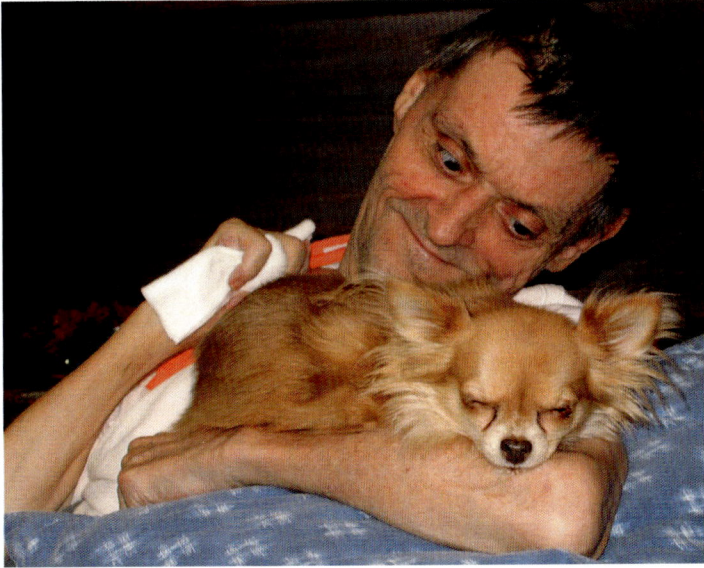

Fig. 9.5 Interacting with a therapy animal does not necessarily require words and can provide unconditional acceptance and support (Photo courtesy of Verein "Tiere als Therapie – TAT", Vienna, Austria)

9.5 Challenges and Ethical Considerations

Mallon et al. (2000) explained "the widespread ardor about the almost universal efficacy of animal-assisted programs has for many years all but obscured any serious questioning of its possible risks" (p. 122). Factors such as physical risk, appropriateness of involving an animal, cultural diversity, health and safety concerns, animal handler competency, and the suitability of the type and personality of the animal all come into play when considering AAT.

No matter how well trained, an animal is never completely predictable. Something as simple as a cat scratching a child, or a dog knocking the client over (Chandler 2005) precludes the need for risk management. Mallon et al. (2000) have written extensively about protocols developed by Green Chimneys Children's Services, which outline concrete methods to minimize risk and guidelines regarding liability and insurance coverage.

Involving an animal does have the potential to negatively affect the therapeutic process. In their review of the challenges within AAT, Chandler (2005) and others offered examples where involving an animal could in fact be counterproductive: if the client perceives rejection from the animal; if the animal becomes ill or even dies; if the client has unreasonable expectations of the animal; or, if the client is not respectful of the animal's safety or well-being.

Cultural belief systems towards animals also influence the interactions (Zamir 2006; Hatch 2007). Consider a person from a ranching background who raises animals for slaughter, versus someone who has only owned animals as pets. Compare these two people with yet another client who is an avid hunter, and with someone who has never interacted with animals, either because of lack of exposure or because animals are considered unclean by their community. As an example of differences amongst various ethnic groups, Chandler (2005) offered the example that some Koreans have expressed discomfort with being close to large dogs as they are often used as aggressive guard animals in their culture, and some Latino communities regard animals as community pets who wander freely, with few restrictions, and are looked after by everyone. In the end, clients and patients bring preconceived notions, expectations and assumptions about animals, which will affect how they interact with an animal. It is the role of the professional to explore this to ensure the involvement of an animal will improve the therapeutic experience (Pichot and Coulter 2007).

No matter where AAT occurs, whether in an office, or outdoor setting, sanitation and the potential for disease must be addressed. Animal inoculations and parasite control, among other things, must be current and clients must also be screened for potential allergies or sensitivities. As well, animals can pass on zoonotic diseases, which are diseases passed between humans and animals (Mallon et al. 2000). Gorczyca et al. (2000) recommended working with dogs and cats older than 9 months as puppies and kittens are more likely to pass on certain parasites to humans. Gorczyca and colleagues listed other diseases, such as acquiring ringworm from cats, or salmonella from cats, birds and horses. They explained that while it is somewhat rare for the transmission of diseases to occur, consideration of the client's immune system is critical and a client at higher risk will require more precautionary measures than someone with a strong immune system. In all cases, if the animals are well-cared for and the client is not exposed to feces or a contaminated environment, the risk should be low.

The United States Department of Agriculture (2012), known as USDA, emphasized cross-species infection can also occur within a farm or between farms, such as the Avian Influenza Virus between chickens and dogs. The department also emphasized viruses can be carried from one farm to another, a consideration when a rural client from a farm or ranch may be involved in therapy on another ranch.

Groups such as EAGALA and Pet Partners emphasize the need for well-trained animals in AAT. However, not all counselling practitioners in AAT agree with this philosophy. For instance, in 2008 we visited a unique facility called the Dreamcatcher Ranch, located in Adrossan, Alberta, Canada where, at that time, they focused on youth with more severe behavioural issues.

According to Amanda Slugoski, one of the counsellors who worked there at that time (2008, personal communication), animals were chosen because of unique behaviours, such as a dog whose brain injury had resulted in severe obsessive compulsive-like behaviours, and a turkey with considerable boundary issues (see Fig. 9.6). Ms. Slugoski took us on a tour of the ranch where we met a horse who, after losing half of her ear from frostbite at her previous home, experienced

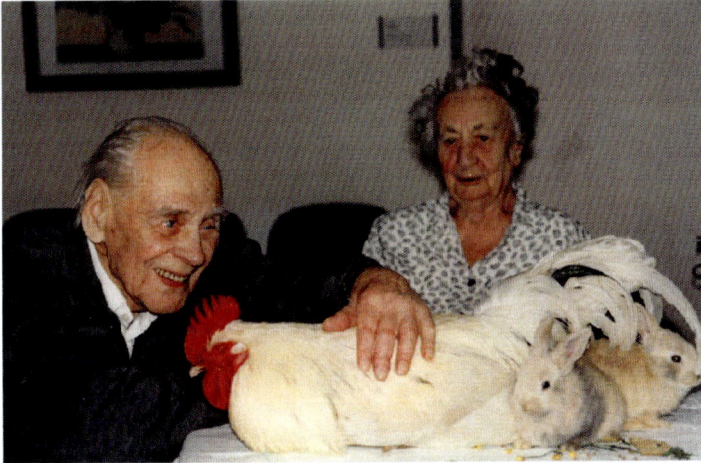

Fig. 9.6 All kinds of animals can be integrated into therapeutic situations, as long as the presence of the animal serves a purpose and is well-supervised, even a rooster. This elderly patient of a nursing home grew up on a farm and especially responded to the animal companions of his childhood (Photo courtesy of Verein "Tiere als Therapie – TAT", Vienna, Austria)

problems with integration into her new environment at the ranch. She explained that the horse's physical disability inadvertently conveyed a threatening pose because of the directionality of her ear, often evoking a negative reaction from some of the other horses. Some youth related well to this animal because of her trouble fitting in, her physical disability and the physical disconnect between her body language and her actual intent. Ms. Slugoski noted that these particular animals mirrored some client problems, which then offered non-contrived opportunities for the youth to consider the challenges others may face when dealing with oppositional or difficult behaviors. She explained the youth engage in managing and problem-solving which in turn encourages insight, empathy and perspective-taking. Working with animals who are less sure of themselves or may not be as trustful of others also created opportunities for youth to practice reading non-verbal behavior to help anticipate problems or potential opportunities, and incited discussions around natural consequences. Ms. Slugoski emphasized the animals' behaviours were predictable to the staff, and therefore their clients were not at risk.

As dogs are frequently included in counselling, it is worth noting the type of breed and size of dog can be an important consideration. For instance, Hart (2000) found that Golden Retrievers, as a breed, are less inclined to be protective than German Shepherds. Hart also noted that the elderly reported feeling safer around smaller breeds because they were afraid of being knocked over by a larger, more rambunctious dog. Chandler (2005), Aanderson (2008), Fredrickson and Howie (2000) and others devised methods that continue to be used to assist practitioners in selecting an appropriate animal for a particular situation.

Current research has begun to examine the physical and mental effects of AAA and AAT on the animals. The Animal Welfare Foundation of Canada (2012), the Canadian Federation of Humane Societies (2012), and the Animal Welfare Council (2012) are three examples of North American organizations focused on the protection and welfare of animals. They promote research, policy making, education, and networking amongst agencies with similar philosophies regarding animal welfare. On their website, the Animal Welfare Council referred to the American Veterinary Medical Association definition of animal welfare as the "human responsibility that encompasses all aspects of animal well-being, including proper housing, management, disease prevention and treatment, responsible care, humane handling, and, when necessary, humane euthanasia" (para. 2). The council advocates for humans to interact with animals in all capacities in a responsible way by ensuring adequate management and care of the animal in each situation. Council participants include rodeos, circuses, animal health organizations, and representation from the entertainment and recreation industries.

With The American Humane Foundation and Pfitzer Health, Macfarland and Ganzert (2012) completed a thorough literature review, interviews and focus groups to gather the most current information on programs and research involving canine therapy with children with cancer (2012). They created in-depth recommendations to guide the research design and questions for a several year study, which will be conducted in pediatric oncology units across the United States. Anecdotal and qualitative accounts consistently reveal positive outcomes on the medical, mental and behavioral outcomes. This study will use various quantitative and qualitative measures to clarify in detail how canine therapy improves overall health in children with cancer and their families and the information will be disseminated through peer reviewed journals to inform best practice in the medical and psychotherapeutic communities.

There is an emergence of universities across North America and Europe providing curriculum, programs and even post-graduate work in the field of AAT. For instance, Purdue University of Veterinary Medicine (2012) through their *Centre for the Human-Animal Bond*, offer curriculum focusing on the care and involvement of animals in therapeutic situations. They promote social, ethical, biological and economical perspectives to improve animal handling, to encourage the development and implementation of policies related to animal welfare, and to advance research.

Other important considerations being addressed in programs focused on AAT are the handler's ability to be attentive to the animal's health and well-being, their working environment and length of their work day (Chandler 2005; Hatch 2007). For example, during our work with alpacas in therapy, we found many clients thought the animal was showing affection when it leaned on them; they enjoyed the feel of its soft fiber and perceived it as almost cuddling. In actual fact, the alpaca was clearly conveying distress. If the handler ignores, or is unaware of, the discomfort the animal is trying to communicate, unexpected and potentially undesirable behaviours from both client and animal may become the focus of the session (see Fig. 9.7).

There are also considerations associated with the environment in which the animal is working, including noise level, excessive feeding by clients, level of risk for injury, adaptability of the animal to the setting, and the type of training the animal

Fig. 9.7 A young friend of ours assists by gently interacting with one of our therapy alpacas on the acreage – continual and calm socialization is critical with therapy farm animals to ensure safety for clients

must endure to fit into the environment (Chandler 2005). Aanderson (2008) offered several considerations, such as choosing the size of animal to fit the space, providing an exercise area, and creating a place for the animal to safely retreat when feeling stressed or uncomfortable. This is particularly important for animals involved in crisis situations where the level of chaos and the state of the environment may be difficult to control.

Finally, consideration must be given to the length of time the animal is expected to work. A working animal may stay in the office all day, travel extensively if they are a part of a mobile operation, or animals on a ranch may normally be able to come and go at their leisure. No matter the situation, without appropriate breaks, an animal that is expected to interact with clients all day may become exhausted and stressed. Depending on the animal, Aanderson (2008) suggested a work session should last no more than one hour at a time, with sizeable downtime away from working which includes exercise, play and rest.

9.6 Directions for Future Research

It is evident in the literature that more comprehensive and solid research is emerging because of increased collaboration between academic institutions, hospitals, veterinary schools, varied medical practitioners, animal advocate organizations, and researchers within the psychotherapeutic community. Collaboration of the medical and psychotherapeutic communities is appealing to potential funders and has encouraged a broader choice of research methodologies and questions. However, problems with inconsistent methodology, confusion around terminology, and difficulties constructing a comprehensive research design continue to make outcome research in AAT complex (Chandler 2005). Additionally, a growing number of AAT programs make claims to therapeutic benefits without the support of substantial research. This leaves the consumer to fend for themselves as they try and sort through which programs follow guidelines set out by organizations such as IAHAIO, The International Association of Human Animal Interaction Organization (2012). IAHAIO promotes rigorous evaluation of AAT by offering viable and valid information to funding bodies, professionals, and the public about the field. Thompson et al. (2012) recently reviewed 115 equine therapy websites and found that there was no evidence-based research offered to support anecdotal claims and testimonials from their happy clients. To protect the consumer, it would be worthwhile to encourage legitimate AAT practitioners to integrate research methodologies appropriate to small, focused studies so that they can contribute to the research pool. In the third edition of Aubrey Fine's AAT handbook, Kazdin (2010) from Yale University offers an extremely comprehensive and practical chapter for practitioners and researchers on the methodological standards and strategies for evidenced-based research. Kazdin addresses overarching concerns such as generating viable questions, control and comparison conditions and sampling and assessment issues. Should AAT be included in mainstream medical treatment, it has the potential to offer a more cost effective method of achieving results such as reducing hospital stay time, treatment compliance, etc. Only commitment to more rigorous research will facilitate this process (Palley et al. 2010).

As AAT-C practitioners, it would be beneficial to understand the efficacy of AAT with various populations. For instance, it would be helpful to know what impact age, gender, ethnicity, or socioeconomic status has; the differences between pet owners versus livestock owners versus those with little experience with animals; or comparisons of the responses of clients with various disabilities. Research investigating if or how much the choice of animal influences the professional/client experience would help guide future practice. The amount or type of involvement of an animal in therapy would also be worth consideration, as well as the influence of the environment and the inclusion of other stimuli in the therapeutic setting including outdoor experiences, plants and so forth.

Further research in AAT could include continued exploration of the influence of animals on human neurophysiology (e.g. oxytocin levels) and the potential benefits of this for improved mental and emotional health. As well, understanding professionals' knowledge of animal welfare issues would be of use in setting policy. Future training needs could be evaluated by a review of professionals' current levels of training, their interest or willingness to access various levels of AAT training, and their perspectives as to what may be required to improve competency. To guide with best practice, it would be helpful to research the efficacy of treatment methods as well as how to best document and evaluate client progress.

9.7 Conclusion

Although there appears to be a number of research studies that support the use of AAT in treatment for both psychological and medical issues, problematic research design leaves many questions for practical application of these results. While it seems intuitively sensible that animals can reduce stress, facilitate therapeutic interaction, etc., more rigorous research needs to be done before it becomes a common practice, especially in light of the numerous ethical and animal welfare issues involved. It is our hope, as practitioners of this approach that this will occur. In the meantime, we will continue to love and respect the animals that have made such an impact on us and our clients.

References

Aanderson KW (2008) Paws on purpose: Implementing an animal assisted therapy program for children and youth, including those with FASD and developmental disabilities. The Chimo Project, Edmonton

Anderson WP, Reid CM, Jennings GL (1992) Pet ownership and risk factors for cardiovascular disease. Med J Aust 157(5):298–301

Animal Welfare Council (2012) Animal Welfare Council.com: welfare vs rights. http://animal-welfarecouncil.com/welfare-vs-rights/. Accessed 27 July 2012

Animal Welfare Foundation of Canada (2012) The Animal Welfare Foundation of Canada. http://www.awfc.ca/english/works.htm. Accessed 10 July 2012

Batson K, McCabe B, Baun MM, Wilson C (1998) The effect of a therapy dog on socialization and physiological indicators of stress in persons diagnosed with Alzheimer's disease. In: Wilson CC, Turner DC (eds) Companion animals in human health. Sage Publications, Inc, Thousand Oaks

Baun MM, McCabe BW (2000) The role animals play in enhancing quality of life for the elderly. In: Fine A (ed) Animal assisted therapy: theoretical foundations and guidelines for practice. Academic, New York

Beck A, Katcher A (1996) Between pets and people: the importance of animal companionship (revised edition). Purdue University Press, West Lafayette

Beck A, Katcher A (2003) Future directions in human-animal bond Research. Am Behav Sci 47(1):79–93

Behling RJ, Haefner J, Stowe M (2011) Animal programs and animal assisted therapy in Illinois long-term care facilities twenty years later (1990–2010). Acad Health Care Manag J 7(2): 109–117

Berget B, Braastad BO (2011) Animal-assisted therapy with farm animals for persons with psychiatric disorders. Ann Ist Super Sanita 47(4):384–390

Braun C, Stangler T, Narveson J, Pettingell S (2009) Animal-assisted therapy as a pain relief intervention for children. Complement Ther Clin Pract 15(2):105–109

Canadian Federation of Humane Societies (2012) The Canadian Federation of Humane Societies. http://cfhs.ca/. Accessed 16 June 2012

Chandler CK (2005) Animal assisted therapy in counseling. Routledge, New York

Chandler CK, Fernando DM, Barrio Minton CA, O'Callaghan DM, Portrie-Bethke TL (2010) Matching animal-assisted therapy techniques and intentions with counseling guiding. J Ment Health Couns 32(4):354

Edwards NE, Beck AM (2002) Animal-assisted therapy and nutrition in Alzheimer's disease. West J Nurs Res 24(6):697–712

Equine-Assisted Growth and Learning Association (2012) The Equine Assisted Growth and Learning Association. http://www.eagala.org/. Accessed 10 June 2012

Fine A (2000) Animals and therapists: incorporating animals in outpatient psychotherapy. In: Fine A (ed) Animal assisted therapy: theoretical foundations and guidelines for practice. Academic, New York

Flom BL (2005) Counseling with pocket pets: using small animals in elementary counseling programs. Prof Sch Couns 8(5):1–5

Fredrickson M, Howie AR (2000) Considerations in selecting animals for animal- assisted therapy: part B. In: Fine A (ed) Animal assisted therapy: theoretical foundations and guidelines for practice. Academic, New York

Friedmann E, Katcher AH, Lynch JJ, Thomas SA (1980) Animal companions and one-year survival of patients after discharge from a coronary care unit. Public Health Rep 95(4):307–312

Gagnon J, Bouchard F, Landry M, Belles-Isles M, Fortier M, Fillion L (2004) Implementing a hospital-based animal therapy program for children with cancer: a descriptive study. Can Oncol Nurs J 14(4):217–222

Gorczyca K, Fine A, Spain CV (2000) History, theory, and development of human-animal support services for people with AIDS and other chronic/terminal illnesses. In: Fine A (ed) Animal assisted therapy: theoretical foundations and guidelines for practice. Academic, New York

Green Chimneys. Green Chimneys International. Who we are. http://www.greenchimneys.org/index.php?option=com_content&view=article&id=44&Itemid=140. Accessed 13 July 2012

Gregg B (2012) Crossing the berm: an occupational therapists' perspective on animal-assisted therapy in a deployed setting. US Army Med Dep J, April–June:55–56

Hamama L, Hamama-Raz Y, Dagan K, Greenfeld H, Rubinstein C, Ben-Ezra M (2011) A preliminary study of group intervention along with basic canine training among traumatized teenagers: a 3-month longitudinal study. Child Youth Serv Rev 33(10):1975–1980

Hart E (2000) Understanding animal behavior, species, and temperament as applied to interaction with specific populations. In: Fine A (ed) Animal assisted therapy: theoretical foundations and guidelines for practice. Academic, New York

Hatch A (2007) The view from all fours: a look at an animal-assisted activity program from the animals' perspective. Anthrozoos 20(1):37–50

Headey B, Grabka M (2007) Pets and human health in Germany and Australia: national longitudinal results. Soc Indic Res 80(2):297–311

Hippotherapy (2012) The American Hippotherapy Association Incorporated: hippotherapy as a treatment strategy. http://www.americanhippotherapyassociation.org/hippotherapy/hippotherapy-as-a-treatment-strategy/. Accessed 4 July 2012

International Association of Human-Animal Interaction Organizations (2012) IAHAIO: about us – mission and goals. http://iahaio.org/pages/aboutus/missiongoals.php. Accessed 5 June 2012

Katcher AH, Wilkins GG (2000) The centaur's lessons: therapeutic education through care of animals and nature study. In: Fine A (ed) Animal assisted therapy: theoretical foundations and guidelines for practice. Academic, New York

Kazdin AE (2010) Methodological standards and strategies for establishing the evidence base of animal-assisted therapies. In: Fine A (ed) Animal assisted therapy: theoretical foundations and guidelines for practice, 3rd edn. Academic, New York

Leimer G (1997) Indication of remedial vaulting for anorexia nervosa. In: Engel BT (ed) Rehabilitation with the aid of a horse: a collection of studies. Barbara Engel Therapy Services, Durango

Levinson BM (1964) Pets: a special technique in child psychotherapy. Ment Hyg 48:243–248

Macfarland, JM, Ganzert, R (2012) Canines and childhood cancer: examining the effects of therapy dogs with childhood cancer patients and their families. http://www.americanhumane.org/assets/pdfs/children/ccc_digitalbook_r19.pdf. Accessed 19 June 2012

Mallon G (1994) Cow as co-therapist: utilization of farm animals as therapeutic aides with children in residential treatment. Child Adolesc Soc Work J 11(6):455–474

Mallon GP, Ross SB, Ross L (2000) Designing and implementing animal- assisted therapy programs in health and mental health organizations. In: Fine A (ed) Animal assisted therapy: theoretical foundations and guidelines for practice. Academic, New York

Marcus D, Bernstein C, Constantin J, Kunkel F, Breuer P, Hanlon R (2012) Animal-assisted therapy at an outpatient pain management clinic. Pain Med 13:45–57

Marr CA, French L, Thompson D, Drum L, Greening G, Mormon J et al (2000) Animal-assisted therapy in psychiatric rehabilitation. Anthrozoos 13:43–47

Marx MS, Cohen-Mansfield J, Riger NG, Dakheel-Ali M, Srihari A, Thein K (2010) The impact of different dog-related stimuli of engagement of persons with dementia. Am J Alzheimers Dis Other Demen 25(1):37–45

Matuszek S (2010) Animal-facilitated therapy in various patient populations: systematic literature review. Holist Nurs Pract 24(4):187–203

McIntosh S (2002) An introduction to equine-facilitated counselling. Self-published, Cremona, Alberta, Canada

Miller SC, Kennedy C, DeVoe D, Hickey M, Nelson T, Kogan L (2009) An examination of changes in oxytocin levels in men and women before and after interaction with a bonded dog. Anthrozoos 22:31–42

Odendaal JSJ (2000) Animal-assisted therapy: magic or medicine? J Psychosom Res 49(4): 275–280

Palley LS, O'Rourke PP, Niemi SM (2010) Mainstreaming animal-assisted therapy. ILAR J 51(3): 199–207

Pet Partners (2012) Pet Partners, Inc: overview. http://www.deltasociety.org/page.aspx?pid=659. Accessed 20 July 2012

Pet Access League Society (2012) http://www.palspets.com/. Accessed 25 July 2012

Pichot T, Coulter M (2007) Animal-assisted brief therapy: a solution-focused approach. The Haworth Press, New York

Prothmann A, Bienert M, Ettrich C (2006) Dogs in psychotherapy: effects on state of mind. Anthrozoos 19(3):265–277

Purdue University of Veterinary Medicine (2012). Purdue University school of veterinary medicine: centre for the human-animal bond. http://www.vet.purdue.edu/chab/index.html. Accessed 10 July 2012

Reichart E (1998) Individual counseling for sexually abused children: a role for animals and storytelling. Child Adolesc Soc Work J 15(3):177–185

Relf PD (2006) Theoretical models for research and program development in agriculture and health care: avoiding random acts of research. In: Hassink J, van Dijk M (eds) Farming for health: green-care farming across Europe and the United States. Springer, Dordrecht

Serpell J (1991) Preventing potential health hazards incidental to the use of pets in therapy. Anthrozoos 4:14–23

Taylor SM (2001) Equine-facilitated psychotherapy: an emerging field. Unpublished master's thesis, University of St. Michael's College, Colchester

Thompson JR, Iacobucci V, Varney R (2012) Giddyup! or whoa Nelly! making sense of the benefit claims on websites of equine programs for children with disabilities. J Dev Phys Disabil 24(4):373–390

United States Department of Agriculture (2012). The United States Department of Agriculture. http://awic.nal.usda.gov/farm-animals/diseases. Accessed 14 June 2012

Urichuk L, Anderson D (2003) Improving mental health through animal-assisted therapy. The Chimo Project, Edmonton

Zamir T (2006) The moral basis of animal-assisted therapy. Soc Anim 14(2):179–199

Chapter 10
Equine-Assisted Therapy: An Overview

Nina Ekholm Fry

10.1 Introduction

Equine-assisted therapy (EAT), an umbrella term for various forms of therapy such as physical therapy, occupational therapy, speech-language therapy and psychotherapy where a horse is part of the treatment team, is an emerging field internationally. The Federation for Horses in Education and Therapy International has members in 49 countries (Federation of Horses in Education and Therapy International (HETI 2012)), and the number of educational programs offering training in this field is growing. In order for equine-assisted therapy to be established as a valid form of treatment and accepted as different from recreational activities involving horses, more research and information is needed. The purpose of this chapter is to provide an overview of equine-assisted therapy including the history of the field and of the human-horse relationship, the current state of research, considerations for a general theoretical framework, and a review of equine-assisted therapy as physical therapy and as psychotherapy.

10.2 History and Background

10.2.1 Defining Equine-Assisted Therapy

Equine-assisted therapy can be defined in the most basic way as the inclusion of a equine, typically a horse, mule, or donkey, in a therapy setting with the purpose of

N. Ekholm Fry, MSSc (✉)
Director, Equine-assisted Mental Health, Associate Faculty, Counseling Psychology,
Master of Arts Program, Prescott College, 220 Grove Avenue, Prescott, AZ 86301, USA
e-mail: nfry@prescott.edu

M. Grassberger et al. (eds.), *Biotherapy - History, Principles and Practice:*
A Practical Guide to the Diagnosis and Treatment of Disease using Living Organisms,
DOI 10.1007/978-94-007-6585-6_10, © Springer Science+Business Media Dordrecht 2013

Fig. 10.1 Equine-assisted therapy is treatment that incorporates equine activities and/or the equine environment (Photo courtesy of Nina Fuller Photography)

enhancing treatment outcomes for the client. The Professional Association of Therapeutic Horsemanship International offers the following definition: "equine-assisted therapy is treatment that incorporates equine activities and/or the equine environment (Fig. 10.1). Rehabilitative goals are related to the patient's needs and the medical professional's standards of practice" (Professional Association of Therapeutic Horsemanship International, PATHIntl). Individuals who may legally provide therapy in their state or country through appropriate license or other applicable credentials can conduct equine-assisted therapy after acquiring training on including horses in the specific form of therapy they practice. It is interesting to note that equine-assisted therapy, although sometimes considered under the general category of animal-assisted therapy, is typically seen as a separate field (Kruger and Serpell 2010).

Equine-assisted therapy should not be seen as a singular, isolated approach, but as a strategy for the practitioner to use within his or her theoretical orientation. Fine (2010) suggests three questions that practitioners should consider when incorporating animal-assisted interventions with their clients. They are modified here for the equine-assisted therapy context:

1. What benefits are the horse and therapeutic equine activities expected to bring to treatment? In general? For a specific client?
2. How can equine-related activities be incorporated into the treatment plan? For instance, will horses be part of every therapy session?

3. How will the therapist adapt his or her clinical work to include equine-assisted therapy? What kind of equine-related activities will be most appropriate?

Equine-assisted therapy can be divided into two general areas: the inclusion of equines in physically oriented therapies such as physical therapy, occupational therapy and speech therapy, and in mental health therapies such as psychotherapy and counseling.

10.2.1.1   Hippotherapy

The term hippotherapy is used when physical aspects of equine-assisted therapy are emphasized (Strauss 1995), It can be defined as a treatment strategy used by a licensed physical therapist with additional training in hippotherapy where the utilization of equine movement is a core component (Deutsches Kuratorium für Therapeutisches Reiten (DKThR), Schweizer Gruppe für Hippotherapie-K®, Österreichisches Kuratorium für Therapeutisches Reiten (OKTR)). In the United States, occupational therapists and speech-language pathologists may also conduct hippotherapy (American Hippotherapy Association). German occupational therapists are offered a separate training for conducting occupational equine-assisted therapy (DKThR). Hippotherapy can literally be translated as 'horse therapy'. Because of this, the term has sometimes been used as an umbrella term, much like equine-assisted therapy. This use of the term is discouraged, as hippotherapy is a well-defined equine-assisted therapy treatment strategy and recognized as such internationally.

10.2.1.2   Equine-Assisted Mental Health

Equine-assisted mental health can be used as an umbrella term for psychotherapeutic services that involve horses and are provided by licensed mental health professionals. Examples of terms commonly used in the United States to describe these services are equine facilitated psychotherapy, equine-assisted psychotherapy and equine-assisted counseling (Hallberg 2008). These sub-terms are not clearly defined; practitioners may choose a term based on personal preference and the equine-assisted mental health training they have received. In Germany, the terms *Psychotherapie mit dem Pferd*; *Psychotherapeutisches Reiten*, and *Pferdegestützte Psychotherapie* are used (Fachgruppe Arbeit mit dem Pferd in der Psychotherapie (FAPP), Münchner Schule für Psychotherapeutisches Reiten, Opgen-Rhein et al. 2011). When including horses in psychotherapy the emphasis is placed on the interactions and relationships between client, horse and therapist. Activities, which may take place on the ground or on horseback (Fig. 10.2), are chosen to support treatment goals set forth by the mental health professional and client (PATHIntl).

Fig. 10.2 Equine-assisted mental health activities can take place on the ground or on horseback (Photo courtesy of Arizona Burn Foundation/Jack Jordan)

10.2.1.3 Discussion

Standardized terminology or, at least, clearly defined terms are important, not only in terms of research and international communication but also for clinicians, clients, and insurance companies. The issues noted in early literature reviews (MacKinnon et al. 1995; Fitzpatrick and Tebay 1998) regarding terminology are still common in the field today.

The misuse of terms is particularly serious when confusion arises over equine-assisted activities that may be legally defined as therapy, and those that may not. Broadly stated, sessions involving horses that are not conducted by licensed therapists with specific training in equine-assisted therapy cannot be considered equine-assisted *therapy*, and the person who is facilitating the session should not present it as such. However, in countries such as Sweden, an education professional with a special education focus may train to become a so-called riding therapist (*ridterapeut*), delivering 'riding therapy' (*ridterapi*) (Intresseföreningen för Ridterapi, IRT). When training programs in equine-assisted therapy are open to professionals in fields outside of traditional therapy professions, such as in the above example, the practitioner should be careful to only provide services within his or her scope of practice as determined by credentials and professional standards.

Equine-assisted activities that cannot be considered therapy are educational in nature. They are lead by professionals who have additional training in how to facilitate learning or coaching sessions that include horses. These activities are different from equestrian pursuits such as riding lessons in that the primary intent of the session is the development of life skills and personal growth through equine interactions and the equine environment (PATHIntl).

Because of the inconsistent use of terms and the different translations employed, systematic reviews of research on various forms of equine-assisted therapy can be challenging. Without clear definitions of terms and thorough descriptions of session activities, studies are difficult to compare and reproduce. An example of a term that has caused confusion internationally is *therapeutic riding* (see also Sect. 10.2.3). Although the word therapeutic may be understood as derived from therapy, it is also used to describe an activity that can be understood as beneficial, without fulfilling the legal requirements of therapy. The term therapeutic riding has been used in Germany since 1970 as an umbrella term for equine activities and therapy in the areas of medicine, education, psychology and riding as a sport for the disabled (DKThR). The term was also formerly used by the Federation for Horses in Education and Therapy International as an all-encompassing description of equine activities and therapy. Today the federation uses equine-assisted activities as an umbrella term for the field (Federation for Horses in Education and Therapy International, HETI). The Horses and Humans Research Foundation defines therapeutic riding in the United States as "mounted activities including traditional riding disciplines or adaptive riding activities conducted by a trained instructor" (Horses and Humans Research Foundation, HHRF), and suggests equine-assisted activities and therapies (EAA/T) as an umbrella term. Therapeutic riding, as conducted in the U.S., is not a form of equine-assisted therapy, despite its name. It is conducted by a certified therapeutic riding instructor with the purpose of contributing positively to the well-being of people with disabilities (PATHIntl). As a result, caution must be taken when comparing services and research studies where the term therapeutic riding is used.

Lack of consistent terminology within the field of equine-assisted therapy is a professional and ethical issue. Practitioners should provide detailed and accurate descriptions of the equine-assisted treatment strategy they provide in order to avoid misrepresentation.

10.2.2 History of the Human-Horse Relationship

The domestication of the horse, which is often considered the starting point of the human-horse relationship, can be traced back to approximately 4000 BCE (Outram et al. 2009). In the past 6,000 years, the versatility of the domestic horse, *equus caballus*, has contributed to the development of civilization in many meaningful ways, most notably in transportation, trade, warfare, and agriculture (Johns 2006). The ability to travel more efficiently with the help of horses made new communications and trade possible, which some scholars believe contributed to the development of modern languages (Anthony 2007). As the western world eventually moved from primarily agrarian to industrial, the horse, an integral part of agriculture and day-to-day life (Fig. 10.3), lost its position in society to motorized vehicles (Budiansky 1997).

Fig. 10.3 The horse has historically been an important part of agriculture and day-to-day life (Photo courtesy of the author)

Horses are still being used in farming, ranch work, and logging around the world, although to a much lesser extent (Brandt and Eklund 2007; Pickeral 2005). With increasing numbers of horses in sport and leisure, and most recently in therapy, it is clear that the role of the horse in society is being redefined (Miller and Lamb 2005).

The horse has been featured in art, legends, and creation stories throughout history. From prehistoric cave paintings to motion pictures, images of the horse can be found across cultures (Pickeral 2008). In the mythology of the Celts, Greeks, Persians and Arabs, among others, the horse has a place of honor and is considered as a messenger of god between the spiritual and material worlds. Norse mythology features Sleipnir, Odin's eight-legged horse, who would travel both in the skies and in the underworld, as well as Skinfaxe and Rimfaxe, the horses of Day and Night, who were ridden across the sky to bring a new day (Frewin and Gardiner 2005; McCormick and McCormick 1997). Pegasus, the winged horse of Greek mythology, is one of the most well known mythological horses (Pickeral 2008). The horse has come to symbolize freedom, guidance, strength, beauty, healing abilities, and fertility, as well as physical and spiritual travel (Kohanov 2001; McCormick and McCormick 1997). Both feminine and masculine qualities can be distinguished in the symbolism surrounding the horse (Root 2000).

The horse relies on perceptual abilities and its place in the herd for survival. The tendency for caution, reactivity and flight, combined with a sense of security and comfort within the social structure of a herd, suggests that the horse is a social animal sensitive to its surroundings (Fig. 10.4). As mammals, horses have brain character-istics that suggest their learning mechanisms may be similar to those in humans (Hahn 2004). Basic learning in both horses and humans takes place through habitu-ation, classic and operant conditioning and trial-and-error. In addition, horses can learn to categorize objects based on physical features as well as generalize stimuli

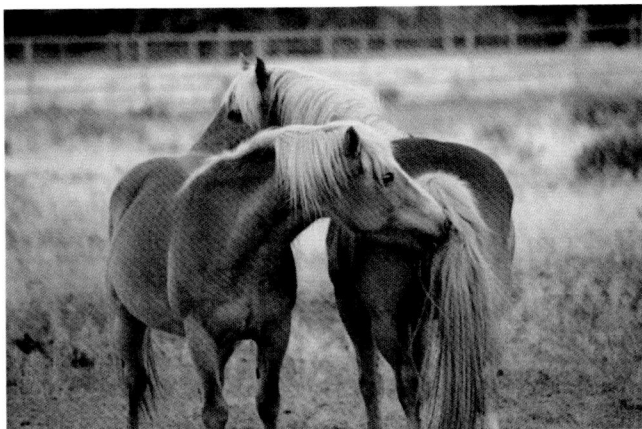

Fig. 10.4 The horse is a social animal (Photo courtesy of Nina Fuller Photography)

(Hänggi 1999; Nicol 2002). There is, however, no evidence at this point to support that horses are capable of abstract thinking, a feature supported by the presence of a large neocortex found in humans. More research is needed to further understand equine cognition and mental states (Hahn 2004).

10.2.3 History of Equine-Assisted Therapy

"That there is no exercise to be compared with horseback-riding is conceded by all well-read physicians" (Magner 1887/2004, p. 261). The positive effects of horseback riding on human health have been noted in medical literature, starting from the second century CE. Physicians such as Galen (129–200), Oribasius (325–403), Hieronymus Merkurialis (1530–1606), Sydenham (1624–1689) and Quellmalz (1697–1758) referred to riding as an excellent form of exercise (Bain 1965; Purjesalo 1991). At the end of the nineteenth century, several works were published in which riding was examined as a possible form of medical treatment. French physician R. Chassaigne reported in 1870 that riding is helpful for patients with specific neurological disorders (Brock 1997). According to Chassaigne, the movement of the horse can influence posture, balance, and flexibility of joints, as well as improve muscle strength in patients. Chassaigne also mentions positive psychological effects (as cited in Durant 1878). The book *Horseback-Riding from a Medical Point of View* was published in 1878 by American physician G. Durant. He recommends riding over other forms of exercise because of its effects on blood circulation, digestion and mental conditions (Durant 1878). In *Magner's Classic Encyclopedia of The Horse*, D. Magner describes the rehabilitative functions of horseback riding, including physical and mental benefits (Magner 1887/2004). It is believed that horses were used in the rehabilitation of wounded soldiers during World War I, but a reliable source for this information is difficult to find.

Fig. 10.5 Lis Hartel was
rehabilitated through
horseback riding and
competed at two Olympic
games

The first organizations and facilities promoting horseback riding for therapeutic purposes were founded in the middle of the twentieth century. The achievements of Danish dressage rider Lis Hartel are typically credited for making the beneficial effects of horseback riding on neuromuscular disorders known to the public. Hartel had contracted polio in one of several epidemics in Scandinavia in the 1940s and had become partially paralyzed in her legs. She rehabilitated herself by riding and went on to win two Olympic silver medals in dressage, in 1952 and 1956 (Fig.10.5).

The first centers for riding as therapy were founded in Olso and Copenhagen during the same period for the treatment of children with disorders such as polio and cerebral palsy (Bain 1965). The British Riding for the Disabled Association and the North American Riding for the Handicapped Association were both founded in 1969. Germany was the first country to develop structured educational standards for professionals in three areas of therapeutic riding: hippotherapy (medicine), Heilpädagogisches Reiten und Voltigieren (education-psychology), and riding for the disabled (sport) (Spink 1993). After the German Kuratorium für Therapeutisches Reiten e.V. was founded in 1970 it was decided that the association would use therapeutic riding as an all-comprising term for the three areas (Weber 1998).

Regarding early development of equine-assisted therapy in Great Britain and the Unites States, Spink (1993) notes:

> (…) the United States and Great Britain appeared to have no structured approach to help guide the early development and organization of therapeutic riding regarding the medical and psychological applications. The initial appeal of using horses to benefit people with disabilities seemed to be greater to horsemen and riding instructors than to professionals in

Fig. 10.6 Interest in the
psychological effects of
therapeutic riding grew
during the 1990s (Photo
courtesy of Nina Fuller
Photography)

medicine, special education, and psychology. Consequently, in these two countries, therapeutic riding developed more as a recreational activity or adapted group sport activity than as a specific medical or remedial treatment method. (pp. 5–6)

The first international congress of therapeutic riding was held in Paris in 1974. Six years later, in 1980, the Federation of Riding for the Disabled International was founded to facilitate international collaboration in the field (HETI). Hippotherapy, developed in Germany, Austria, and Switzerland, was introduced to the United States in the 1980s (American Hippotherapy Association, AHA). In the late 1980s therapeutic riding programs existed in countries such as Australia, Canada, Belgium, France and Italy (Purjesalo 1991). An adapted form of hippotherapy called developmental riding therapy emerged in the United States around the same time (Spink 1993).

While hippotherapy and therapeutic riding were gaining acceptance, the idea of including horses, or any kind of animal, in psychotherapy had still been largely unexplored. In an early paper, American psychotherapist Boris Levinson (1961) described the use of a dog as therapeutic agent in counseling with children, based on his own experiences in this practice. In the book that followed, *Pet-Oriented Child Psychotherapy*, Levinson (1969) provided examples of ways in which animals could enhance therapy. These publications are typically credited as the first purposeful descriptions of animals as part of psychotherapy in the role of a co-therapist (Fig. 10.6).

The increasing interest in the psychological effects of therapeutic riding, as well as in the inclusion of horses in psychotherapeutic services lead to the 1996 founding of the Equine Facilitated Mental Health Association in the Unites States, and the publication of a special edition journal *Die Arbeit mit dem Pferd in Psychiatrie und Psychotherapie* in 1993 by the German Kuratorium für Therapeutisches Reiten (Klüwer 2009; Hallberg 2008).

In the two decades that have passed since the early beginnings of the field, an ever-increasing number of individuals and associations wishing to promote, practice, and provide training for areas of equine-assisted therapy have emerged. Statements regarding the effectiveness of equine-assisted therapy need to be supported by research in order for the field to gain validity.

10.2.4 Current State of Research in Equine-Assisted Therapy

Despite an exponential growth of the equine-assisted therapy field in the past decade the quality of research is considered moderate at best (Selby 2009). Challenges noted by MacKinnon et al., in 1995, such as small sample sizes, unsound methodology and lack of appropriate measures are still present in studies today. Another prevalent issue in current research and literature is when unreasonable statements are made about the "power" of these forms of treatment (Fine and Mio 2010, p. 569). It is important for clinicians to communicate day-to-day observations in order to promote further understanding of the field, but an issue of validity arises when unstructured observations are communicated as broad, definite statement about the ever-present positive effects and central elements of equine-assisted therapy. Unsupported statements are perpetuated in the literature, likely due to the wealth of observations of that equine-assisted therapy 'works', but a lack of research to support it. In other words, there needs to be a balance between the engagement required to advance a new field and the risk of overstating its effectiveness

Recent reviews of research studies in hippotherapy were completed by Bronson et al. (2010), Snider et al. (2007), Sterba (2007) and Zadnikar and Kastrin (2011). Selby (2009) has provided an extensive review of studies where a horse is a part of psychotherapeutic treatment. Conclusions that can be drawn from reviews in both areas are that equine-assisted therapy, as hippotherapy or as a psychotherapeutic treatment, shows promise as a treatment strategy but that more evidence is needed to support its clinical effectiveness.

Fine and Mio (2010) mention the sometimes uneasy relation between clinicians and researchers and the need for collaboration between the two. In interviews with stakeholders in the fields of equine facilitated mental health and equine facilitated learning conducted by the Certification Board for Equine Interaction Professionals, some respondents were concerned about the increasing importance of research. These respondents feared that research investigation may ultimately create requirements which reduce practice to a number of pre-set ways, diminishing the value of intuitive skill within the profession (Certification Board for Equine Interaction Professionals, CBEIP 2008). However, research should not be seen as restriction but as a way to ensure longevity and effective treatment. It is also important in terms of justifying the additional costs of including a horse in treatment, including travel and use of facilities.

Qualitative studies exploring experiences and themes, as well as case studies, are important for insight into and development of clinical practice. The use of randomized control trials, regarded as the gold standard of evidence-based practice, in addition to sound methodological design and large sample sizes, is encouraged in order to establish an evidence base for animal-assisted therapies (Kazdin 2010).

10.3 Principles

10.3.1 Considerations for a Theoretical Framework of Equine-Assisted Therapy

Despite the lack of an evidence-based framework for equine-assisted therapy, a number of theoretical assumptions of why it is useful are commonly found in the literature. What is presented here should be considered broad, potential foundations of equine-assisted therapy. Section 10.5 contains descriptions of additional elements specific to each treatment strategy.

10.3.1.1 Biophilia

The biophilia hypothesis is defined as an innate human tendency to focus on other living beings and lifelike processes (Wilson 1984). It does not suggest that humans have a natural affection for animals but that attention to, and knowledge of, environmental cues have been an important part of human evolution (Kruger and Serpell 2010). There is a large body of data supporting the stress-reducing effects of simply watching landscapes or the movement patterns of animals in calm states (see Katcher and Wilkins 1993). In other words, events in nature may be associated with a sense of safety or a sense of danger in humans. The biophilia hypothesis is sometimes used to support the notion that low-intensity interaction with a relaxed horse can have a calming, de-arousing effect, but it is likely that other factors, such as using rhythmic touch, may have a stronger effect in reducing arousal in the client. Research on neurobiological aspect of equine interactions has recently been given funding (National Institute of Child Health and Human Development (NICHD); Beetz et al. 2011). More research is needed regarding the effects of equine interactions on neurobiology.

10.3.1.2 Touch

The use of touch is common in physical and body-based therapies but considered controversial in psychotherapy (Phelan 2009). Yet, bodily touch is considered a significant part of the human experience (Montague 1971). Equine-assisted therapy practitioners frequently comment how the body of the horse, with large areas of contact, presents many opportunities for touch (Esbjorn 2006; McConnell 2010). Physical closeness with a horse may provide a non-threatening opportunity for touch for clients with a background of abuse. Sokolof (2009) presents a case study in which the client, through sensory exploration of the horse's coat, found a renewed connection with her physical self. Brooks (2006) notes how the horse can provide a client with the experience of being held in a safe environment. She also comments on how the client's use of touch with an animal can be part of diagnostic

Fig. 10.7 Equine-assisted therapy offers many opportunities for touch and movement (Photo courtesy of Nina Fuller Photography)

assessment (Fig. 10.7). As proposed by Schulz (1999), rhythm, touch and skin contact, naturally present in equine-assisted therapy, are essential elements of human development.

10.3.1.3 Therapeutic Environment

The environment in which equine-assisted therapy takes place can be considered an important element of treatment (Bizub et al. 2003). Vidrine, Owen-Smith and Faulkner (2002) propose:

> Part of the farm experience's therapeutic potential lies in its unpredictability. It is by no means a sterile, controlled environment. Groups had to problem-solve around the weather and horse issues such as illness or missing shoes. The 'controlled chaos' of the farm invited creativity and trial and error. It provides a 'good enough' environment filled with incidental learning opportunities such as a visit from the horse dentist, vet, or blacksmith (p. 600).

Based on her own clinical experiences, Hallberg (2008) proposes that the barn environment may contribute to the client disclosing thoughts and feelings more easily. Hanna et al. (1999) suggest that there is an increased likelihood of self-disclosure in environments outside of the counseling office with defiant and aggressive adolescents. In addition being to a non-traditional setting for therapy, the equine-assisted therapy environment enhances sensory stimulation through numerous interactions within a natural setting. Research is needed to establish which, if any, elements of the outdoor therapy environment contribute to or detract from treatment. In a recent review of the mental health effects of participating in physical activity in an outdoor natural environment as compared to indoors, Thompson-Coon et al. (2011) found promising effects on self-reported mental wellbeing immediately following exercise in nature but cautions readers that the interpretation of the findings is hampered by the low methodological quality of some of the studies included.

10.3.1.4 Movement

Movement is an inherent part of any equine-assisted therapy, regardless of physical or psychological treatment focus. In hippotherapy, movement is controlled and manipulated according to goals of treatment and central to the treatment strategy. Movement is also present in equine-assisted mental health through both mounted work and activities on the ground with the horse. In addition to general physical benefits of exercise for all populations, research indicates a beneficial effect of physical activity on depression (Brosse et al. 2002; Mata et al. 2012; Rethorst et al. 2009).

10.3.1.5 Relationship

A major feature of therapies incorporating a horse is the opportunity for the client to form a relationship with an other-than-human being. In physical therapy, added social and emotional aspects through the relationship with the horse may contribute to a more enjoyable and dynamic treatment session for the client (Strauss 1998). In equine-assisted mental health, exploring the client-horse relationship is a central aspect of treatment (Esbjorn 2006; Hayden 2005). It is also believed that animals may assist in building rapport between client and therapist (Kruger and Serpell 2010).

10.3.1.6 Client Motivation for Therapy

Lack of client motivation for treatment is a multifaceted concept in both physically and psychologically oriented therapy (Maclean and Pound 2000; Bachelor et al. 2007). Statements regarding increased client motivation for therapy in equine-assisted treatments typically refer to motivation to attend and engage in the therapy situation (Cook 1997; Heine 1997). Macauley and Gutierrez (2004) found that hippotherapy improves the motivation of children to attend and participate in speech-language therapy. However, due to a very small sample size (n = 3) the results cannot be generalized. In a qualitative study by McConnell (2010), a number of programs reported that it is the horse that draws the client to treatment. It has been theorized that the presence of a horse may help the client endure physical or emotional pain in treatment session (Kruger and Serpell 2010), and thus increase motivation for participation. For instance, the horse might help divert attention from uncomfortable physical experiences (client petting the horse while stretching in hippotherapy), or be perceived as a source of nonjudgmental support and warmth (client leaning against the horse while sharing a traumatic experience in equine-assisted mental health, Fig. 10.8).

Fig. 10.8 Equines may represent an additional source of support to the client in the therapy process (Photo courtesy of Nina Fuller Photography)

10.3.2 Ethical Considerations

The lack of an evidence-based framework for equine-assisted treatment strategies can lead to practical, ethical and safety issues for practitioners and clients alike. An awareness of ethical considerations that exist when including a horse as part of treatment is of crucial importance. Recognizing that a moral and philosophical discussion underlies the entire concept of using animals in human therapy (Zamir 2006), what is presented here are general issues of ethical concern in equine-assisted therapy.

10.3.2.1 The Therapist

Hallberg (2008) identified three areas of ethical concern for practitioners in the fields of equine-assisted mental health and learning: lack of competence regarding the practical application of equine-assisted services, lack of equine knowledge and experience, and lack of safety awareness. These issues can be generalized to the whole field of equine-assisted therapy.

Therapist training in the field of equine-assisted therapy is paramount. There are a number of training programs available and the professional must consider whether a brief exposure, such as a workshop, is enough in order to incorporate horses into

Fig. 10.9 The therapist is responsible for the equine-assisted therapy session (Photo courtesy of PATH Intl., Denver, CO)

treatment and still ethically provide services within his or her scope of practice. Needless to say, a well-trained therapist will seek lengthy training and supervision when incorporating a new treatment strategy into his or her work. He or she will also strive to keep informed through continuing education. It is difficult for clients to make an informed decision about the experience of a therapist providing equine-assisted therapy as training programs may offer a certificate after only a brief seminar. The therapist should provide a detailed account of training and experience to the client prior to the start of treatment, as part of general ethical practice.

Therapists wishing to incorporate a form of equine-assisted treatment strategy into their practice should have sufficient horse experience to be able to ensure a safe and ethical environment for both client and horse. Whether the treatment session includes safety support assistants, which is standard practice in hippotherapy, or so-called equine specialists, which is recommended by some training programs in equine-assisted mental health, the therapist is ultimately responsible for the session (Fig. 10.9). It can be argued that practicing equine-assisted therapy without equine knowledge or experience is an ethical violation related to scope of practice.

Considerations for reducing risk of harm in therapy are especially important when including a large animal in the treatment situation. The inherent physical risks of equine activities should be explained in the client consent document and the therapist should maintain good risk management practices. This includes the engagement of experienced horse handlers, assistants and co-therapists, when applicable. Hippotherapy should always be conducted in a team (Strauss 1995). The therapist

Fig. 10.10 An understanding of equine behavior, mental processes and learning is necessary in order to maintain the wellbeing of the therapy horse (Photo courtesy of Nina Fuller Photography)

needs to assess the suitability of any assistant in the treatment situation and ensure that issues around role and confidentiality are resolved prior to client work.

In order to reduce risk, the activities used in equine-assisted treatment, especially in equine-assisted mental health, should be carefully chosen as to not intentionally trigger a fight/flight/freeze response in the horse. Transmission of zoonotic disease, an area of risk typically addressed in the literature on animal-assisted therapy (Fine 2010), has not been discussed much in the context of equine-assisted therapy.

10.3.2.2 The Horse

Ethical considerations related to the therapy horse concern welfare and care, as well as suitability and preparedness for the activities it will take part of in equine-assisted treatment. Currently, few published (Gehrke et al. 2011; Kaiser et al. 2006) and unpublished (Pyle 2006; Suthers-McCabe and Albano n.d.) studies can be found on the effects of equine-assisted therapy on the horse. A well-defined work role for the therapy horse will help the practitioner in the selection and assessment for suitability. For instance, if the horse is included as part of physical therapy, its conformation, quality of gait, and obedience are characteristics central to its work role (Spink 1993).

An understanding of equine behavior, mental processes and learning is imperative in order to maintain the wellbeing of the therapy horse and to avoid the emergence of unwanted behaviors such as aggression (Fig. 10.10). Considerations for equine welfare should go beyond biological functioning and the absence of pain (Heleski and Anthony 2012). In addition, an awareness of the human tendency for inaccurate anthropocentric and anthropomorphic descriptions of the horse (such as the horse being 'kind', 'bad' or 'lazy') is important (McGreevy and McLean 2010).

In equine-assisted mental health, where a variety of mounted and ground work activities can be used with clients, the practitioner might design activities meant to challenge the horse so that the client may visually perceive how he or she is affecting the environment. Kruger and Serpell (2010) note that animals are commonly thought to expedite the rapport-building process between therapist and client in mental health. Thus, it is hard to imagine that therapeutic alliance, considered a consistent predictor of therapy outcome (Martin et al. 2000) where trust and respect, among other things, are central (Ackerman and Hilsenroth 2003), could be fostered in an environment where the equine is intentionally distressed. Careful consideration of current practices is needed to ensure that high-levels of stress are not induced in the therapy horse.

10.3.2.3 The Client

Equine-assisted therapy is not indicated for all client populations. The therapist needs to be knowledgeable about specific precautions and contraindications for the equine-assisted treatment strategy they employ and assess the client for suitability prior to entering treatment. The client should have an understanding of the nature of the treatment strategy as well as the inherent risks of horses prior to treatment in order to give informed consent. In addition, issues related to maintaining confidentiality due to the nature of the treatment setting should be clearly communicated. Finally, regardless of the therapist's own interest in equine-assisted therapy, he or she must recognize when a client might benefit more from a different treatment strategy and make an appropriate referral.

Hallberg (2008) comments that clients may perceive the client-therapist relationship in a more informal way when in the equine environment. The therapist is likely casually dressed and it is possible that treatment is taking place in equine facilities at the therapist's home. The opportunity for the client to see the therapist outside a conventional office setting may increase the likelihood of the client misperceiving the nature of the therapeutic relationship. Therapists need to be aware of the potential of this client perception and maintain healthy professional boundaries at all times.

It is common, even recommended, that the therapist has a pre-existing relationship with the therapy horse in equine-assisted mental health (Yrjölä 2009). Practitioners must explore feelings and projections in their personal relationship with the therapy horse before seeing clients. This is necessary so that the therapist does not become personally affected by what the client projects onto the therapy horse, especially when projections are experienced as negative or in conflict with those of the therapist (Scheidhacker 2009).

10.4   Practice

The physical and psychotherapeutic treatment strategies in equine-assisted therapy share common theoretical assumptions about why they are beneficial (see Sect. 10.3.1). In addition, each treatment strategy emphasizes specific aspects of the interaction between client and horse. The way the therapist utilizes an equine-assisted treatment strategy in clinical practice should support and augment his or her theoretical orientation (Fine 2010). A brief description of the two main areas of equine-assisted therapy is provided here, including nature of treatment, role of horse and therapist, as well as indications and contraindications for clients.

10.4.1   Hippotherapy

Hippotherapy can be defined as a treatment strategy in which the utilization of equine movement is a core component. The therapist who includes hippotherapy in his or her practice is trained and licensed as a physical therapist, occupational therapist or speech-language pathologist with additional training in hippotherapy (AHA, DKThR). The purpose of this treatment strategy is to address impairments, functional limitations, and disabilities in clients with neuromusculoskeletal dysfunction (AHA).

10.4.1.1   Nature of Treatment Strategy

The three-dimensional movement of the horse at a walk provides input to the client closely resembling that of human walk (Uchiyama et al. 2011). The effect is very similar in terms of lateral pelvic tilt, direction of displacement from center of gravity, and temporal sequence of stride (Fleck 1997). Lateral pelvic shift is larger when riding, which indicates a greater effect on trunk and spine (Heine 1997). The rhythmical impulses from the horse cause continuous muscle flexion and provide an opportunity for the client to practice equilibrium and righting reactions. According to the American Hippotherapy Association (n.d.a, b):

> The effects of equine movement on postural control, sensory systems, and motor planning can be used to facilitate coordination and timing, grading of responses, respiratory control, sensory integration skills and attention skills. Equine movement can be used to facilitate the neurophysiologic systems that support all of our functional daily living skills.

Hippotherapy usually takes place at the walk in an enclosed area. If the equine facility has a trail course within its premises it can be used in the hippotherapy session. The client is mounted on the therapy horse throughout the session but does not control the horse. The client may also take part in grooming the horse prior to the session to support functional skills and bonding with the horse. Adaptive equipment, such as a surcingle with handles or safety stirrups, is chosen depending on the client's needs and functional level. A saddle is typically not used in order for the

Fig. 10.11 Hippotherapy is
always conducted in a team
(Photo courtesy of PATH
Intl., Denver, CO)

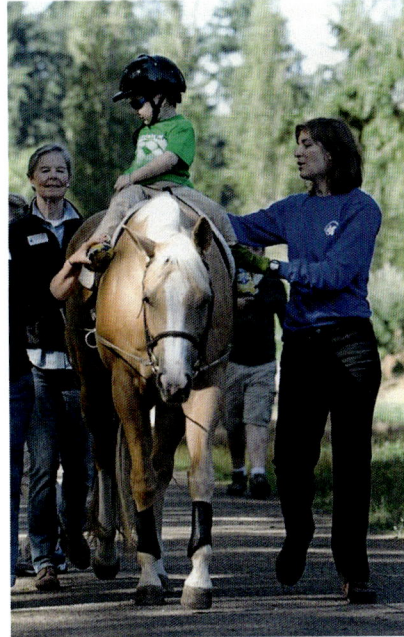

client to receive maximum input from the movement of the horse. The session
includes a horse handler, who leads or long-lines the horse (this can also be done by
the therapist), and side walkers if needed. Side walkers are positioned on either side
of the horse to ensure that the unstable client is not at risk for falling and may also
assist the client in carrying out movements as directed by the therapist (Fig. 10.11).
Activities in the session include changes in length of stride, tempo and direction;
patterns such as serpentines or figure eights, modified client positions on the back
of the horse, and the use of equipment such as balls or rings for exercises (Heine
1997; Spink 1993).

10.4.1.2   Role of the Horse

The therapy horse provides the movements that are central to the treatment strategy.
Unlike equipment used in traditional physical therapy, the horse may also be a
source of client motivation and bring additional social and emotional benefits to the
session (Strauss 1998). The therapy horse is carefully chosen for quality of gait and
overall conformation. He or she is continuously trained to maintain muscle sym-
metry and to ensure compliance with cues from the therapist and the horse handler.
Additionally, soundness needs to be carefully monitored and an assessment should
be done prior to each session. It is important that the therapy horse has been desensitized to various elements of the hippotherapy situation by a knowledgeable

equine professional in order to reduce the risk of stress and stress-related behaviors. A thorough screening and training system, such as the one proposed by Spink (1993) should be in place to ensure quality of treatment and ethical practice with regards to the therapy horse.

10.4.1.3 Role of Therapist

The therapist trained in hippotherapy will not only need to assess the client and make decisions for treatment but also evaluate the therapy horse, the assistant(s) and the facility (Strauss 1995). An important part of maintaining a professional and therapeutic setting, which is the responsibility of the therapist, is to have appropriately trained assistants in the hippotherapy session. Assistants should be familiar with the treatment strategy and therapy-specific requirements such as confidentiality and professional boundaries. This will reduce the risk of assistants inadvertently disrupting the therapeutic process through affecting the horse or the client in a negative or inappropriate way. The therapist should also monitor his or her own actions and use of language with regards to the horse and team members.

The dynamic nature of a hippotherapy session requires the therapist to constantly monitor the client and modify activities. It is the skill and creativity of the therapist that determine the quality of treatment. When matching a client with a therapy horse the therapist considers size and gait characteristics of the equine. The therapist also chooses the appropriate adaptive equipment for the client, which may change as treatment progresses. Finally, therapists must remember that adequate insurance is needed when conducting hippotherapy as part of their practice (Strauss 1995).

10.4.1.4 Indications and Contraindications

Each client needs to be individually evaluated by the referring physician for hippotherapy. The indications and contraindications presented here are general recommendations.

Hippotherapy is indicated for individuals with the following conditions: cerebral palsy, multiple sclerosis, head and brain trauma, developmental neurological conditions, post-traumatic neurological conditions, post-inflammatory neurological conditions and degenerative neurological conditions (Strauss 1995). Impairments include: abnormal muscle tone, impaired balance responses, impaired coordination, impaired communication, impaired sensorimotor function, postural asymmetry, poor postural control, decreased mobility, limbic system dysfunction related to arousal and attention skills (AHA). (Fig. 10.12)

Hippotherapy is contraindicated if the movement stimulus of the horse aggravates the client's neurological symptoms. Bone and joint changes of an inflammatory nature, inadequate cushioning between the vertebrae of the spine, atlanto-axial instability, danger of hip dislocation when sitting astride the horse, severe osteoporosis and severe scoliosis are considered contraindicative. In addition, the adult client

Fig. 10.12 The movements of the horse are beneficial for individuals with a number of impairments (Photo courtesy of PATH Intl., Denver, CO)

must be able to sit independently with sufficient head control to qualify for treatment. Contraindications unrelated to movement disorders are severe allergies, phobias related to the therapy situation, respiratory diseases, a severe heart condition and obesity (Strauss 1995).

10.4.2   Equine-Assisted Mental Health

When a horse is included in psychotherapy, a number of different terms are used internationally. In this chapter, the umbrella term equine-assisted mental health is used to encompass equine-assisted psychotherapeutic services provided by licensed mental health professionals who may legally practice psychotherapy or counseling in their country of residence.

10.4.2.1   Nature of Treatment Strategy

Horses can be included in psychotherapy in a variety of ways (McConnell 2010). There is no commonly accepted theoretical framework for equine-assisted mental health, and therapists should, in fact, include horses in a way that informs and enhances their theoretical approach and training (Fine 2010).

The format (individual, family or group) and the activities used for the equine-assisted mental health sessions are determined by the therapist, based on the client's treatment plan. Therapy can take place in a number of equine settings (stable, pasture, arena, riding trail), and a variety of activities can be incorporated to support treatment goals. Activities such as grooming, leading, longing, vaulting, driving and riding are modified for the purpose of therapeutic gain when included in treatment

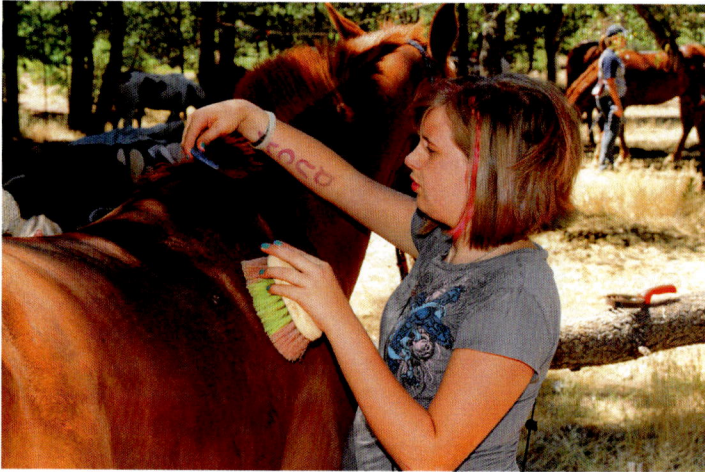

Fig. 10.13 Activities in equine-assisted mental health, such as grooming, are modified for the purpose of therapeutic gain (Photo courtesy of Arizona Burn Foundation/Jack Jordan)

(Fig. 10.13). Clients may also interact with horses in a number of non-traditional ways, such as painting the horse or observing equine behaviors. Challenge-based activities in a game format, based on experiential learning principles, are sometimes used (Hallberg 2008).

The difference between equine-assisted mental health activities and those done in a traditional equestrian setting is the focus on therapeutic goals. Whereas learning equine skills can be part of treatment it is not the main reason for interactions with the horse in therapy. The client's interpretation of an equine interaction is considered more important than the horse's actual behavior, as it may offer insight into beliefs the client has about him or herself and others. The therapist uses therapeutic metaphors when facilitating activities in order to help generalize the client's experiences to the rest of his or her life.

Karol (2007, p. 80) suggests, based on her own clinical experiences, that the following six aspects may be conducive to psychotherapeutic work including horses: (a) the actual experience; (b) the unique experience of being in relation to the horse; (c) the experience of the therapeutic relationship with the practitioner; (d) nonverbal experiences in communication with the horse; (e) preverbal/primitive experience such as contact, comfort, touch and rhythm; and (f) the therapeutic use of metaphor.

Recent qualitative investigations have aimed to describe essential features of the equine-assisted mental health session from both the client (Hayden 2005; Whitely 2009) and therapist (Esbjorn 2006; Frame 2006; McConnell 2010) point of view. It is still unclear which specific elements in the treatment session with the horse that need to be present for treatment to be effective (such as a positive relationship with the horse, or the perception of risk), and which populations are indicated for the best therapeutic outcome. In addition, each individual will likely benefit from a unique combination of elements in the equine-assisted mental health session.

Anthropomorphism, defined as the tendency to ascribe human characteristics to animals or innate objects, is central to the therapeutic experience in equine-assisted therapy (McConnell 2010). The fact that humans anthropomorphize animals (as being 'good', 'bad', 'nonjudgmental', 'accepting' and so forth) makes it possible for the clients to interpret the horse's behaviors within their own emotional world; to project feelings onto the horse; to feel a sense of bonding or attachment to the horse; and to accept interactions with the horse as metaphors for other relationships.

Many practitioners report that the horse 'mirrors' the client, or can 'read' the client's emotions. When these statements are being used in a therapeutic context for the benefit of the client, they can be helpful. For instance, when the horse reacts to the actions of the client, the therapist can ask the client to interpret the behavior or may offer possible explanations as part of insight-oriented work. Horses react to a number of stimuli in their environment and there is currently no support in the scientific literature for an inherent equine ability to 'read' human emotional states (McGreevy and McLean 2010). Practitioners have an ethical responsibility not to overstate what can be expected from this treatment strategy.

Brooks (2006) describes two models for working in equine-assisted mental health: a triangle model, which consists of the client, the horse and the therapist; and a diamond model, which also includes an equine professional as an integral part of treatment (Fig. 10.14). The role of the equine professional, often referred to as an equine specialist in the United States, is to monitor equine behavior in the session and to offer equine-specific observations to the client. The therapist may also include assistants in the session, to lead the horse or to support a client when riding.

10.4.2.2 Role of the Horse

The therapy horse is typically seen as a source of support and relationship, as well as a facilitator of insight and change (Esbjorn 2006; Frame 2006; McConnell 2010). Additionally, the horse has been described as a co-facilitator, an assistant, and sometimes as a therapeutic tool in the therapy session (Esbjorn 2006). Since the horse may have different functions depending on client goals, therapist orientation, and activities used in the session, a training protocol for the therapy horse in equine-assisted mental health is difficult to outline. Pain, fear and high-levels of stress will increase the risk for behaviors that might results in the client or horse becoming injured. The therapy horse should be trained for client interactions, such as riding and specific groundwork activities. The horse should also be familiar with tools and props used in sessions. It is important that the therapy horse receives consistent training and handling outside the client situation to reinforce desired behaviors and reduce the probability of stress (McGreevy and McLean 2010).

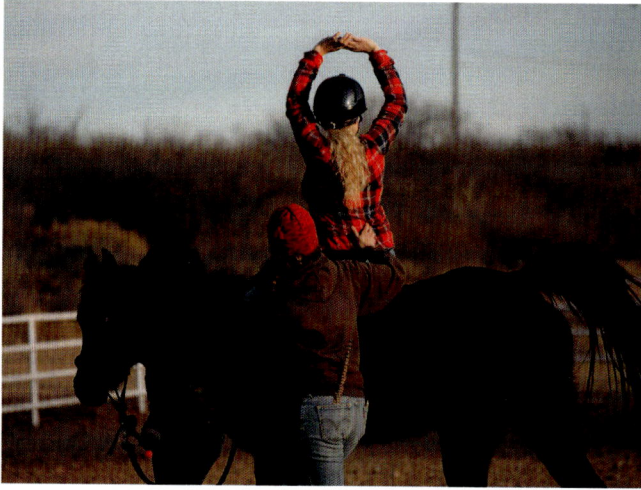

Fig. 10.14 The difference between equine-assisted mental health activities and those taking place in a traditional equestrian setting is the focus on therapeutic goals (Photo courtesy of Nina Fuller Photography)

10.4.2.3   Role of Therapist

The nature of psychotherapeutic work with horses requires the therapist to not only be aware of the dynamic between him or herself and the client but also what is taking place between the client and the horse. As equine-assisted mental health does not rely on a specific psychotherapeutic approach, the therapist can interpret the interactions and what the client offers within the context of his or her theoretical orientation. The inclusion of a living being into the therapy session opens up new possibilities for treatment but also for the risk of things not going as planned. The therapist should be prepared for events such as the death of a therapy horse and the impact it might have on the client (Cohen 2010).

In determining the client's suitability for equine-assisted treatment, the therapist must consider a number of factors specifically related to the nature of this treatment strategy. In addition to pondering the three questions presenter earlier in this chapter, the therapist should ask the client about previous equine experiences and explore whether there are cultural or religious beliefs that may be counterintuitive to treatment. Traumatic experiences involving horses, or a history of animal abuse should be known by the therapist ahead of time. These client experiences are not necessarily contraindicative but the level of severity will guide the therapist in making sound clinical decisions regarding treatment.

A practitioner with a strong interest in equine-assisted mental health might wish to consciously (or unconsciously) 'convert' clients who do not seem interested in having horses as part of treatment. These and other potential issues related to the therapist should be discussed in supervision (Scheidhacker 2009).

The therapist's responsibility to maintain a therapeutic and professional treatment setting extends to determining the suitability of assistants that are part of the session. In addition, the therapist should recognize when it might be advisable to co-facilitate sessions with a colleague, such as when conducting group therapy with horses included in the session.

Prior to seeing clients, the therapist needs to acquire the different types of insurances needed for equine-assisted work. Finally, the therapist must remember that when including a therapy horse in treatment, he or she is not only responsible for the safety of the client but also that of the horse.

10.4.2.4 Indications and Contraindications

Indications and contraindications for equine-assisted mental health have rarely been discussed in the literature (McConnell 2010). A number of studies have been conducted with adolescents (see Selby 2009) but there is not enough research to recommend or discourage treatment for specific populations or diagnoses.

It has been suggested that clients who are contraindicated for equine-assisted therapy are actively dangerous to self or others, actively psychotic, hold severe delusions involving horses, are medically unstable or in active substance abuse (PATHIntl). A history of animal abuse or fire setting is sometimes considered contraindicative because of the need to protect the therapy horses. However, the level of risk needs to be carefully assessed as individuals with such histories may benefit from equine-assisted treatment. Contraindications related to the general nature of treatment are severe allergies, phobias related to the therapy situation, respiratory diseases, a severe heart condition and severe obesity (Strauss 1995).

10.5 Conclusion

An international field of study and practice based on an interest in horses as therapeutic agents has emerged during the past decades. Currently, professionals in a number of countries utilize equine-assisted therapy, the inclusion of horses in physical or psychotherapeutic treatment. This chapter has provided a brief overview of equine-assisted therapy, its history, terminology, ethical considerations, research and practice. The inclusion of horses in therapy is set apart from other horse-related activities by the involvement of licensed/credentialed professionals and the focus on therapeutic goals based on the client's treatment plan. While some progress has been made to clarify this distinction through the use of appropriate terminology, there are still many issues that need to be addressed in order for equine-assisted therapy to be recognized as a valid strategy in treating both physical and psychological disorders.

For an emerging field to move forward, enthusiasm and engagement are undoubtedly needed. Practitioners and others interested in the development of equine-assisted therapy should, however, be careful not to overstate the benefits of this treatment strategy. As aptly inquired by the late Barbara L. Heine, an American pioneer in the field of hippotherapy: "[is it better] to be exuberant, or to be correct"? (AHA). More research is needed to establish the effectiveness of equine-assisted therapy, as the two most fundamental questions for this field are still unanswered: does including a horse in treatment improve the effectiveness of therapy? And if it does, what is it about the equine interaction that makes a difference? The future of the field will depend on the continued collaboration between practitioners, researchers, and other stakeholders in answering these questions.

References

Ackerman S, Hilsenroth MJ (2003) A review of therapist characteristics and techniques positively impacting the therapeutic alliance. Clin Psychol Rev 23:1–33

American Hippotherapy Association (AHA) (n.d.a) Hippotherapy as a treatment strategy. Retrieved from http://www.americanhippotherapyassociation.org/hippotherapy/hippotherapy-as-a-treatment-strategy. Accessed 23 Dec 2012

American Hippotherapy Association (AHA) (n.d.b) History of hippotherapy and AHA inc. Retrieved from http://www.americanhippotherapyassociation.org/hippotherapy/history-of-hippotherapy. Accessed 23 Dec 2012

Anthony DW (2007) The horse, the wheel, and language. Princeton University Press, Princeton

Bachelor A, Laverdière O, Gamache D, Bordeleau V (2007) Clients' collaboration in therapy: self-perceptions and relationships with client psychological functioning, interpersonal relations, and motivation. Psychotherapy 44(2):175–192

Bain AM (1965) Pony riding for the disabled. Physiotherapy 51(8):263–265

Beetz A, Kotrschal K, Uvnäs-Moberg K, Julius H (2011) Basic neurobiological and psychological mechanisms underlying therapeutic effects of Equine Assisted Activities (EAA/T). HHRF grant 2011 – public report. Retrieved from http://www.horsesandhumans.org/Research_AwardedProjects.html. Accessed 12 Oct 2012

Bizub AL, Joy A, Davidson L (2003) "It's like being in another world": demonstrating the benefits of therapeutic horseback riding for individuals with psychiatric disability. Psychiatr Rehabil J 26(4):377–384

Brandt N, Eklund E (2007) Häst – människa – samhälle: Om den nya hästhushållningens utveckling i Finland. Notat 1/2007, Forskningsinstitutet SSKH. Helsingfors: Helsingfors Universitet. Retrieved from http://sockom.helsinki.fi/info/notat/notat107.pdf. Accessed 12 Oct 2012

Brock B (1997) Therapy on horseback: psychomotor and psychological change in physically disabled adults. In: Engel BT (ed) Rehabilitation with the help of a horse: a collection of studies. Barbara Engel Therapy Services, Durango, pp 107–116

Bronson C, Brewerton K, Ong J, Palanca C, Sullivan S (2010) Does hippotherapy improve balance in persons with multiple sclerosis: a systematic review. Eur J Phys Rehabil Med 46(3):347–353

Brooks SM (2006) Animal-assisted psychotherapy and equine-facilitated psychotherapy. In: Webb N (ed) Working with traumatized youth in child welfare. Guilford Press, New York, pp 196–219

Brosse AL, Sheets ES, Lett HS, Blumenthal JA (2002) Exercise and the treatment of clinical depression in adults: recent findings and future directions. Sports Med 32(12):741–760

Budiansky S (1997) The nature of horses: exploring equine evolution, intelligence and behavior. Free Press, New York

Certification Board for Equine Interaction Professionals (CBEIP) (2008) State of the art report. Retrieved from http://www.cbeip.com/. Accessed 12 Oct 2012

Cohen SP (2010) Loss of a therapy animal: assessment and healing. In: Fine AH (ed) Handbook of animal-assisted therapy: theoretical foundations and guidelines for practice. Academic, San Diego, pp 441–456

Cook R (1997) The effects of developmental riding therapy on a client with post-traumatic brain injury: a single-case study. In: Engel BT (ed) Rehabilitation with the help of a horse: a collection of studies. Barbara Engel Therapy Services, Durango, pp 249–272

Deutsches Kuratorium für Therapeutisches Reiten (DKThR) (n.d.) Bereiche im Therapeutischen Reiten. Retrieved from http://www.dkthr.de/dkthrfakten.php?n2=therapie. Accessed 23 Dec 2012

Durant G (1878) Horse-back riding from a medical point of view. Cassell, Fetter & Galpin, New York. Retrieved from http://www.archive.org/stream/horsebackridingf00durarich/horse-backridingf00durarich_djvu.txt. Accessed 23 Dec 2012

Esbjorn RJ (2006) When horses heal: a qualitative inquiry into equine facilitated psychotherapy. Unpublished doctoral dissertation, Institute of Transpersonal Psychology, Palo Alto

Fachgruppe Arbeit mit dem Pferd in der Psychotherapie (FAPP) (n.d.) Geschichte und entwicklung der FAPP. Retrieved from http://www.fapp.net/treff1.htm. Accessed 23 Dec 2012

Federation of Horses in Education and Therapy International (HETI) (2012) Current members. Retrieved from http://www.frdi.net/membership_list.html. Accessed 12 Oct 2012

Federation of Horses in Education and Therapy International (HETI) (n.d.) Equine-assisted activities. Retrieved from http://www.frdi.net/EAA.html. Accessed 12 Oct 2012

Fine AH (2010) Incorporating animal-assisted therapy into psychotherapy: guidelines and suggestions for therapists. In: Fine AH (ed) Handbook of animal-assisted therapy: theoretical foundations and guidelines for practice. Academic, San Diego, pp 169–191

Fine AH, Mio JS (2010) The role of AAT in clinical practice: the importance of demonstrating empirically oriented psychotherapies. In: Fine AH (ed) Handbook of animal-assisted therapy: theoretical foundations and guidelines for practice. Academic, San Diego

Fitzpatrick JC, Tebay JM (1998) Hippotherapy and therapeutic riding: an international review. In: Wilson C, Turner D (eds) Companion animals in human health. Sage Publications, Thousand Oaks, pp 41–58

Fleck CA (1997) Hippotherapy: mechanics of human walking and horseback riding. In: Engel BT (ed) Rehabilitation with the help of a horse: a collection of studies. Barbara Engel Therapy Services, Durango, pp 155–176

Frame DL (2006) Practices of therapists using equine facilitated/assisted psychotherapy in the treatment of adolescents diagnosed with depression: a qualitative study. Unpublished doctoral dissertation, New York University School of Social Work, New York

Frewin K, Gardiner B (2005) New age or old sage? a review of equine assisted psychotherapy. Aust J Couns Psychol 6:13–17

Gehrke E, Baldwin A, Schiltz P (2011) Heart rate variability in horses engaged in equine-assisted activities. J Equine Vet Sci 31:78–84

Hahn C (2004) Behavior and the brain. In: McGreevy P (ed) Equine behavior: a guide for veterinarians and equine scientists. Saunders, London, pp 55–85

Hallberg L (2008) Walking the way of the horse: exploring the power of the horse-human relationship. iUniverse, Bloomington

Hänggi EB (1999) Categorization learning in the horse (Equus caballus). J Comp Psychol 113(3):1–10

Hanna FJ, Hanna CA, Keys SG (1999) Fifty strategies for counseling defiant, aggressive adolescents: reaching, accepting, and relating. J Couns Dev 77:395–404

Hayden A (2005) An exploration of the experiences of adolescents who participated in equine facilitated psychotherapy: a resiliency perspective. Unpublished doctoral dissertation, Alliant International University, San Diego

Heine B (1997) Hippotherapy. A multisystem approach to the treatment of neuromuscular disorders. Aust J Physiother 43:145–149

Heleski CR, Anthony R (2012) Science alone is not always enough: the importance of ethical assessment for a more comprehensive view of equine welfare. J Vet Behav Clin Appl Res 7(3):169–178

Intresseföreningen för Ridterapi (IRT) (n.d) Ridterapi. Retrieved from http://www.irt-ridterapi.se/ridterapi.html. Accessed 12 Oct 2012

Johns C (2006) Horses: history, myth, art. Harvard University Press, Cambridge, MA

Kaiser LK, Heleski CR, Siegford J, Smith KA (2006) Stress-related behaviors among horses used in a therapeutic riding program. J Am Vet Med Assoc 228:39–45

Karol J (2007) Applying a traditional individual psychotherapy model to equine-facilitated psychotherapy (EFP): theory and method. Clin Child Psychol Psychiatry 12(1):77–90

Katcher A, Wilkins G (1993) Dialogue with animals: its nature and culture. In: Kellert SR, Wilson EO (eds) The biophilia hypothesis. Island Press, Washington, DC

Kazdin AH (2010) Methodological standards and strategies for establishing the evidence base of animal-assisted therapies. In: Fine AH (ed) Handbook of animal-assisted therapy: theoretical foundations and guidelines for practice. Academic, San Diego, pp 563–578

Klüwer C (2009) Introduction. In: Fachgruppe Arbeit mit dem Pferd in der Psychotherapie (ed) Equine facilitated psychotherapy: case studies and international reports. FNverlag, Warendorf

Kohanov L (2001) The tao of equus. New World Library, Novato

Kruger KA, Serpell JA (2010) Animal-assisted interventions in mental health: definitions and theoretical foundations. In: Fine AH (ed) Handbook of animal-assisted therapy: theoretical foundations and guidelines for practice. Academic, San Diego, pp 33–48

Largo-Wright E (2011) Cultivating healthy places and communities: evidenced-based nature contact recommendations. Int J Environ Health Res 21(1):41–61

Levinson B (1961) The dog as a "co-therapist". Ment Hyg 46:59–65

Levinson B (1969) Pet-oriented child psychotherapy. Charles C. Thomas, Springfield

Macauley BL, Gutierrez KM (2004) The effectiveness of hippotherapy for children with language-learning disabilities. Commun Disord Q 25:205–217

MacKinnon JR, Noh S, Laliberte D, Lariviere J, Allan DE (1995) Therapeutic horseback riding: a review of the literature. Phys Occup Ther Pediatr 15:1–15

Maclean N, Pound P (2000) A critical review of the concept of patient motivation in the literature on physical rehabilitation. Soc Sci Med 50(4):495–506

Magner D (1887/2004) Classic encyclopedia of the horse. Book Sales Inc, New York

Martin DJ, Garske JP, Davis K (2000) Relation of the therapeutic alliance with outcome and other variables: a meta-analytic review. J Consult Clin Psychol 68(3):438–450

Mata J, Thompson RJ, Jaeggi SM, Buschkuehl M, Jonides J, Gotlib IH (2012) Walk on the bright side: physical activity and affect in major depressive disorder. J Abnorm Psychol 121(2):297–308

McConnell PJ (2010) National survey on equine assisted therapy: an exploratory study of current practitioners and programs. Unpublished doctoral dissertation, Walden University, Minneapolis

McCormick A, McCormick M (1997) Horse sense and the human heart. Health Communications, Deerfield Beach

McGreevy PD, McLean AD (2010) Equitation science. Wiley-Blackwell, Oxon

Miller RM, Lamb R (2005) The revolution in horsemanship and what it means to mankind. The Lyons Press, Guilford

Montague A (1971) Touching: the human significance of the skin. Columbia University, New York

Münchner Schule für Psychotherapeutisches Reiten (n.d.) Definition – Entwicklung der Münchner schule für psychotherapeutisches reiten. Retrieved from http://www.psychotherapeutisches-reiten.de/Frameset.html. Accessed 23 Dec 2012

National Institute of Child Health and Human Development (NICHD) (n.d) NICHD funding by state: Washington for 2011 (Patricia Pendry). Retrieved from http://www.nichd.nih.gov/about/budget/Pages/fundstate.aspx. Accessed 12 Oct 2012

Nicol CJ (2002) Equine learning: progress and suggestions for future research. Appl Anim Behav Sci 78(2):193–208

Opgen-Rhein C, Klaeschen M, Dettling M (2011) Pferdegestuetzte Therapie bei psychischen Erkrankungen. Schattauer GmbH, Stuttgart

Österreichischen Kuratorium für Therapeutisches Reiten (OKTR) (n.d.) Hippotherapie. Retrieved from http://www.oktr.at/pub/hippo/hippo_base.php?tab=11. Accessed 12 Oct 2012

Outram AK, Stear NA, Bendrey R, Olsen S, Kasparov A, Zaibert V, Thorpe N, Evershed RP (2009) The earliest horse harnessing and milking. Science 323(5919):1332–1335

Phelan JE (2009) Exploring the use of touch in the psychotherapeutic setting: a phenomenological review. Psychother Theory Res Pract Train 46(1):97–111

Pickeral T (2005) The encyclopedia of horses and ponies. Parragon, London

Pickeral T (2008) The horse: 30,000 years of the horse in art. Merrell, London

Professional Association of Therapeutic Horsemanship (PATHIntl.) (n.d.) EAAT definitions. Retrieved from http://www.pathintl.org/resources-education/resources/eaat/193-eaat-definitions. Accessed 23 Dec 2012

Purjesalo K (1991) Ratsastus hoito- ja kuntoutusmuotona. Tampereen Yliopisto, Tampere

Pyle AA (2006) Stress responses in horses used for hippotherapy. Unpublished master's thesis, Texas Technical University, Lubbock. http://repositories.tdl.org/ttu-ir/handle/2346/11442. Accessed 23 Oct 2012

Rethorst CD, Wipfli BM, Landers DM (2009) The antidepressive effects of exercise: a meta-analysis of randomized trials. Sports Med 39(6):491–511

Root A (2000) Equine-assisted therapy: a healing arena for myth and method. Unpublished doctoral dissertation, Pacifica Graduate Institute, Carpinteria

Scheidhacker M (2009) Searching for the whole (healing) center. In: Fachgruppe Arbeit mit dem Pferd in der Psychotherapie (ed) Equine facilitated psychotherapy: case studies and international reports. FNverlag, Warendorf

Schulz M (1999) Remedial and psychomotor aspects of the human movement and its development: a theoretical approach to developmental riding. Sci Educ J Ther Riding 7:44–57

Schweizer Gruppe für Hippotherapie-K® (n.d.) Was ist Hippotherapie-K®? Retrieved from http://www.hippotherapie-k.org/index.php?id=39. Accessed 23 Dec 2012

Selby A (2009) A systematic review of the effects of psychotherapy involving equines. Unpublished master's thesis, University of Texas, Austin

Snider L, Korner-Bitensky N, Kamman C, Warner S, Saleh M (2007) Horseback riding as therapy for children with cerebral palsy: is there evidence of its effectiveness? Phys Occup Ther Pediatr 27(2):5–23

Sokolof M (2009) Treating abuse with the help of a horse. In: Fachgruppe Arbeit mit dem Pferd in der Psychotherapie (ed) Equine facilitated psychotherapy: case studies and international reports. FNverlag, Warendorf

Spink J (1993) Developmental riding therapy: a team approach to assessment and treatment. Therapy Skill Builders, Tucson

Sterba JA (2007) Does horseback riding therapy or therapist-directed hippotherapy rehabilitate children with cerebral palsy? Dev Med Child Neurol 49(1):68–73

Strauss I (1995) Hippotherapy: neurophysiological therapy on the horse (trans: Takeuchi M). Ontario Therapeutic Riding Association (original work published in 1991), Toronto

Strauss I (1998) Hippotherapy: its unique position within physiotherapy. In: Deutsches Kuratorium für Therapeutisches Reiten (ed) Therapeutic riding in Germany: selected contributions from the special brochures of the DKThR. DKThR, Warendorf, pp 15–17

Suthers-McCabe HM, Albano L (n.d.) Evaluation of stress response of horses in equine assisted therapy programs. Center for Animal Human Relationships. Virginia-Maryland Regional College of Veterinary Medicine. Study results available at http://www.petpartners.org/Document.Doc?id=268. Accessed 23 Oct 2012

Thompson-Coon J, Boddy K, Stein K, Whear R, Barton J, Depledge MH (2011) Does participating in physical activity in outdoor natural environments have a greater effect on physical and mental well-being than physical activity indoors? A systematic review. Environ Sci Technol 45(5):1761–1772

Uchiyama H, Ohtani N, Ohta M (2011) Three-dimensional analysis of horse and human gaits in therapeutic riding. Appl Anim Behav Sci 135(4):271–276

Vidrine M, Owen-Smith P, Faulkner P (2002) Equine-facilitated group psychotherapy: applications for therapeutic vaulting. Issues Ment Health Nurs 23(6):587–603

Weber A (1998) Hippotherapy for patients suffering from multiple sclerosis. In: Deutsches Kuratorium für Therapeutisches Reiten (ed) Therapeutic riding in Germany: selected contributions from the special brochures of the DKThR. DKThR, Warendorf, pp 22–39

Whitely R (2009) Therapeutic benefits of equine assisted psychotherapy for at-risk adolescents. Unpublished doctoral dissertation, Texas A&M University, College Station

Wilson EO (1984) Biophilia. Harvard University Press, Cambridge

Yrjölä M-L (2009) The horse as a good object in long-term psychotherapy. In: Fachgruppe Arbeit mit dem Pferd in der Psychotherapie (FAPP) (ed) Equine facilitated psychotherapy: case studies and international reports. FNverlag, Warendorf

Zadnikar M, Kastrin A (2011) Effects of hippotherapy and therapeutic horseback riding on postural control or balance in children with cerebral palsy: a meta-analysis. Dev Med Child Neurol 53(8):684–691

Zamir T (2006) The moral basis of animal assisted therapy. Soc Anim 14(2):179–198

Chapter 11
Canine Olfactory Detection of Human Disease

Claire Guest

11.1 Introduction

For centuries, it has been understood that many human diseases have a characteristic odor. Physicians often came to recognise the odor associated with conditions such as lung infection, diabetes, and typhoid. Dogs have been used by man for their olfactory abilities for many years and are used in a wide variety of disciplines, including the detection of drugs, explosives and currency. In recent years it has been recognised that dogs may be able to assist in the early detection of human disease, notably epileptic seizures (Dalziel et al. 2003; Kirton et al. 2008). However, it is unclear in this instance whether the dogs are detecting oncoming seizures by smell or by some other means.

The first indication that cancer cells may release abnormal volatile substances, and that dogs may be able to smell them, came in a case report published in *The Lancet* by Williams and Pembroke (1989). The authors described how a woman sought medical advice because her dog had been taking an inordinate interest in a mole on her leg. When the mole was removed, it was found to be a malignant melanoma.

The report attracted the attention of an orthopaedic surgeon, Dr. John Church, who, over the intervening years, unearthed another 16 similar stories, one of which he also published as a case report in *The Lancet*. Most of these incidents related again to skin cancer, but several involved internal cancers, including breast cancer (Church and Williams 2001). Prior to clinical evidence of the ability of dogs to detect cancer, there were a number of anecdotes reported. Some of these are presented below.

C. Guest, B.Sc., M.Sc., DSc. (✉)
Medical Detection Dogs, 3 Millfield, Greenway Business Park,
Great Horwood, Milton Keynes MK 17 0NP, UK
e-mail: claire.guest@medicaldetectiondogs.org.uk

M. Grassberger et al. (eds.), *Biotherapy - History, Principles and Practice:*
A Practical Guide to the Diagnosis and Treatment of Disease using Living Organisms,
DOI 10.1007/978-94-007-6585-6_11, © Springer Science+Business Media Dordrecht 2013

Fig. 11.1 Maureen with her
dog Max, a red Collie cross

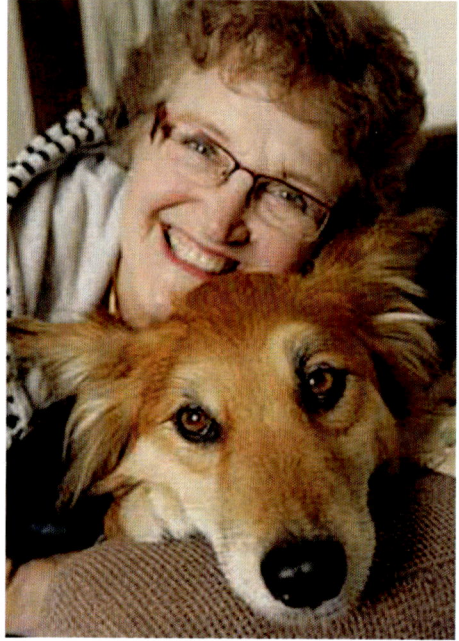

In 1978, magazine editor Gillian noticed that her pet Dalmatian named Trudi was paying constant attention to what she thought was a freckle on her leg. Gillian did not pay attention to a small brownish-yellow spot on her skin until the time when her dog kept sniffing this area. When Gillian went to a physician, the freckle was removed and tests showed that it was a malignant melanoma.

In 2006, Maureen Burns found a lump in her breast during self-examination. A few months earlier, her dog Max, a red Collie cross aged nearly 10 (Fig. 11.1), began showing unusual signs and became less playful, wouldn't jump on Maureen's lap, share her bed or sit at her feet, and his eyes were sad and dull. Max would from time to time come up to Maureen and touch the lump area and back off very unhappily. Also, he kept sniffing her breath on numerous occasions. Maureen was confused and wondered if Max's strange behaviour was due to his advancing years, or indicated that something was really wrong with her.

When she consulted her general practitioner (GP), she was quickly was referred to hospital where numerous mammograms and scans did not show anything extraordinary. Maureen went in for further testing and after having two biopsies, her cancer was finally confirmed. However, she told the nurse that she already knew she had cancer as her dog already had "told" her. Soon thereafter, the lump and four lymph nodes were surgically removed. It was diagnosed that the patient had an invasive lobular carcinoma (grade 1), the size of which was 2.5 cm, while the four lymph nodes were not infiltrated. When Maureen returned home from the surgery, she was greeted by Max, and his behaviour was totally different. He was overjoyed and doing all the things he had done before the diagnosis. The patient began to receive routine radiotherapy and the prognosis was excellent.

11.2 Training of Cancer Detection Dogs

In 2002, following several months of training with many dogs and dog trainers involved, a spaniel called Tangle sniffed up and down a line of patients' urine samples and identified the one that had been provided by a cancer patient. The previous stages of training had gone well, with Tangle proving to be a reliable dog, however there was still uncertainty as to whether the dogs would be able to tell the difference between those half ml samples that had been given by patients with cancer as opposed to those given by patients with other diseases. However, in a training environment, Tangle had demonstrated that the theory was in fact a reality: dogs could be trained to detect cancer. Further studies were now required.

In collaboration with Dr. Church, and together with a research team of scientists and doctors from Amersham Hospital the author started a thorough investigation to see whether there was a scientific basis to these anecdotal stories. It was decided to conduct a simple, but stringent proof of principle experiment, the first of its kind, to see whether dogs could be trained to detect bladder cancer from the odor of urine.

Over a period of 7 months, a training program was developed with six dogs, all of whom were gradually taught to "key into" the specific smell for cancer and to ignore all of the other odors present in the urine, including those associated with other diseases and conditions. Samples were frozen to minus 35 °C. A contribution was given by someone who knew their life may be nearing the end but who hoped to help the team make crucial advances in the fight against cancer – a fight that would continue well after their death. It was agreed to start with dogs that were representative of dogdom itself, rather than an elite subsection of specially picked "scent experts". After all, this was a proof of principle study to see if dogs in general could be trained to detect cancer. All those involved suspected that the best dogs for the job would be the breeds typically associated with working disciplines, but wished to make no assumptions at this stage. Finally, the band of six dogs was selected and consisted of a Papillon, a Labrador, three Cocker Spaniels and a Mongrel.

The dogs were then tested in a rigorously controlled, blinded trial in which they were required to select cancer urines from a range of 'controls' obtained from age-matched, diseased and healthy subjects. All the samples were entirely new to them. Their combined accuracy rate of 41 % was statistically significant, as the results proved unequivocally that dogs can be trained to detect cancer by the sense of smell. Two of the most accurate dogs who performed in this first test had a 56 % accuracy rate. One of the dogs was Tangle as discussed earlier (Fig. 11.2).

The results of the study were published in the *British Medical Journal* (Willis et al. 2004) and were widely reported across the world, both in the press and on television. The dogs have shown that human cancer cells release odorous substances of diagnostic importance. The scientists related to the study above, and other international research groups, now wished to further utilize the dog's olfactory capabilities to help develop diagnostic instruments for clinical use.

Fig. 11.2 Tangle is not
alerting he is screening
cancer samples

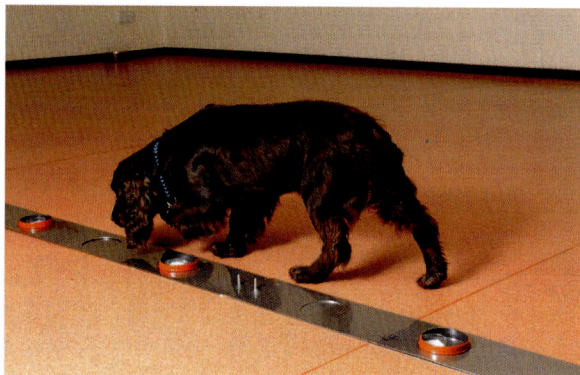

11.3 Development of "Medical Detection Dogs"

During 2005, work continued but progressed in a new direction. It was essential that dogs in training were regularly tested in blinded trials run by hospital personnel in order to assess their accuracy rate and progress. These blind runs were lengthy and needed to be run during the working day, since hospital staff and researchers were not in a position to work in the evenings. It was obvious that the project needed some new initiative in order to succeed. It had become apparent over an extended period of time that to progress in this difficult field, a really intense concentrated effort was required. The staff would have to be freely available to do the work, and premises would be required which enabled the dogs to work in optimal conditions. Additionally, there would need to be more rigorous selection of dogs as well as multi-disciplinary teamwork. A decision was made to take action on a conclusion previously reached, that the best way forward was to form a new charity organization with fund-raising capabilities that would have its own premises and with its own fully employed staff.

"Cancer and Bio-Detection Dogs" was formed as a not-for-profit company in October 2007 and charitable status was obtained in June 2008. Dr. Church became a trustee and the Honorary Medical Director of the organization. Since the development of this not-for-profit organization, the members have now come to understand that dogs are able to detect a variety of diseases and debilitating conditions by identifying changes in the odors of the body or breath. In 2011, the name of the charity was changed to "Medical Detection Dogs" as this described the work more accurately.

The charity is now working in conjunction with Professor Karol Sikora, a leading UK oncologist, as well as mass spectrometry scientists and medical statisticians in the detection of cancer from human breath and urine. Currently, the charity's main focus is the detection of prostate cancer. There are powerful reasons for such a study. Prostate Cancer (PCa) is a major killer in male human populations. Current testing such as the prostate specific antigen test (PSA test) is unreliable, and as a result, many GPs are reluctant to use it. If dogs can sniff prostate cancer from a urine sample, chances are high that results from the dogs' sniffing research can allow for a test to be developed that is superior than the PSA test. Results would indicate the existence of a potential odor signature of PCa that may correspond to one or, more likely, multiple Volatile Organic Compounds (VOCs). These molecules should then be assessed by specific gas chromatography/mass spectrometry analysis.

Society can benefit tremendously from the work undertaken by Medical Detection Dogs. First and foremost, the work can advance research into the early diagnoses of cancer. It is well known that early diagnosis would save countless lives and would benefit the public significantly. However, research is warranted when there is sufficient motivation to find an answer and a reasonable belief that a particular line of investigation can provide answers. Both of these premises are clear with this work. If answers cannot be determined, i.e. that dogs cannot in practice help doctors with early diagnoses, the research that is conducted will nevertheless be beneficial to overall research into cancer.

11.4   Research with Cancer Detection Dogs

Since the early *British Medical Journal* article in 2004, there have been a number of supporting studies from around the world.

11.4.1   Prostate Cancer

In a recent study Cornu et al. (2011) evaluated the efficacy of prostate cancer (PCa) detection by a trained dog (a Belgian Malinois) on human urine samples. After a learning phase and a training period of 2 years, the dog's ability to discriminate PCa and control urine was tested in a double-blind procedure. The dog correctly designated the cancer samples in 30 of 33 cases. Of the three cases wrongly classified as cancer, one patient was rebiopsied and a PCa was diagnosed. The sensitivity and specificity were both 91 %. The authors concluded that dogs can be trained to detect the odour signature of PCa by smelling urine with a significant success rate. They further suggest that identification of the VOCs involved could lead to a potentially useful screening tool for PCa.

11.4.2   Colorectal Cancer

A study by Sonoda et al. (2011) on colorectal cancer using breath and faecal samples showed equally promising results. Sensitivity was at 0.97 with a specificity of 0.99 for faecal samples, while sensitivity was at 0.91 with a specificity of 0.99 for breath samples. The accuracy of canine scent detection was high even for early cancer. Canine scent detection was not confounded by current smoking, benign colorectal disease, inflammatory disease or the presence of human haemoglobin or transferrin. The authors conclude that a specific cancer scent does indeed exist and that cancer-specific chemical compounds may be circulating throughout the body. These odour materials may become effective tools in colorectal cancer screening.

11.4.3 Lung Cancer

McCulloch et al. (2006) evaluated the ability of trained dogs to distinguish, by scent alone, exhaled breath samples of 55 lung cancer patients from those of 83 healthy controls. Among lung cancer patients and controls, overall sensitivity of canine scent detection compared to biopsy-confirmed conventional diagnosis was 0.99 (95 % confidence interval [CI], 0.99, 1.00) and overall specificity 0.99 (95 % CI, 0.96, 1.00). The authors state that in a matter of weeks, ordinary household dogs with only basic behavioral "puppy training" were trained to accurately distinguish breath samples of lung and breast cancer patients from those of controls.

Additional evidence was published by Ehmann et al. (2012) who found that sniffer dogs can identify lung cancer with an overall sensitivity of 71 % and a specificity of 93 %. The authors anticipated that a robust and specific volatile organic compound (or pattern) is present in the breath of patients with lung cancer.

11.4.4 Breast Cancer

In the same study from 2006, McCulloch et al. also tested the dog's ability to distinguish exhaled breath samples of 31 breast cancer patients from those of the 83 healthy controls. Dog handlers and experimental observers were blinded to the identity of breath samples, obtained from subjects not previously encountered by the dogs during the training period. Among breast cancer patients and controls, sensitivity was 0.88 (95 % CI, 0.75, 1.00) and specificity 0.98 (95 % CI, 0.90, 0.99).

11.4.5 Ovarian Carcinoma

Horvath et al. (2008) trained a dog to distinguish different histopathological types and grades of ovarian carcinomas, including borderline tumours, from healthy control samples. With double-blind tests showing 100 % sensitivity and 97.5 % specificity the authors clearly demonstrated that human ovarian carcinoma tissues can be characterized by a specific odour, detectable by a trained dog. In a subsequent study Horvath et al. (2010) examined whether the cancer-specific odour can also be found in the blood. Using two specially trained dogs both, ovarian cancer tissues and blood from patients with ovarian carcinoma, were tested. As a result the tissue tests showed sensitivity of 100 % and specificity of 95 %, while the blood tests showed sensitivity of 100 % and specificity of 98 %. In accordance with these results the study strongly suggests that the characteristic odour emitted by ovarian cancer samples is also present in blood drawn from patients with the disease.

11.4.6   Melanoma

In a small, more anectdotal study Pickel et al. (2004) describe two dogs that initially demonstrated reliable localization of melanoma tissue samples hidden on the skin of healthy volunteers. Subsequently one dog "confirmed" clinically suspected (and later biopsy-proven) diagnoses of melanoma in five patients. Interestingly, in a sixth patient, the dog "reported" melanoma at a skin location for which initial pathological examination was negative (despite clinical suspicion). More thorough histopathological examination in this individual then confirmed melanoma in a fraction of the cells.

Whilst all the above-mentioned studies showed huge promise there is still a long way to go. Much of this work has been with few dogs and on a very limited number of samples. In some cases there has been repeated use of controls in testing and controls have been from younger healthy individuals and not from age and symptom matched individuals. This would result in much higher accuracy levels from the dogs that may not be replicated in a more detailed clinical trial. Further research is required with a larger group of dogs, much larger sample sizes and robust double blind testing with age and symptoms matched controls. This can be done but does require funding as ethics and patient consent must be passed and sample collection can be a lengthy process.

11.4.7   Volatile Organic Compounds (VOCs)

The scientific basis of the ability of dogs to detect cancer is believed to be linked to VOCs produced by malignant cells. It has been established that during tumor growth, protein changes in malignant cells lead to peroxidation of the cell membrane components, which produce VOCs that are detected in the headspace of the cell cultures (Brunner et al. 2010). The interest in the biomarkers of cancer has risen steadily over the past 20 years and a variety of approaches have been used including gas chromatography/mass spectrometry, so called eNoses, and trained sniffer dogs.

A recent study completed by the charity and colleagues at the Buckinghamshire NHS trust was recently published in the Journal *Cancer Biomark* entitled, "Volatile organic compounds as biomarkers of bladder cancer: Sensitivity and specificity using trained sniffer dogs" (Willis et al. 2010). This study aimed to evaluate

Fig. 11.3 Samples being prepared

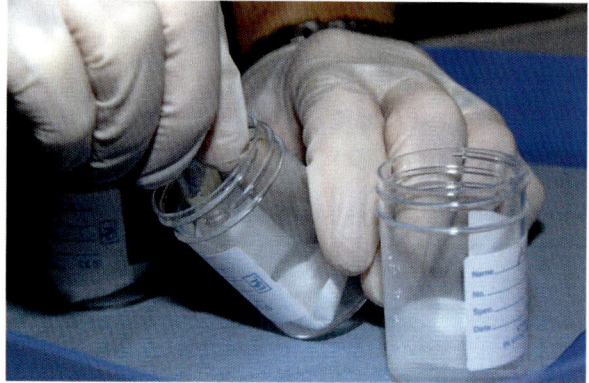

Fig. 11.4 (a, b) Trained sniffer dog Daisy working on the test carousel to identify samples from patients with bladder cancer

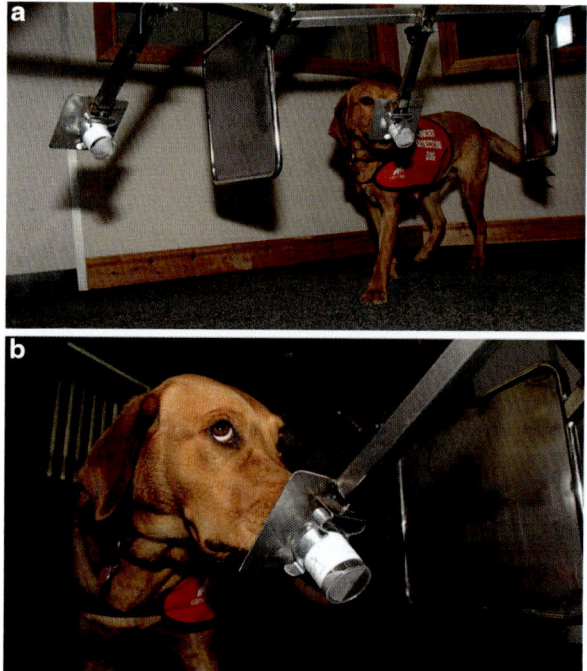

the sensitivity and specificity that can be achieved by dogs in a series of 30 double-blind tests runs, each consisting of one cancer urine placed alongside six controls (see Figs. 11.3 and 11.4).

The highest sensitivity achieved by the best performing dog was 73 % overall. This particular dog scored 100 % on grade 1 tumors, 60 and 61 % on Grades 2 and 3, respectively. The sniffer dog group as a whole identified cancer samples in 64 % of the cases. Specificity ranged from 92 % from healthy, young volunteers down to

56 % for those taken from older patients with non-cancerous urological disease. The sensitivities achieved collectively by the dogs for G1, G2, and G3 tumors in the study were 75 % (with one dog achieving 100 % accuracy, 61 and 60 % respectively). Odds ratio comparisons confirmed a significant decrease in performance as the extent of urine dipstick abnormality and/or pathology amongst the control population increased. More importantly, statistical analysis indicated covariates such as smoking, gender, age, blood protein and/or leucocytes did not alter the odds of indication. This means that it is highly unlikely that the dogs were using a confounder for their detection.

These results give further evidence of volatile biomarkers for cancer. In addition, they indicate that the performance of the dogs is higher on early stages of a tumor, though this is yet to be statistically determined. This also increases the possibility that these kinds of works may lead to an early non-invasive way of diagnosis.

Published studies in relation to canine detection of cancer biomarkers show that there is considerable variability between the sensitivity and specificity reported (Lippi and Cervellin 2012). For example, there is a lower level of accuracy rates for canines on bladder cancer than those reported by other groups for different tumor types. There are a number of possible reasons for this difference. First, urine is a very complex medium, which contains numerous compounds.

The study by Cornu et al. (2011) on the canine olfactory detection of prostate cancer, with reported accuracy of 91 %, used warmed samples, which may have improved relative proportions. In this study however, a repeated use of controls was used. This may well have lead to a higher success rate of the dogs as they might have been able to learn the signature of the individuals who were "non-reward individuals". Dogs are extremely adept at this and may use this to 'short cut' the detection if it results in a successful outcome.

It is also likely that the constrictions on the training protocol as set by clinicians in the study by Willis et al. (2010) in relation to rewarding the dogs, was detrimental to performance. This study used the most complicated controls following relatively few training samples. These controls, which were presented as double blinds came from individuals with a variety of other medical conditions, some of which the dogs had not been trained to ignore. In addition, the dogs were not rewarded at all during test runs during the double blind phase. This inevitably resulted in a reduction in motivation and confidence in the dogs. However, the work of the dogs raises the possibility that modern headspace analysis could be used for a diagnosis of bladder and other cancers. Regression analysis also indicated that the dog's ability to detect cancer volatiles was not due to the recognition of confounders.

A recent publication by Weber et al. (2011) evaluating a gas sensor array for the identification of bladder cancer from urine, reported that specificity and sensitivity for cancer resulted in lower accuracy rates when the sensor was compared against diseased controls. Accuracy for the eNose was still slightly lower as compared to using the dogs when more demanding controls were used. The current overall accuracy of the gas sensor was reported at 70 % with a score of 70 % sensitivity and 70 % specificity. Accordingly, this method is not considered sufficient for a diagnostic test.

However, this study opens the door of VOC detection for cancer diagnosis and furthermore suggests that the conditioned dog should be used in the near future to validate candidate molecules emerging from metabolomic screening. The results provide a new insight into the field while additional work with more dogs is now required to further investigate and validate this work.

11.5   Medical Alert Dogs

The use of dogs to assist people living with diabetes is a recent development. For many years, some diabetic individuals have anecdotally reported that their companion animals have reacted to a decrease in their blood glucose (Wells et al. 2008). O'Connor et al. (2008) reported case of hypoglycaemia in a non-diabetic patient being detected by a dog.

The organisation "Medical Detection Dogs" has begun to formally train dogs for this purpose. These dogs are trained to live and work alongside people to provide early alerts of changing glucose levels, prompting clients to take appropriate action. The dogs provide vital early warning signals to individuals who suffer from brittle or aggressive diabetes and/or poor hypoglycaemic (low blood sugar) awareness. This early warning enables the client to better regulate their own blood sugar levels, thereby reducing the incidence of hypoglycaemic episodes and also preventing harmful side effects. For someone living with diabetes, hypoglycaemia, or the avoidance of it can be a daily problem. This condition is frightening and also very distressing; symptoms can vary from confusion to seizures and comas, and can also be life-threatening. Medical Assistance Dogs can be invaluable companions for individuals in this situation. Preliminary studies of the current partnerships suggest that the trained dogs facilitate pronounced reductions in hypoglycaemic episodes and significantly reduce recipients' requirement for paramedic assistance and hospitalization. Self-management of their condition using evidence-based methods can help people with diabetes to live longer, healthier lives and experience improved quality of life.

Until now, Medical Assistance Dogs have been trained to assist individuals who must manage their complex medical conditions on a day-to-day basis. The dogs are taught to identify the odour changes that are associated with certain medical events. The charity has now placed 40 of these blood sugar detection assistance dogs with people living with so called brittle or unstable diabetes. Trained to be highly sensitive to glucose levels, they warn the individual when levels deviate from the normal range, and can indicate changes within one or two millimoles of blood sugar. Normally the dogs are trained to alert blood sugar levels below 4.5 (but above 3) millimoles per litre and also alert to high blood sugars at the level chosen by the client (normally above 10–12 millimoles per litre).

Additional benefits of the placement of a blood sugar detection dog include:

- Increases in individuals' confidence, reducing the tendency to maintain high blood sugar levels as a method of preventing hypoglycaemia;

- Reduction of prolonged periods of hyperglycaemia (high blood sugar level), which has a major benefit on the long-term health and well-being of individuals with diabetes;
- Reduction in symptoms in patients with high blood sugar is the cause of many long-term complications normally associated with diabetes such as sight loss, severe ulcerated wounds, amputation of limbs and kidney damage.

There is some evidence that diabetics with assistance dogs might have improved HbA1c (glycosylated haemoglobin; a long term marker for average blood glucose levels) or average blood glucose levels (Korljan-Babić et al. 2011). This fact alone has a significant impact on the lives of those with diabetes or hypoglycaemia.

11.5.1 Training Medical Alert Dogs

Training a blood sugar detection dog requires both training in odour identification and a reliable alert, which enables the dog to communicate blood level changes to the individual. When outside the normal range, dogs, once trained, can warn before the symptoms are felt. Depending on their 'owners' needs, the dogs will alert in a variety of ways; for example, with jumping up, licking or pawing. They can also be trained to push alarm buttons. These dogs are also trained to fetch the medical bag/blood testing kit if asked or when, due to the cognitive confusion caused by low blood sugars, the person does not respond to the alert.

For many individuals with unstable diabetes who get no warning of such an episode, these dogs are truly lifesavers. The dogs are also trained to warn when blood sugars become too high. Although this is rarely an emergency situation there are numerous serious and potentially life-threatening side effects of hyperglycaemia, as previously discussed. Medical Alert Dogs quickly recognise these signs and are taught to bring vital medical supplies and to summon help. Below are some examples.

11.5.2 Rebecca and Shirley's Story

Six-year-old Rebecca had been hospitalised eight times in 2 years due to her unstable diabetes resulting in unpredictable changes in her blood sugar level. To support the family, the charity paired Rebecca with a Labrador cross Golden Retriever named Shirley (Fig. 11.5). Within just 3 weeks following her placement, the dog was alerting successfully. Shirley wakes the mother at night when Rebecca is experiencing a dangerous change in her blood sugar levels, meaning prompt actions can be taken at home and avoidance of trips to the hospital. Rebecca was previously collapsing three or four times a week, unable to feel any change in her sugar levels. With the arrival of Shirley, the fear of collapsing has eased. As a hypo-alert dog trained to sense changes in odour when the sugar levels drop or increase, Shirley can

Fig. 11.5 Rebecca and
Shirley, a Labrador cross
Golden Retriever trained to
detect unpredictable changes
in her blood sugar level

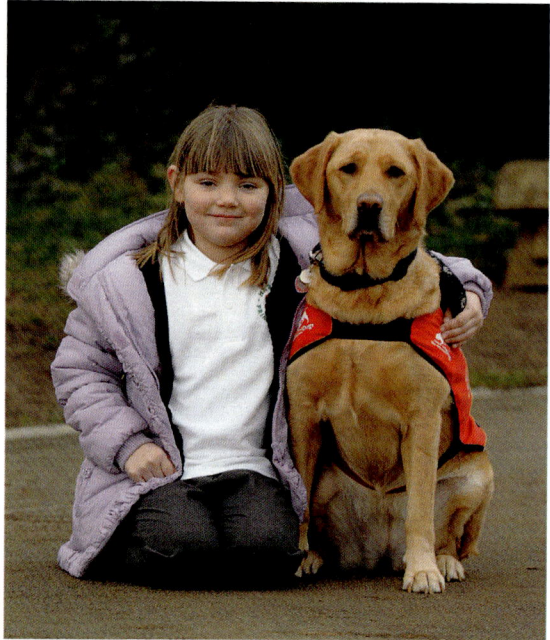

warn Rebecca by licking her hands, sitting on her lap and bringing a sugar-level testing kit to her.

Since the arrival of Shirley, the young girl regained confidence, and it has been a relief to the whole family. Shirley accompanies Rebecca wherever she goes, and there has not been a single hospital admission since the dog arrived. On the request of the Headmaster and Governors, the dog now attends school with Rebecca and alerts the teachers when the young girl's blood sugar levels become high or low. This is the first case where an assistance dog has been permitted to attend a mainstream primary school in the UK.

11.5.3 Claire and Kiska's Story

Claire was diabetic for 40 years, since she was a child. She had always tried to manage her diabetes but for the last 16 years had very little or no hypo awareness, with her sugar levels often dropping as low as 1.1 (values below 4.0 are defined as hypoglycaemic). She was experiencing regular diabetic seizures and collapsing, often on a daily basis.

As a direct consequence of long-term diabetes, she was diagnosed with early stage retinal maculopathy, which is likely to result in loss of sight. Ten years ago she also experienced a heart attack followed by a second, 18 months later. At this point,

Fig. 11.6 Claire and Kiska, her medical alert dog in training

she was collapsing on an almost daily basis, sometimes even several times a day from severe hypos. Her family were constantly worrying and were unable to leave her alone. Research has shown the link of heart attacks to repeating severe diabetic hypoglycemic episodes (Frier et al. 2011). Claire now has a complex heart condition and takes daily medication, some of which masks low blood sugar levels. Her doctors have warned that another severe hypo could cause a further heart attack, which may be fatal.

The charity has placed a dog named Kiska into training as a Medical Alert Dog with Claire (Fig. 11.6), which has already made a difference to the whole family's lives. Although still in training, the dog is accurately alerting on a daily basis, meaning Claire can take appropriate action before her blood sugar levels drop too low. The last severe hypo attack she experienced was in April 2010, the longest she had survived without collapsing since being diagnosed with diabetes. Her retinal maculopathy has also stabilised and her sugar levels are much more balanced, which in turn has assisted in her diabetes related complications.

11.5.4 Neil, Jack and Roots

Roots, a working Cocker Spaniel has become the first Medical Alert Dog to alert two Type 1 diabetics to dangerous drops in their blood sugar levels. Neil and his

Fig. 11.7 Neil and his son Jack with Roots, a trained Cocker Spaniel able to detect dangerous drops in their blood sugar levels

8-year old son Jack are the first to share a Medical Alert Dog. Roots has been specially trained to sniff out drops in both Neil and Jack's blood sugar levels, long before they begin to experience any ill-effects, so that they can treat themselves before their condition escalates to a medical emergency (Fig. 11.7).

Both Neil and Jack were diagnosed with Type 1 diabetes when they were just 2 years old and have been relying on Neil's wife, Sarah, to spot the signs of an oncoming hypo. Although they tried to lead a normal life, they always had to be careful due to their unawareness. This placed a great deal of pressure on Sarah as she was on constant alert. Before Roots moved in with the family, father and son would both collapse with very little warning. On occasions, Sarah would find her husband unconscious downstairs during the night. Since the arrival of their medical alert dog, Neil and Jack are no longer collapsing, because Roots is alerting when one of them has a drop in blood sugar, while retrieving the test kit for either one of them. This is a great relief to the whole family, giving them greater confidence when they leave the house together with Roots.

11.5.5 Medical Alert Dogs for Diabetics – Further Research

In conjunction with scientists and endocrinologists, the charity is now working to further establish a dog's sensitivity skills and to learn exactly what it is that the dogs are detecting. This study will explore the reliability of the dogs' response, and the factors, which may affect this response. It will also be possible to determine the range of benefits as perceived by the clients.

The organization is empirically assessing the effectiveness of dogs as an intervention modality for people living with diabetes by comparing the range of blood glucose levels of the individuals before and after allocating trained dogs. HbA1c levels and rates of medical emergencies reported are compared before and after placement of trained dogs. In addition, the health and psychosocial benefits of alert dogs are being assessed by means of established interview methodologies and questionnaires.

The National Health Service (NHS) is currently spending £1 million every hour on treating diabetes and its complications, accordingly Medical Alert Dogs have the potential to save money and resources related to these services.

11.5.6 Addison's Disease Alert Dog

Soon after the training of the first blood sugar detection dogs, the charity began to recognize that dogs were able to indicate the changes in odor that occurred with other medical emergencies. In 2009, it was discovered that the same principles applied to people living with Addison's disease, a chronic condition resulting from an adrenal cortex hypofunction/dysfunction with a deficient production of glucocorticoids, mineralocorticoids and androgens. One of the most significant consequences of Addison's disease is the body's failure to adapt to stress and, in the absence of adequate steroid cover, this may result in a state of shock, known as an Addison's crisis requiring urgent treatment. Usually, lifelong replacement therapy with gluco- and mineralocorticoids is required for patients with Addison's Disease.

The charity "Medical Detection Dogs" has trained what is believed to be the world's first dog to reliably detect and alert an oncoming episode of Addision's Crisis. The dog reliably alerts Karen when her cortisol levels drop dangerously, thereby allowing sufficient time for her to self medicate and preventing an emergency hospital admission (Fig. 11.8). Karen, after having regular hospital visits including admissions to intensive care, is now able to manage her condition at home. At the present moment, there are a number of Medical Assistance Dogs in training for individuals with this condition.

Fig. 11.8 Karen with her
dog Coco, trained to detect
episodes of low cortisol
levels

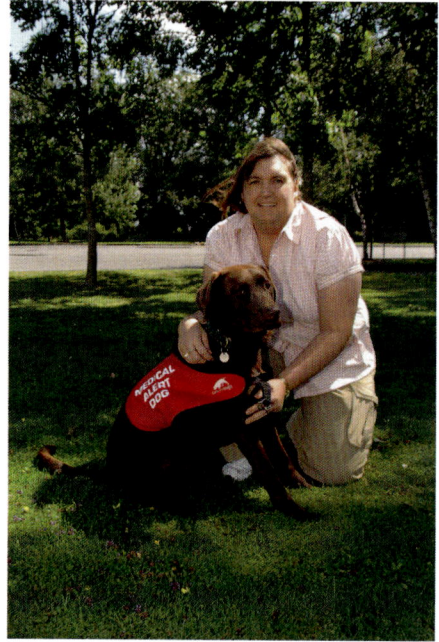

11.6   Future Research

The charity continues to investigate and train dogs for individuals with other debilitating and life threatening conditions, which dogs have the potential to detect
through odour. These include severe pain induced seizure and allergic reactions, as
well as narcolepsy (a malfunction of the sleep/wake regulating system, which causes
sleep attacks and paralysis). As yet, no rigorous data exist as to whether seizure prediction by Seizure-Alert Dogs is better than chance, and what false positive and
negative prediction rates might be (Brown and Goldstein 2011; Kirton et al. 2008).

Experienced medical practitioners and those who have worked in countries
where there are high incidences of tuberculosis have reported that infected patients
frequently have a distinctive breath odour, which distinguishes them from uninfected individuals. This enables an initial diagnosis of such patients. Rapid screening of individuals affected with tuberculosis has developed to a more sophisticated
level in southern Africa by the APOPO group. This group has shown that specially
trained African giant pouched rats (initially trained to detect land mines) can also be
trained to perform reliable screening of human sputum samples, where the reward
for successful identification of Tuberculosis, are morsels of food. Published results
suggest that pouched rats are a valuable adjunct to, and may be a viable substitute
for, sputum smear microscopy as a tuberculosis diagnostic in resource-poor countries (Mahoney et al. 2012).

As assistance animals must reside in our homes and accompany us in social situations, clearly dogs are the species of choice. However, what is clear is that we are just scratching the surface. Dogs evidently have the ability to detect small odor changes that occur as a result of illness. Training of a dog is currently underway to support a young lady with severe narcolepsy and also a woman with a severe life threatening auto-immune/allergic condition. There is great optimism that life-saving preventative cares and measures for these diseases will be successfully provided. All of the training regimens and methods have developed based on anecdotal reports, and the list continues to grow. Many of the anecdotal stories describe dogs displaying fear or an avoidance response to a medical event.

Moreover, it is clear that medical dog assistance care dramatically increases an overall sense of added security and comfort, thereby decreasing anxiety in the affected individuals and limiting potential dangerous situations or episodes. All Medical Detection Dogs have the potential to help and save countless lives. For those living with life threatening and disabling health conditions, having a Medical Detection Dog can make all the difference. Not only can they reduce both the cost of health care and hospital admissions, but more importantly they provide owners with a better quality of life, freedom, and independence.

Whilst the work of medical assistance dogs is overall well accepted by the medical establishment, the use of cancer detection dogs is greeted with scepticism (Lippi and Cervellin 2012). The author of this chapter was inspired to continue this part of the charity's work by her own personal experience and story. In 2009, one of her cancer detection dogs, Daisy, started to behave anxiously around her. Daisy is trained to detect bladder, prostate and renal cancer from small urine samples. However, 1 day she jumped against the author and caused her to feel what felt like a deep bruise in her left breast. Soon thereafter she felt a small lump, which was investigated and diagnosed as a harmless cyst. However, a mammogram and subsequent biopsies revealed a deep cancer well behind this cyst. Through surgery, lymph node removal and radiotherapy, the author came to appreciate firsthand the importance of early diagnosis. The surgeon told her how fortunate she was that the tumour had been detected early, as due to the deep location it would have been detected at least 1–2 years later. A routine mammogram was not necessary for the next 5 years, and the prognosis today would have been very different, if Daisy had not detected it early.

References

Brown SW, Goldstein LH (2011) Can Seizure-Alert Dogs predict seizures? Epilepsy Res 97(3):236–242

Brunner C, Szymczak W, Höllriegl V, Mörtl S, Oelmez H, Bergner A, Huber RM, Hoeschen C, Oeh U (2010) Discrimination of cancerous and non-cancerous cell lines by headspace-analysis with PTR-MS. Anal Bioanal Chem 397(6):2315–2324

Church J, Williams H (2001) Another sniffer dog for the clinic? Lancet 358:930

Cornu JN, Cancel-Tassin G, Ondet V, Girardet C, Cussenot O (2011) Olfactory detection of prostate cancer by dogs sniffing urine: a step forward in early diagnosis. Eur Urol 59(2):197–201

Dalziel DJ, Uthman BM, Mcgorray SP, Reep RL (2003) Seizure-alert dogs: a review and preliminary study. Seizure 12(2):115–120

Ehmann R, Boedeker E, Friedrich U, Sagert J, Dippon J, Friedel G, Walles T (2012) Canine scent detection in the diagnosis of lung cancer: revisiting a puzzling phenomenon. Eur Respir J 39(3):669–676

Frier BM, Schernthaner G, Heller SR (2011) Hypoglycemia and cardiovascular risks. Diabetes Care 34(Suppl 2):S132–S137

Horvath G, Järverud GA, Järverud S, Horváth I (2008) Human ovarian carcinomas detected by specific odor. Integr Cancer Ther 7(2):76–80

Horvath G, Andersson H, Paulsson G (2010) Characteristic odour in the blood reveals ovarian carcinoma. BMC Cancer 10:643. http://www.biomedcentral.com/1471-2407/10/643. Accessed 26 Dec 2012

Kirton A, Winter A, Wirrell E, Snead OC (2008) Seizure response dogs: evaluation of a formal training program. Epilepsy Behav 13(3):499–504

Korljan-Babić B, Barsić-Ostojić S, Metelko Z, Car N, Prasek M, Skrabić V, Kokić S (2011) Impact of a guide dog on glycemia regulation in blind/visually impaired persons due to diabetes mellitus. Acta Clin Croat 50(2):229–232

Lippi G, Cervellin G (2012) Canine olfactory detection of cancer versus laboratory testing: myth or opportunity? Clin Chem Lab Med 50(3):435–439

Mahoney A, Weetjens BJ, Cox C, Beyene N, Reither K, Makingi G, Jubitana M, Kazwala R, Mfinanga GS, Kahwa A, Durgin A, Poling A (2012) Pouched rats' detection of tuberculosis in human sputum: comparison to culturing and polymerase chain reaction. Tuberc Res Treat 2012:716989. doi:10.1155/2012/716989

McCulloch M, Jezierski T, Broffman M, Hubbard A, Turner K, Janecki T (2006) Diagnostic accuracy of canine scent detection in early- and late-stage lung and breast cancers. Integr Cancer Ther 5(1):30–39

O'Connor MB, O'Connor C, Walsh CH (2008) A dog's detection of low blood sugar: a case report. Ir J Med Sci 177(2):155–157

Pickel D, Manucy GP, Walker DB, Hall SB, Walker JC (2004) Evidence for canine olfactory detection of melanoma. Appl Anim Behav Sci 89(1–2):107–116

Sonoda H, Kohnoe S, Yamazato T, Satoh Y, Morizono G, Shikata K, Morita M, Watanabe A, Morita M, Kakeji Y, Inoue F, Maehara Y (2011) Colorectal cancer screening with odour material by canine scent detection. Gut 60(6):814–819

Weber CM, Cauchi M, Patel M, Bessant C, Turner C, Britton LE, Willis CM (2011) Evaluation of a gas sensor array and pattern recognition for the identification of bladder cancer from urine headspace. Analyst 136(2):359–364

Wells DL, Lawson SW, Siriwardena AN (2008) Canine responses to hypoglycemia in patients with type 1 diabetes. J Altern Complement Med 14(10):1235–1241

Williams H, Pembroke A (1989) Sniffer dogs in the melanoma clinic? Lancet 1:734

Willis CM, Church SM, Guest CM, Cook WA, McCarthy N, Bransbury AJ, Church MRT, Church JCT (2004) Olfactory detection of human bladder cancer by dogs: proof of principle study. BMJ 329:712

Willis CM, Britton LE, Harris R, Wallace J, Guest CM (2010) Volatile organic compounds as biomarkers of bladder cancer: sensitivity and specificity using trained sniffer dogs. Cancer Biomark 8:145–153

Index

A

AAA. *See* Animal-assisted activities (AAA)
Aanderson, K.W., 241, 247, 249
Abdulrahman, M., 136
Abedon, S., 207, 212, 223, 224
Abuladze, A.S., 44
Addison's Disease, 299
Adolapin, 85, 87–89
Alexander, H., 9
Allergies, 18, 82, 127, 226, 246, 275, 279
Allergy test, 84
American Apitherapy Society (AAS),
    80, 121
Andereya, S., 51
Anderson, W.P., 241
Animal-assisted activities (AAA), 235, 236,
    239, 248
Animal-assisted interventions, 256
Animal-assisted therapy, viii, 2, 233–251,
    256, 270
Anti-*E. coli* phage cocktail, 192, 225
Anti-fungal/anti-bacterial & anti-viral
    properties, 88, 99–100
Anti-inflammatory properties, 51, 87, 89, 97,
    100–101, 138
Apamin, 85, 87, 88, 94, 100
Apis mellifera, 79–81, 91, 126, 147
Apitherapy, vii, 1, 2, 7–107, 113–142, 147
Application of a leech, 2, 31–34, 36,
    41–43, 45, 48–49, 51, 58–60,
    65, 67–69
Armstrong, D.G., 15
Arthritis, vi, vii, 50–52, 77–79, 82, 85, 89–91,
    94–97, 210
Asthma, 127, 138, 142, 178

Atopic dermatitis, 148, 163
Avicenna, 33
Azab, K.S., 141

B

Bacterial infections, vii, 2, 132, 165–166, 168,
    182, 191, 211
Bacteriophage biology, 193–198
Bacteriophages, vii, 2, 191, 193–203, 208,
    211–214, 221
Bacteriophage T4, 194, 197, 202
Baer, W.S., 6, 9, 11, 15, 16
Balneotherapy, 45, 50, 51
Banks, B.E., 100
Barnaulov, O.D., 63
Barnes, K.M., 15
Barroso, P.R., 137
Batson, K., 241
Bayne-Jones, S., 218, 219
Beck, A.M., 234, 235, 241, 244
Beck, B.F., 79, 80
Bee Bread, 26, 125
Beekeeper, M., 80
Bee products, vii, 2, 114–115, 117,
    126, 128, 136
Bee sting therapy (BST), 78, 81–84,
    102, 104, 107
Bee venom, vi, vii, 2, 77–79, 82–86, 89–104,
    106, 107, 114, 115, 117–122
Bee venom therapy (BVT), 77–107, 117,
    119–122
Behling, R.J., 235
Belliveau, J., 102
Bell, T., 64

M. Grassberger et al. (eds.), *Biotherapy - History, Principles and Practice:*
A Practical Guide to the Diagnosis and Treatment of Disease using Living Organisms,
DOI 10.1007/978-94-007-6585-6, © Springer Science+Business Media Dordrecht 2013

Printed by Publishers' Graphics LLC
MLSI130618.15.15.184